Golden Notes

Biochemistry

Rapid Revision

Handwritten Notes

New Topics in Third Edition

- More than 100 topic-wise clinical cases
- ORS in treatment of dehydration
- Marker enzymes of cellular organelles
- ABC transport proteins
- Glucose transporters
- Cori cycle and glucose-alanine cycle
- Hereditary fructose intolerance
- Glucose tolerance curve
- Lipid storage disorders
- Lipid profile
- Triacylglycerol metabolism and role of HSL
- Diagnostic approach to hyperlipoproteinemias
- Amines and polyamines
- Obesity and role of leptin
- Eukaryotes—regulation of gene expression
- Vitamins and minerals revision chart
- Laboratory interpretations
- DNA fingerprinting, blotting techniques
- High yield MCQs, reasons, viva questions

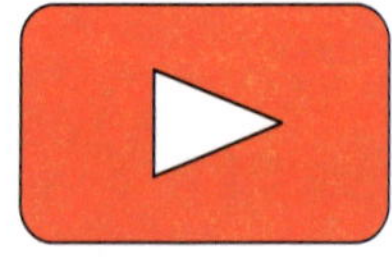

Based on

YouTube Channel

Biochemistry by Dr Rajesh Jambhulkar

458,000 Subscribers and 41,486,524 views till date

Golden Notes

Biochemistry Rapid Revision

Handwritten Notes

Third Edition

Rajesh Jambhulkar
MBBS MD DNB (Biochemistry)
Associate Professor
Department of Biochemistry
Government Medical College
Jalgaon, Maharashtra, India

JAYPEE BROTHERS MEDICAL PUBLISHERS
The Health Sciences Publisher
New Delhi | London

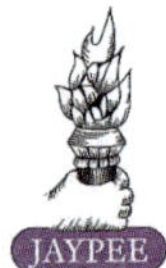

Jaypee Brothers Medical Publishers (P) Ltd

Headquarters
EMCA House
23/23-B, Ansari Road, Daryaganj
New Delhi 110 002, India
Landline: +91-11-23272143, +91-11-23272703
+91-11-23282021, +91-11-23245672
Email: jaypee@jaypeebrothers.com

Corporate Office
4838/24, Ansari Road, Daryaganj
New Delhi 110 002, India
Phone: +91-11-43574357
Fax: +91-11-43574314
Email: jaypee@jaypeebrothers.com

Overseas Office
J.P. Medical Ltd
83 Victoria Street, London
SW1H 0HW (UK)
Phone: +44 20 3170 8910
Email: info@jpmedpub.com

EU GPSR Authorised Representative
Logos Europe, 9 rue Nicolas Poussin
17000, La Rochelle, France
Phone: +33 (0) 6 67 93 73 78
E-mail: Contact@logoseurope.eu

Website: www.jaypeebrothers.com

Website: www.jaypeedigital.com

Inquiries for bulk sales may be solicited at: jaypee@jaypeebrothers.com

Golden Notes—Biochemistry Rapid Revision Handwritten Notes

Third Edition: 2025, **Reprint:** 2025, **2026**

ISBN: 978-93-5696-578-2

Printed at: Samrat Offset Pvt. Ltd.

CONTENTS

CELL MEMBRANE

MCI COMPETENCY BI 1.1

Important Diagram:- 1. fluid mosaic model of cell membrane
2. Mitochondria

Important LAQ:- Transport across the cell membrane
SAQ:- Cellular organelles with their functions

MCQ

Zellweger syndrome [cerebrohepatorenal syndrome]

peroxisome biogenesis disorder

- due to mutation in genes involved in the biogenesis of peroxisomes.
- PEX family genes – Peroxins
- Absence of functional peroxisomes [Protein targeting defect]
- Oxidation of very long chain fatty acid is affected
- Elevated levels of very long chain fatty acids in brain causes profound neurologic impairment, hypotonia, vision loss, hearing loss.

Zellweger spectrum disorders

Most severe form:- Zellweger syndrome
Intermediate form:- Neonatal adrenoleukodystrophy
Mild form:- Infantile Refsums disease.

Fluid Mosaic Model - Plasma membrane

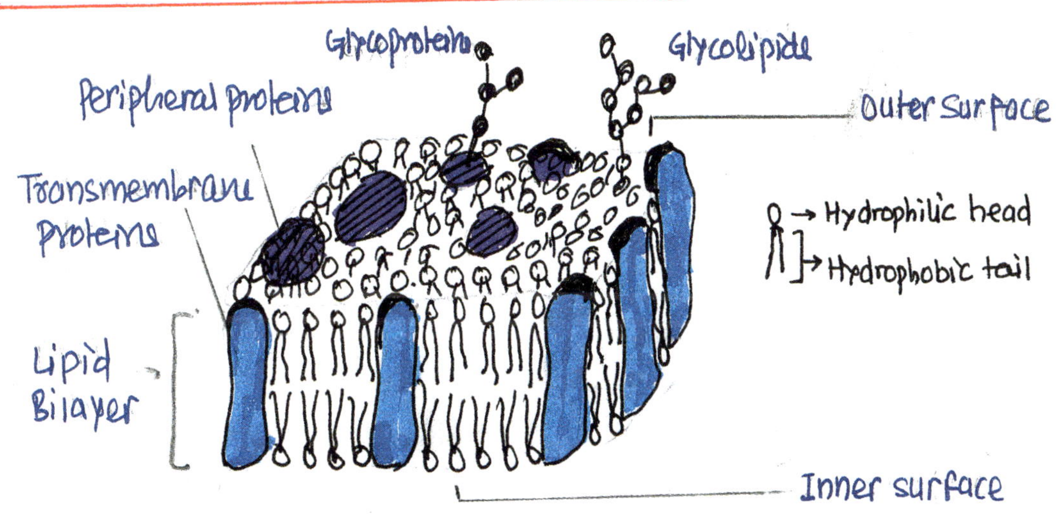

Components of plasma membrane

1. Lipid → phospholipid bilayer - composed of hydrophilic head and hydrophobic tails.
   Cholesterol :- contributes to membrane stability & fluidity.

2. proteins → Integral proteins - embeded within the bilayer and serves as channels, transporters, receptors, enzymes.
   Integral proteins facilitates transport across the membrane.
   peripheral proteins - essential role in cell adhesion, shape, recognition of self & non-self cells.

3. Carbohydrates :- As glycoproteins & glycolipids

factors affecting fluidity of the membrane

1. Temperature → increases fluidity → phospholipids move rapidly
2. Length of FA chain in the phospholipid
3. Cholesterol → Prevent phospholipids from packing too tightly.

MCQs

Marker enzymes for cellular Organelles

1. Plasma membrane : 5' Nucleotidase, Adenyl cyclase, Na^+-K^+ ATPase
2. Mitochondria :- ATP synthase
3. Lysosomes :- Cathepsin
4. Endoplasmic reticulum :- Glucose 6 phosphatase
5. Golgi Apparatus :- Galactosyl transferase

ATP-binding cassette transporter - MCQ

Associated diseases

ATP 7A → Absorption of copper from intestine → Menke's disease

ATP 7B → Incorporation of copper into ceruloplasmin & excretion in the bile acid → Wilson's disease

ABC A1 → Transport of cholesterol from tissues to HDL → Tangier disease

ABC C2 → MRP-2 [Multidrug resistance associated Protein-2] → Dubin Johnson's syndrome

CFTR [cystic fibrosis transmembrane regulator protein] → chloride channel protein → cystic fibrosis

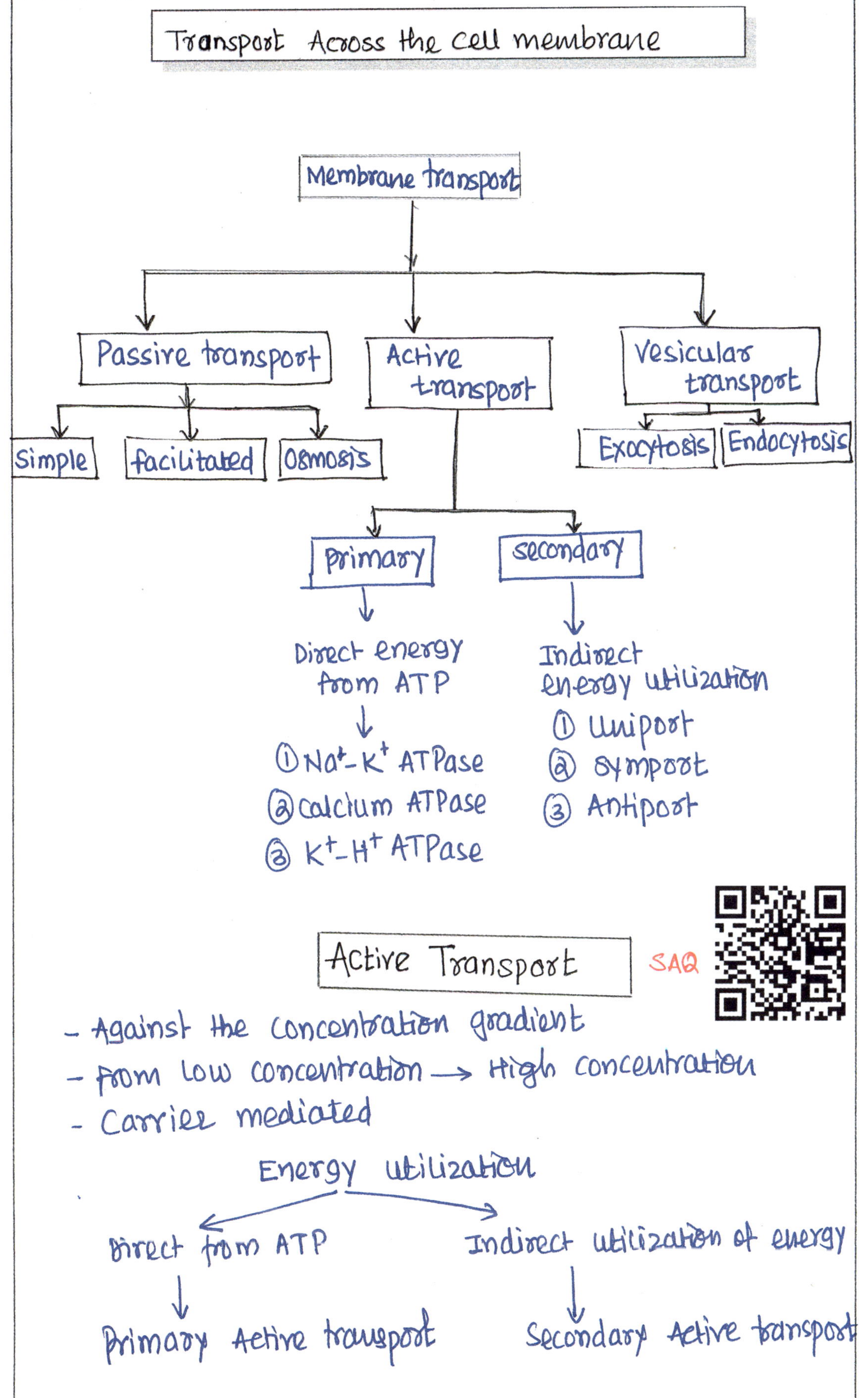
Transport Across the cell membrane
Membrane transport
Passive transport
Active transport
Vesicular transport
Simple
facilitated
Osmosis
Exocytosis
Endocytosis
Primary
Secondary
Direct energy from ATP
① Na+-K+ ATPase
② calcium ATPase
③ K+-H+ ATPase
Indirect energy utilization
① Uniport
② Symport
③ Antiport
SAQ
Active Transport
- Against the concentration gradient
- From low concentration → High concentration
- Carrier mediated
Energy utilization
Direct from ATP
Primary Active transport
Indirect utilization of energy
Secondary Active transport

Primary Active transport

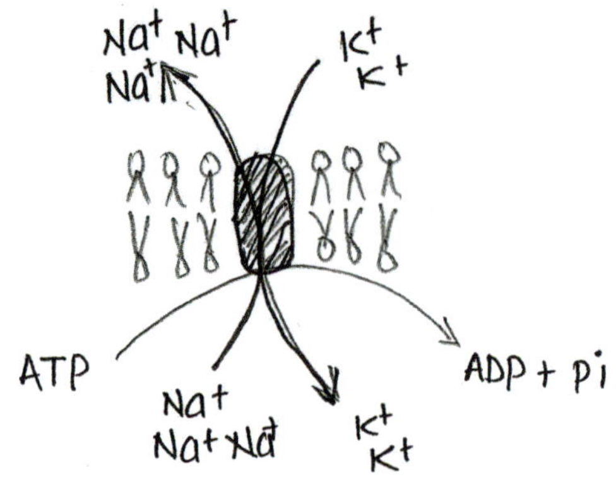

Examples - ① Na^+-K^+ ATPase
② calcium ATPase
③ K^+-H^+ ATPase

Secondary Active transport

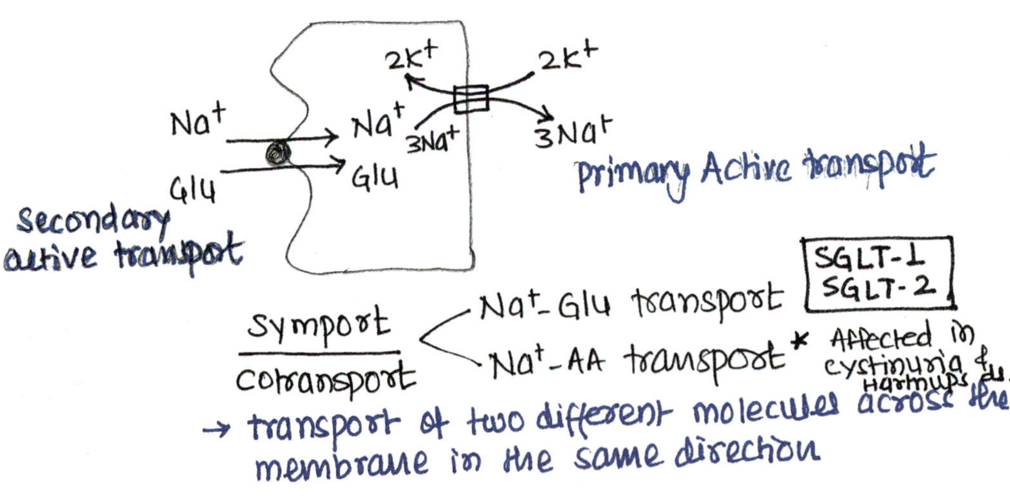

Symport / Cotransport
- Na^+-Glu transport — SGLT-1, SGLT-2
- Na^+-AA transport* — *Affected in cystinuria & Hartnups dis.

→ transport of two different molecules across the membrane in the same direction

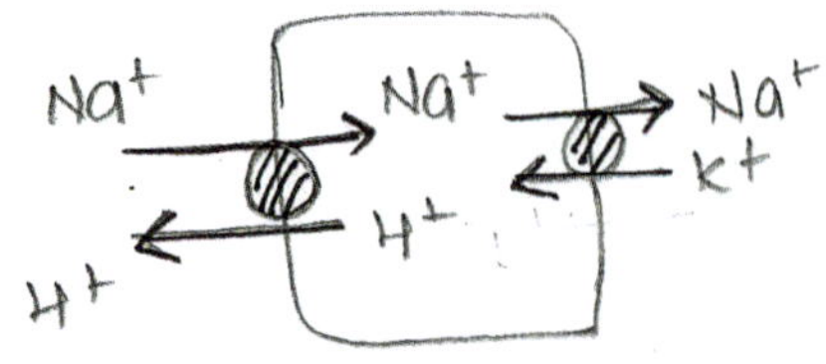

Antiport / Counter transport
① Na^+-H^+ exchanger
② Ca^{2+}-H^+ exchanger

→ transport of two different molecules across the membrane in the opposite direction

ORS :- Oral rehydration Solutions

ORS – used for treatment of dehydration due to diarrhea or vomiting.

ORs helps to replenish lost fluids.

components → Glucose → 13.5 grams
Sodium chloride → 2.6 grams
Potassium chloride → 1.5 grams
Trisodium citrate dehydrate → 2.9 grams.

Glucose → facilitates absorption of sodium and hence water in the intestine [symport - cotransport]

Na^+-Glu cotransporters [SGLT-1] in the small intestine is important for absorption of glucose & sodium and it is major mode of glucose absorption from intestine.

Na^+/K^+ → to replenish the lost electrolytes

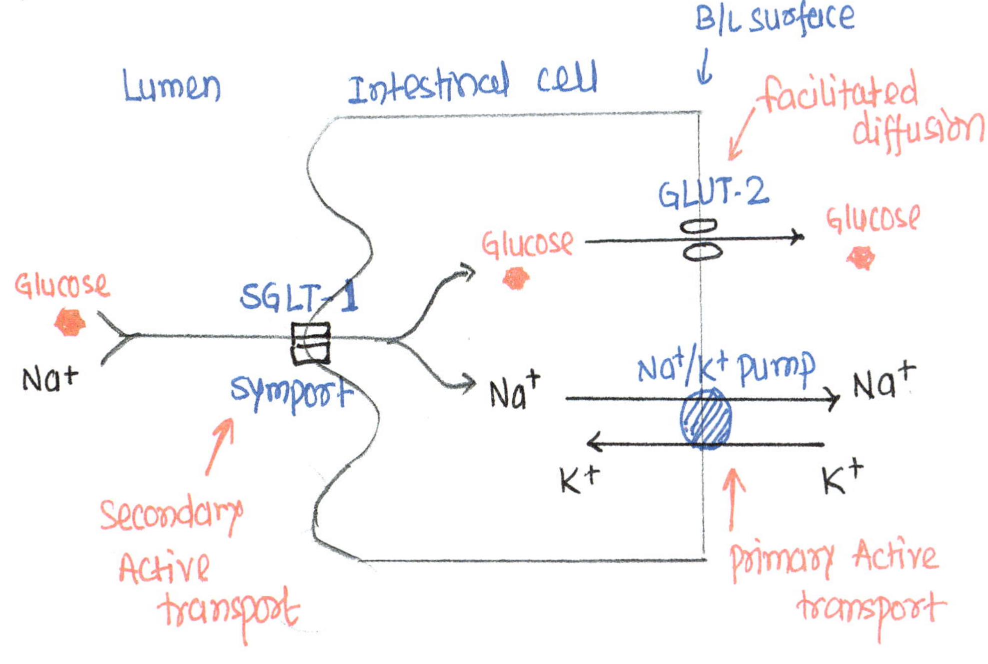

Mode of Absorption of Glucose + Na^+ from Intestine

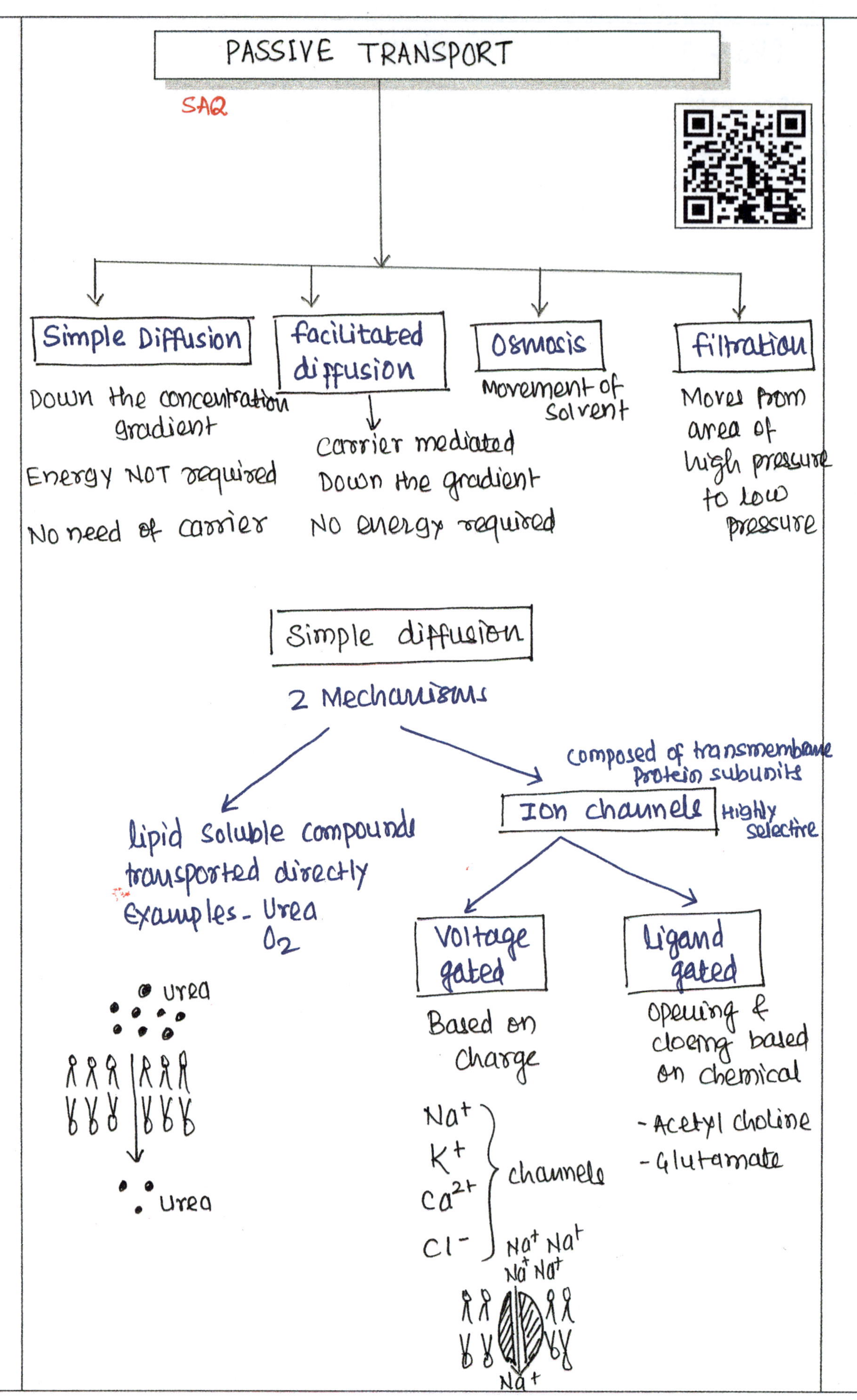

PASSIVE TRANSPORT
SAQ
Simple Diffusion
Down the concentration gradient
Energy NOT required
No need of carrier
facilitated diffusion
carrier mediated
Down the gradient
No energy required
Osmosis
Movement of solvent
filtration
Moves from area of high pressure to low pressure
Simple diffusion
2 Mechanisms
lipid soluble compounds transported directly
Examples - Urea
O_2
Ion channels
composed of transmembrane protein subunits
Highly selective
Voltage gated
Based on charge
Na^+
K^+
Ca^{2+}
Cl^-
channels
Ligand gated
opening & closing based on chemical
- Acetyl choline
- Glutamate
Urea
Urea
Na^+ Na^+
Na^+ Na^+
Na^+

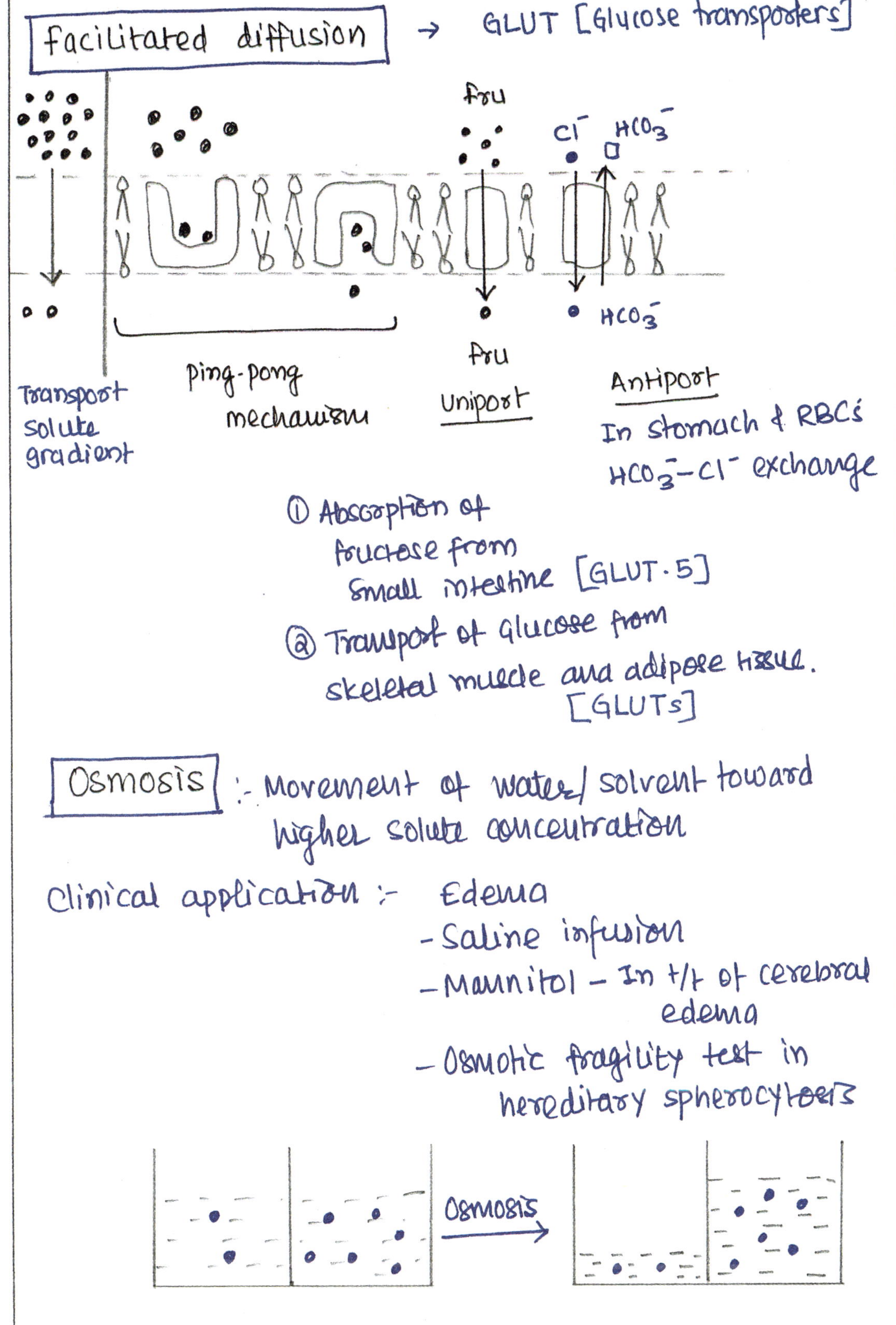

Facilitated diffusion → GLUT [Glucose transporters]

Antiport — In stomach & RBCs HCO_3^-–Cl^- exchange

① Absorption of fructose from small intestine [GLUT-5]

② Transport of glucose from skeletal muscle and adipose tissue. [GLUT5]

Osmosis

Osmosis :- Movement of water/solvent toward higher solute concentration

Clinical application :- Edema

- Saline infusion
- Mannitol – In t/t of cerebral edema
- Osmotic fragility test in hereditary spherocytosis

Vesicular Transport

Exocytosis

Secretion of molecules outside the cell via a vesicle fusing to a membrane

vesicle migrates to the cell membrane, binds & releases its content out of the cell

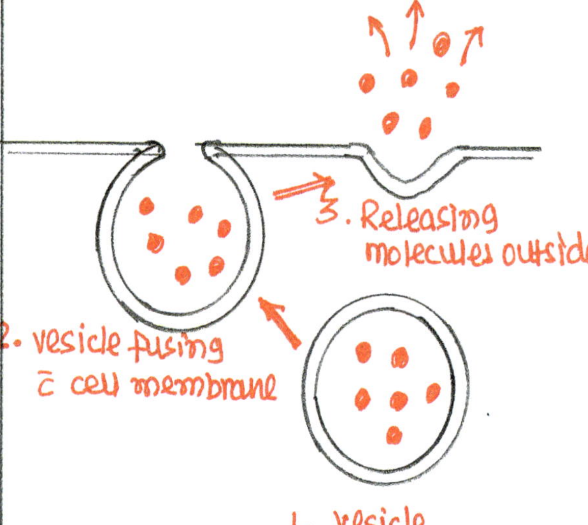

Example

Insulin, PTH & catecholamines are packaged in granules and released from the cells on appropriate stimulation.

Endocytosis

Engulfing of molecules inside the cell via vesicle formation

- uptake of molecule from extracellular space into the cell

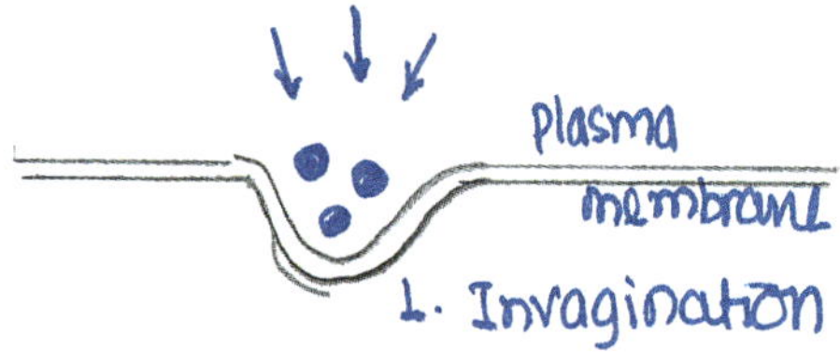

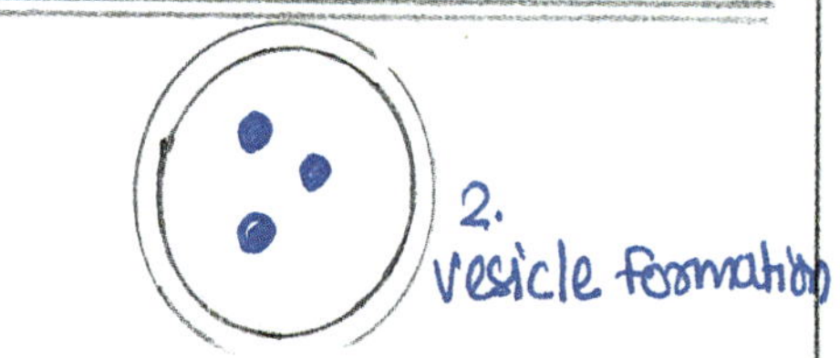

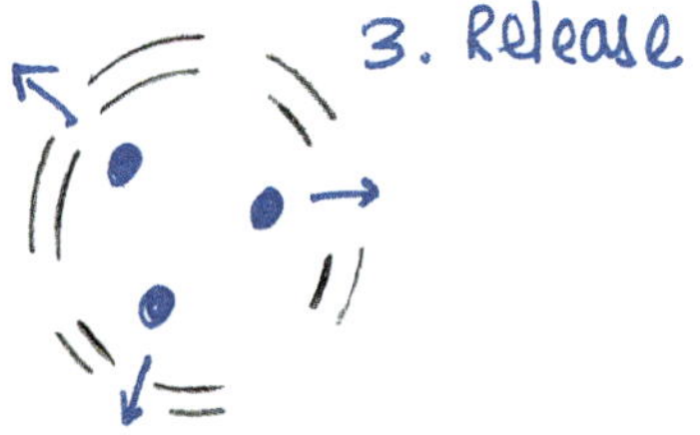

1. Phagocytosis → cell eating
   e.g. → Bacteria
2. Pinocytosis → cell drinking
   e.g. → small molecules or fluids
3. Receptor mediated endocytosis
   - clathrin mediated
   e.g. → LDL uptake MCQ

Cellular organelles and their functions.

1. **Nucleus**
   Contains genetic material - DNA
   DNA replication and transcription occurs in the nucleus.

2. **Endoplasmic reticulum** - Marker enzyme → glucose 6 phosphatase MCQ
   Membrane enclosed spaces
   Two types - Smooth and Rough ER
   Synthesis of proteins, glycoproteins, lipoproteins
   Metabolism of drugs

3. **Lysosomes**
   It is a bag of digestive enzymes of the cell
   Degradation of proteins, carbohydrates, lipids and NTs occurs in the lysosomes.

4. **Mitochondria** - power house of cell
   ETC, ATP synthesis, TCA cycle, FA oxidation, ketone body production occurs in mitochondria.
   part of Heme synthesis, Urea cycle, gluconeogenesis and Pyrimidine synthesis occurs in mitochondria.

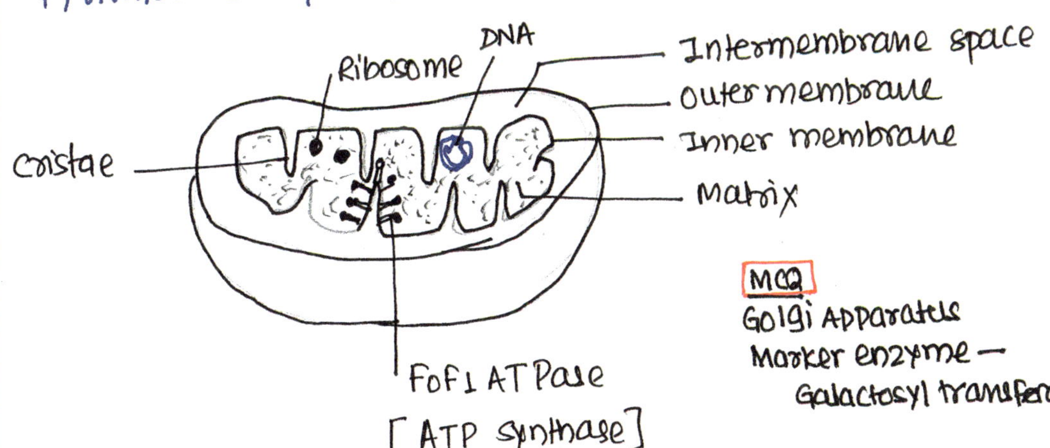

MCQ
Golgi Apparatus
Marker enzyme —
Galactosyl transferase

5. **Golgi bodies** - It is a network of flattened membranes & vesicles involved in protein packaging, sorting and secretion.

6. **Peroxisomes** - contains peroxidases and catalases.
   Oxidation of long chain FA occurs in peroxisomes.

7. **Ribosomes** - Factories of protein synthesis, found on ER.

8. **Cytosol** - Gel like matrix, surrounded by cell membrane
   - site for protein synthesis, glycolysis, glycogen metabolism, fatty acid synthesis.
   - part of heme synthesis, Urea cycle, Pyrimidine synthesis occurs in the Cytosol.

ENZYMES

MCI COMPETENCY BI 2.1-2.7

Important LAQs

1. Define and classify enzymes c̄ suitable examples.
2. Mechanism of enzyme action
3. factors affecting enzyme activity
4. Different types of enzyme inhibition c̄ therapeutic role
5. Diagnostic importance of enzymes & Isoenzymes.
6. Regulation of enzyme activity

Important SAQs

① Cardiac markers
② Enzyme pattern in liver diseases.
③ coenzymes c̄ functions
④ Isoenzymes
⑤ Therapeutic uses of competitive inhibition
⑥ Normal levels of SGOT, SGPT, ALP and interpretation of laboratory data

Clinical cases

① Myocardial infarction - chest pain, ECG changes, ↑ CK-MB, ↑ Troponins

② Hepatic Jaundice - SGPT ↑↑↑↑, Direct bilirubin ↑↑
SGOT ↑↑↑↑, Indirect bilirubin ↑↑
Slight increase in ALP

③ Obstructive Jaundice - ALP ↑↑↑↑, SGPT ↑, SGOT ↑
Direct bilirubin ↑↑, ↑↑ 5'NT, ↑↑ GGT

④ Acute Pancreatitis - Amylase ↑↑, lipase ↑↑
Abdominal pain, fever, nausea, vomiting

Enzymes

LAQ. Define and classify enzymes with suitable examples.

Biocatalyst (Accelerate the rate of reaction without undergoing any change)

C - Colloidal

T - Thermolabile

P - Protein

→ Except Ribozyme *MCQ

Peptidyl transferase, SnRNA, RNAse H, Ribonuclease P

S - Specific in action

Classification and Enzyme Commission number

EC Number :- 4 digit number
Given by IUBMB [International Union of Biochemistry & Molecular Biology] in their Nomenclature & Classification

Example ⇒ 2.6.1.1 → SGOT
2.6.1.2 → SGPT
3.1.3.1 → ALP

3 → Class
1 → Subclass
3 → Sub.subclass
1 → Individual enzyme number

Classification

Mnemonic **OTHLIL-T**

EC 1 → Oxidoreductases
EC 2 → Transferases
EC 3 → Hydrolases
EC 4 → Lyases
EC 5 → Isomerases
EC 6 → Ligases
EC 7 → Translocases [Newly added class in 2018] *MCQ

EC 1 - Oxidoreductases

- Catalyses oxidation-reduction reactions
- Transfer of H^+ and electrons.
- This class includes Dehydrogenases, Oxidases, Peroxidases

Example - Alcohol dehydrogenase EC 1.1.1.1

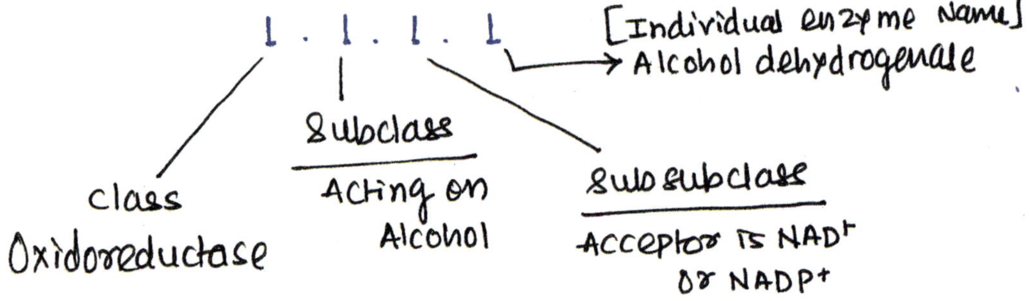

Lactate dehydrogenase EC 1.1.1.27

$$\text{Lactate} \xrightleftharpoons[\text{LDH}]{NAD^+ \rightarrow NADH + H^+} \text{Pyruvate}$$

EC 2 - Transferases

Catalyses transfer of group other than H^+ ions.

Example - SGOT (AST), EC 2.6.1.1 } transamination reaction
SGPT (ALT), EC 2.6.1.2 } transfer of aminogroup

$$\text{Aspartate} + \alpha\text{ ketoglutarate} \xrightleftharpoons[\text{PLP}]{\text{SGOT}} \text{Glutamate} + \text{oxaloacetate}$$

$$\text{Alanine} + \alpha\text{ ketoglutarate} \xrightleftharpoons[\text{PLP}]{\text{SGPT}} \text{Pyruvate} + \text{Glutamate}$$

Hexokinase → transfer of Ⓟ from ATP to glucose

$$\text{Glucose} \xrightarrow[\text{ATP} \rightarrow \text{ADP}]{\text{Hexokinase}} \text{Glucose 6-Ⓟ}$$

EC 3 - Hydrolases

These enzymes catalyses the hydrolysis of various bonds by addition of water molecules

Most of the digestive enzymes belongs to this class

Subclassified on the basis of bond cleaved

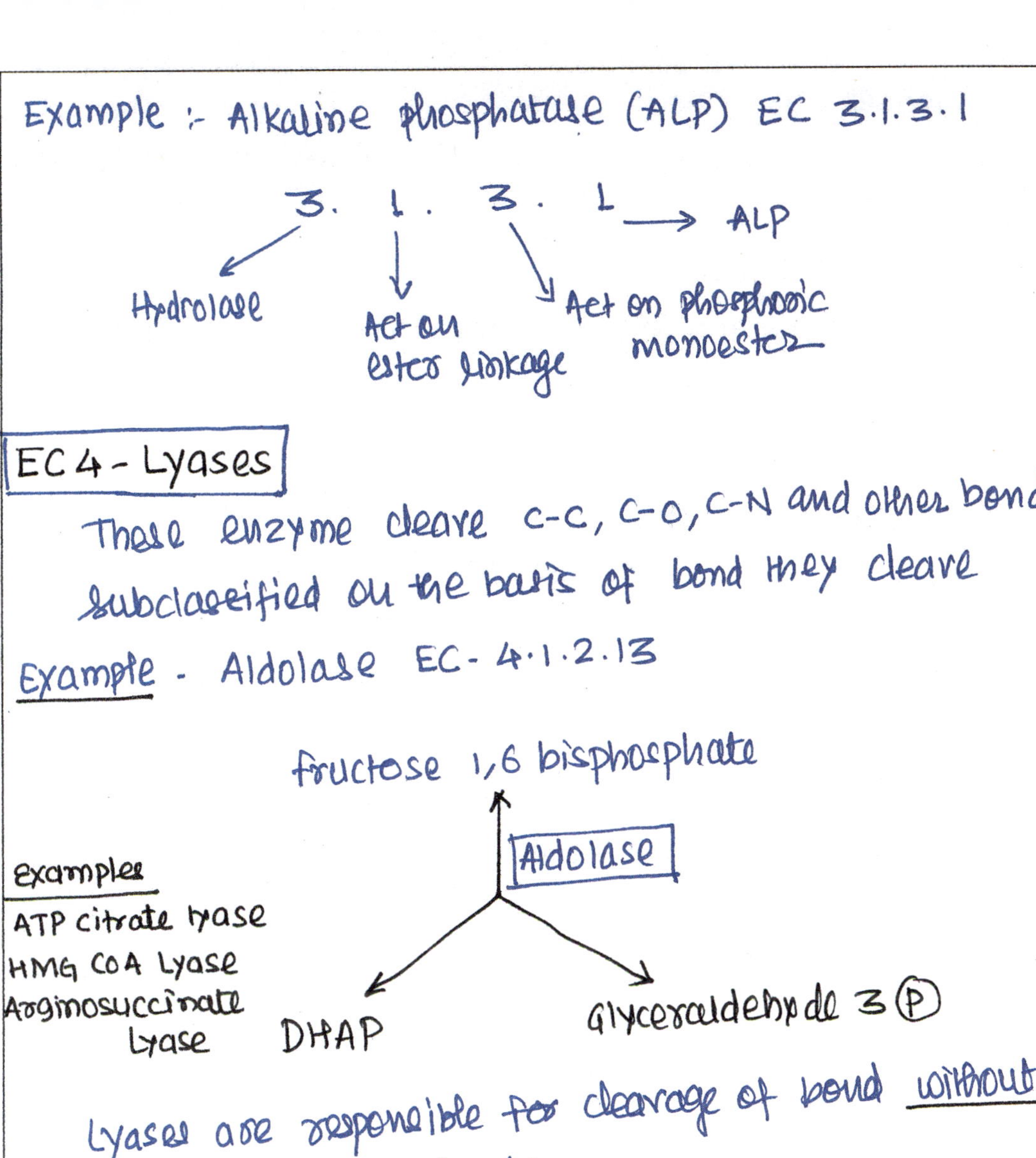

Example :- Alkaline phosphatase (ALP) EC 3.1.3.1

EC 4 - Lyases

These enzyme cleave C-C, C-O, C-N and other bonds subclassified on the basis of bond they cleave

Example - Aldolase EC- 4.1.2.13

examples
ATP citrate lyase
HMG CoA Lyase
Arginosuccinate Lyase

Lyases are responsible for cleavage of bond without addition of water molecule.

EC 5 - Isomerases

Responsible for transfer of intramolecular atoms
Responsible for interconversion of isomers

Includes - Racemases
Epimerase
Cis-trans isomerase
Mutase etc.

Example - Triose phosphate isomerase EC 5.3.1.1

DHAP $\xrightleftharpoons[\text{Isomerase}]{\text{Phosphotriose}}$ Glyceraldehyde 3 Ⓟ

Glucose 6 Ⓟ $\xrightleftharpoons[\text{Isomerase}]{\text{Phosphohexose}}$ fructose 6 Ⓟ

EC6- Ligases

catalyses the joining of two molecules with utilization of ATP

Example ① DNA ligase EC 6.5.1.1

It is a DNA joining enzyme. Two nucleotides are joined by 3'- 5' phosphodiester linkage.

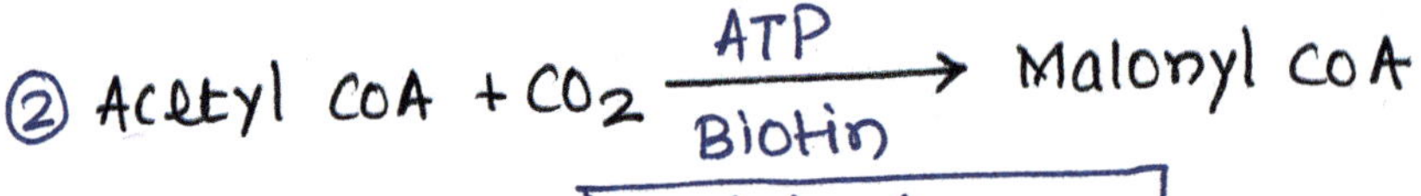

Acetyl CoA carboxylase

③ propionyl CoA carboxylase

④ pyruvate carboxylase

EC 7- Translocases

Catalyses the translocation of ions across the membrane. Subclassified on the basis of component transferred. Sub sub classified on the basis of energy used.

Example: ① ATPase → Na^+/K^+ exchanging ATPase
EC 7.2.2.13

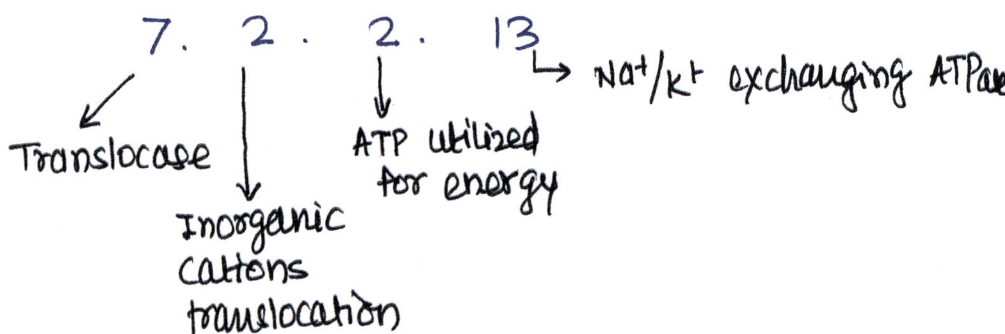

② Carnitine translocase

③ Ornithine translocase

Mechanism of Enzyme action

LAQ :- Describe the mechanism of enzyme action. Add a note on different theories of enzyme-substrate complex formation.

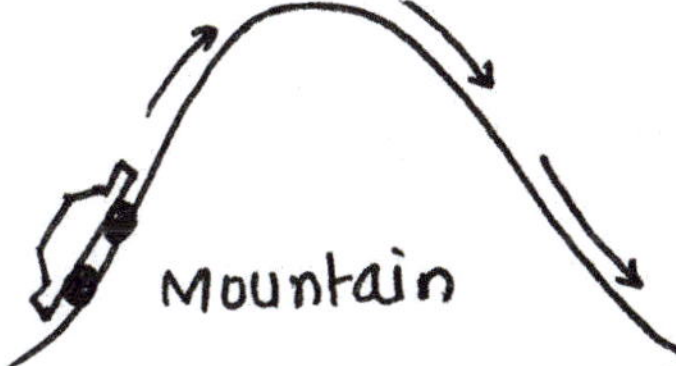

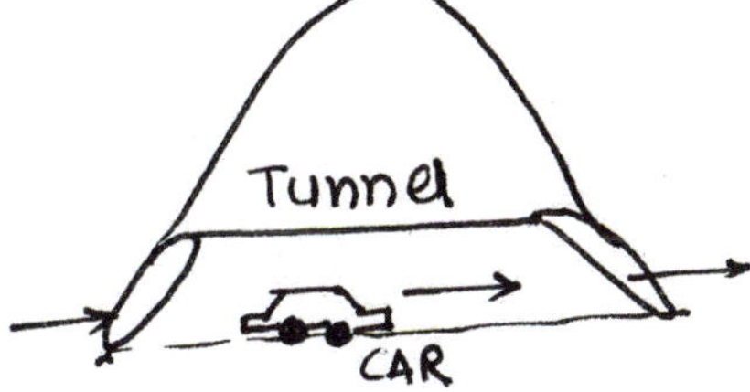

Think about energy requirement to climb the mountain

↑

Reaction without enzyme

Now think about energy requirement to pass through the tunnel made in the mountain

↑

Enzymatic reaction (Here tunnel is enzyme)

1. Enzyme lowers the Activation energy

Activation energy :- energy required to convert the substrate from ground state to transition state (High energy state)

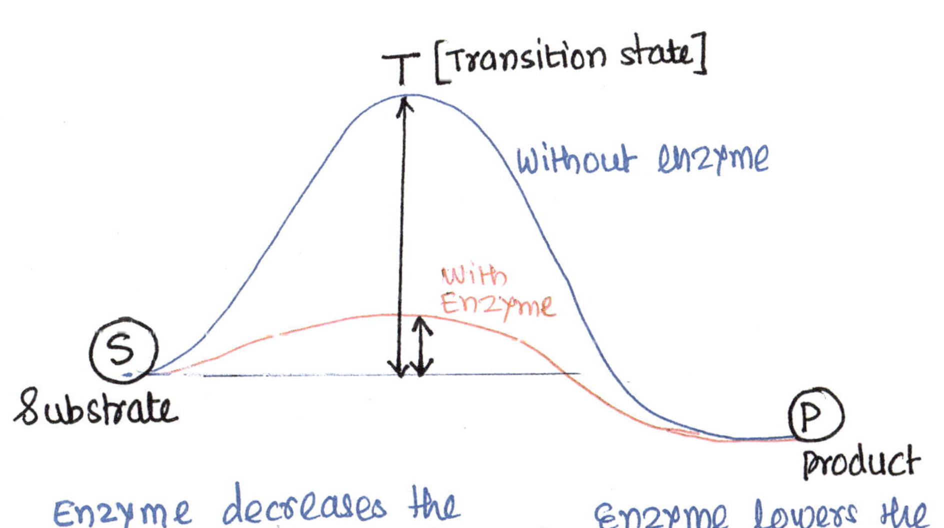

Enzyme decreases the energy barriers in the reaction ⟹ Enzyme lowers the Activation energy

2. **Transition state**

It is a Intermediate state and high energy state between substrate and product

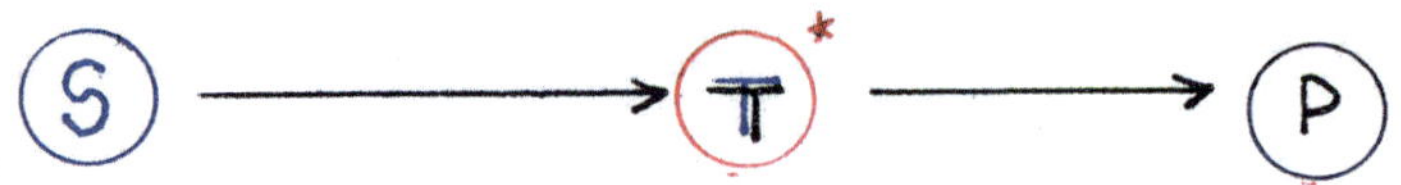

Enzyme favours the formation of transition state

3. **Enzyme - substrate complex formation**

ⓐ Lock and key model [Fischer Template theory]

- Active site of an enzyme is rigid & preshaped
- Only specific substrate can bind to the active site of an enzyme & forms enzyme-substrate complex
- This theory failed to explain the enzyme kinetics and obsolete now.

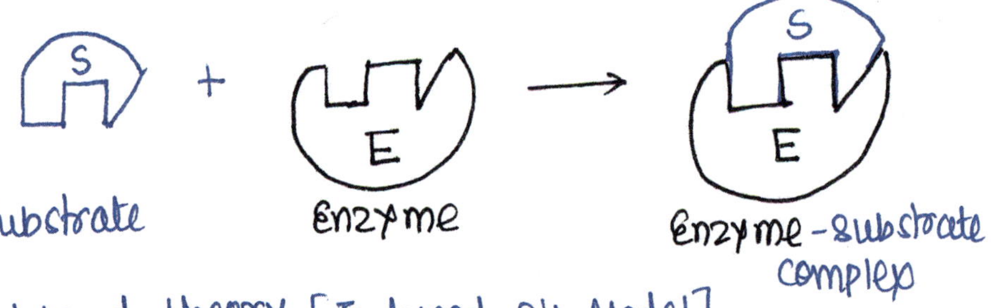

ⓑ Koshland theory [Induced fit Model]

- Active site of an enzyme is flexible [Not rigid]
- Substrate induces a conformation change in the active site of an enzyme and favours the formation of E-S complex

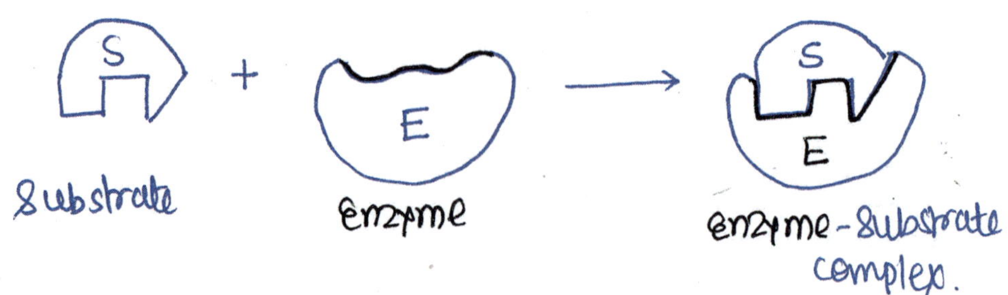

(c) Substrate strain theory

- Active site is flexible
- Substrate brings about conformational change in the active site of an enzyme
- Substrate also get strained, this results in cleavage of the bonds present in substrate molecule.

4. Mechanism of catalysis

(a) Acid-base catalysis

- There are acidic and basic groups at the active site
- These groups attack the substrate
- favours formation of E-S complex and catalysis.

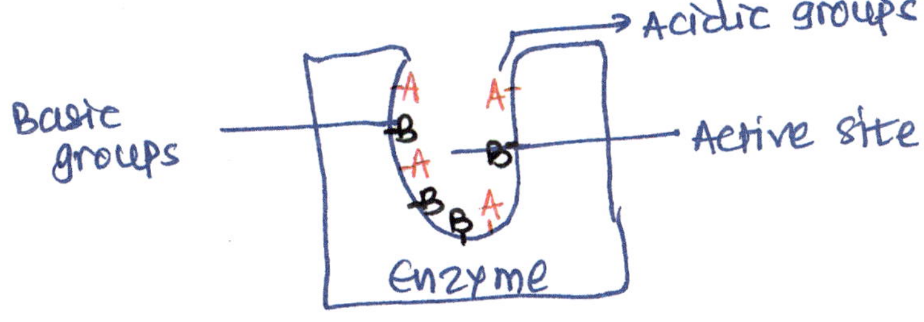

(b) Covalent catalysis

- There are positively and negatively charged groups in the active site
- Reacts with the substrate, E-S complex formation and catalysis occurs.

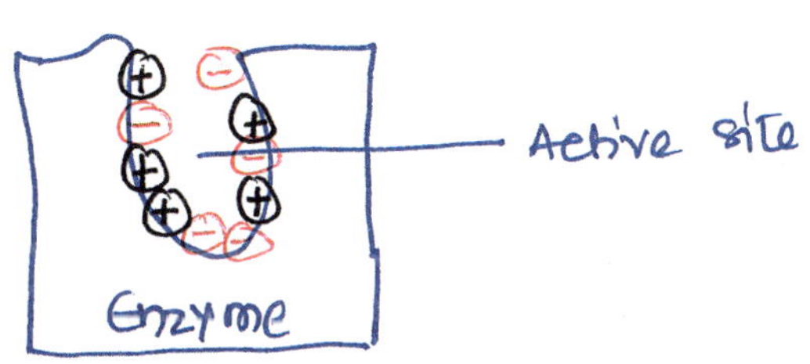

(c) Proximity effect [Entropy effect]

- If concentration of [S] & [E] is more, they will encounter with each other, they will be in bond forming distance, that results in E-S formation and catalysis of reaction

(d) Substrate strain :- Simillar to substrate strain theory.

COENZYME

SAQ:- Coenzymes with suitable examples.

SAQ:- Differentiate between coenzyme, cofactor & prosthetic group. Give suitable example of each.

BAQ:- Enumerate coenzymes derived from vit B.

BAQ:- Enumerate any four non-vitamin coenzymes.

Definition :- These are nonprotein, small molecules or metal ions that participate directly in substrate binding and catalysis.

Also known as Cosubstrate / secondary substrate

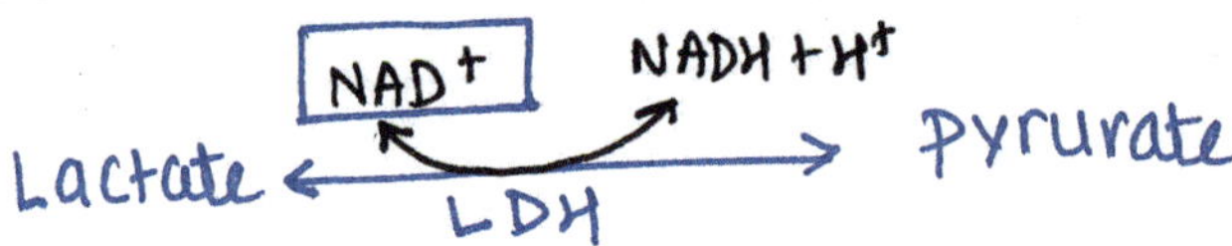

Holoenzyme [functional part] = Apoenzyme [protein] + Coenzyme / cofactor / prosthetic group [Non protein]

These are small molecules or metal ions associated with enzyme action.

Coenzyme

N Nonprotein
O Organic
L Low molecular weight
D Dialyzable [easily dissociate] substance associated c̄ enzyme action

Prosthetic group

- usually metal ions
- tight & stable association
- with apoenzyme

These enzymes are called as Metalloenzymes. Examples:

Zn – Superoxide dismutase, Carbonic anhydrase
Mo – Xanthine oxidase
Se – Glutathione peroxidase

Cofactor

- weak & transient association with apoenzymes
- These enzymes are called as Metal activated enzymes.

example :- ATPase [Mg^{2+} & Ca^{2+}]

Classification of Coenzymes.

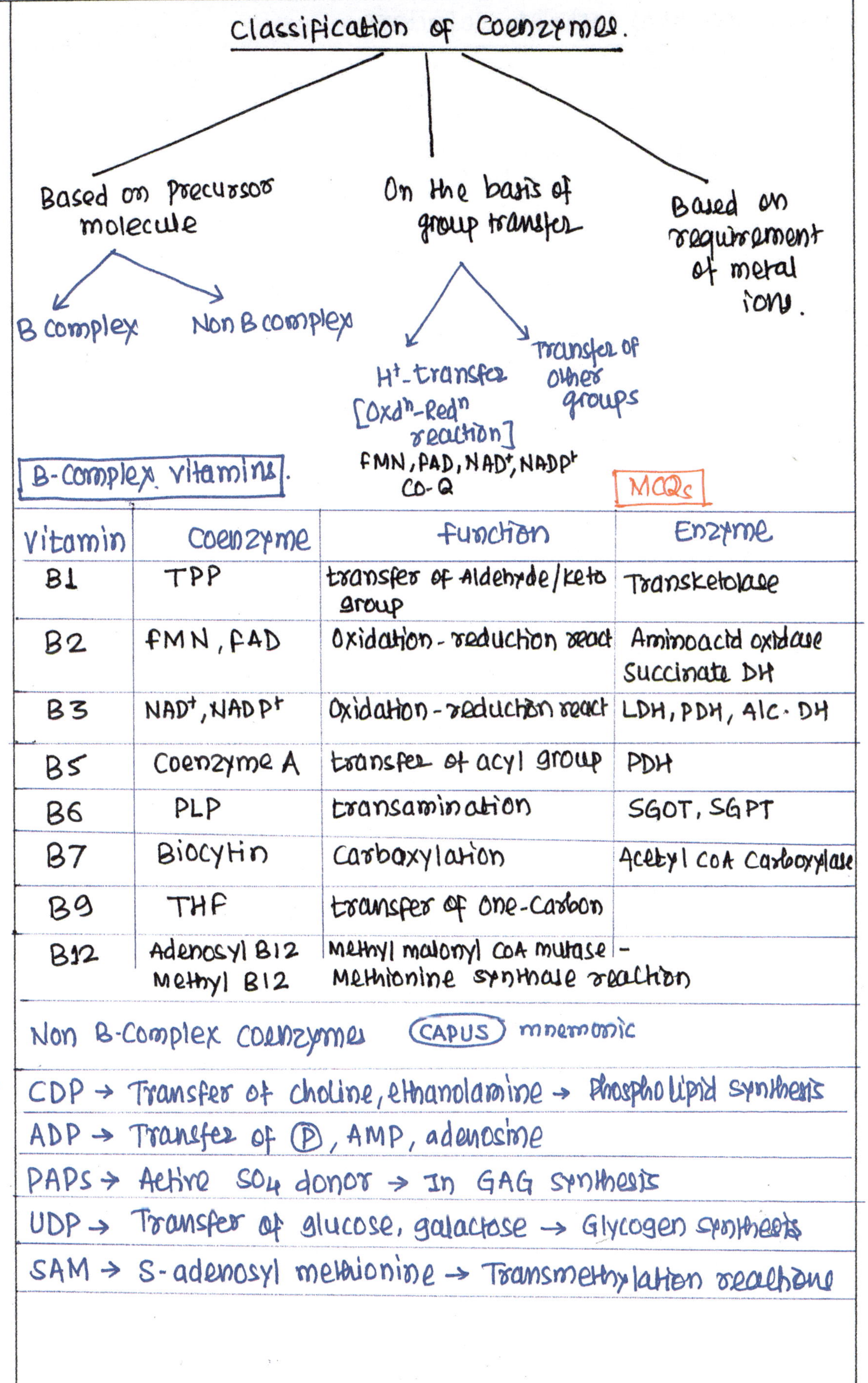

B-Complex vitamins.

MCQs

Vitamin	Coenzyme	function	Enzyme
B1	TPP	transfer of Aldehyde/keto group	Transketolase
B2	FMN, FAD	Oxidation - reduction react	Aminoacid oxidase Succinate DH
B3	NAD^+, $NADP^+$	Oxidation - reduction react	LDH, PDH, Alc. DH
B5	Coenzyme A	transfer of acyl group	PDH
B6	PLP	transamination	SGOT, SGPT
B7	Biocytin	Carboxylation	Acetyl CoA Carboxylase
B9	THF	transfer of one-Carbon	
B12	Adenosyl B12 Methyl B12	Methyl malonyl CoA mutase - Methionine synthase reaction	

Non B-Complex coenzymes (CAPUS) mnemonic

CDP → Transfer of choline, ethanolamine → Phospholipid synthesis

ADP → Transfer of Ⓟ, AMP, adenosine

PAPS → Active SO_4 donor → In GAG synthesis

UDP → Transfer of glucose, galactose → Glycogen synthesis

SAM → S-adenosyl methionine → Transmethylation reactions

Metal ions that act as Cofactors

MCQs*

Metal	Enzymes.
Zn	Carbonic anhydrase, Superoxide dismutase, LDH Alc DH, ALP, RNA Polymerase
Cu	Cytochrome oxidase, Tyrosinase, Lysyl oxidase Superoxide dismutase
Se	Glutathione peroxidase, Deiodinase, Thioredoxin Reductase
Mo	Xanthine oxidase
Mg	Kinase, Enolase, Glu 6 ⓟase, HK, PFK
Mn	mt. SOD, Arginase, Kinase, Enolase

ACTIVE SITE

Active site is the region of an enzyme where substrate bind and undergo a chemical reaction

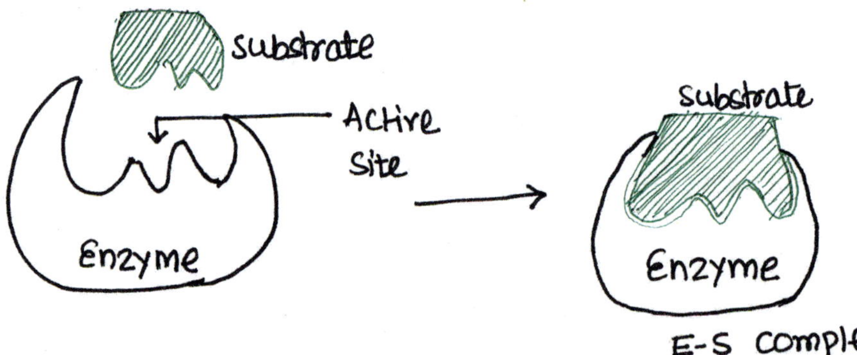

3D cleft or crevice, formed by groups from different parts of amino acid sequence.

Generally, possesses a substrate binding site and catalytic site. Substrate bind at the active site by weak noncovalent bonds.

Amino acids forming the active site are away from each other in a linear sequence.

MC aminoacid at the active site is Serine

AA forming Active site

Other Amino acids

FACTORS AFFECTING ENZYME ACTIVITY

LAQ: Describe the important factors affecting enzyme activity.

1. Enzyme concentration

If substrate is available, rate of reaction is directly proportional to enzyme concentration.

Velocity $\propto$ [E]

This property is used to estimate concentration of various enzyme levels in serum.

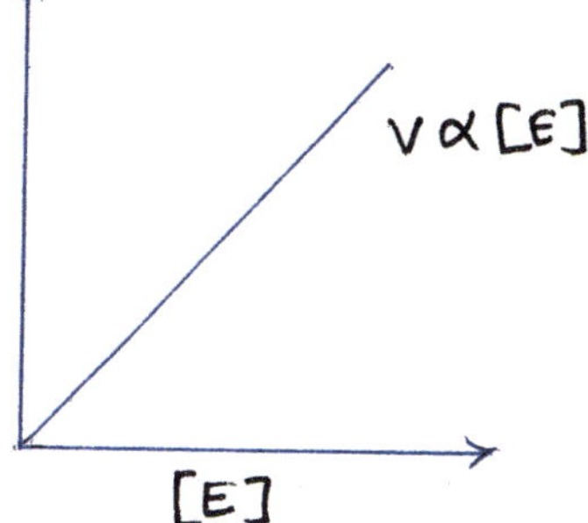

2. Substrate concentration

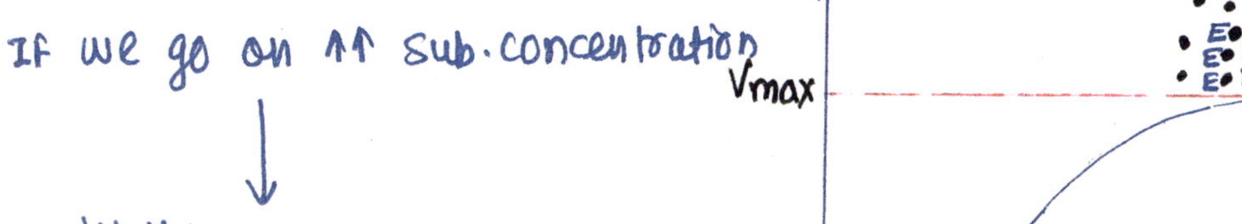

If we go on ↑↑ sub. concentration

↓

Initially,

velocity $\propto$ [S]

first order kinetic

↓

Enzyme exhausted

& plateau in graph – zero order kinetic

- Rate of reaction is directly proportional to the ~~enzyme~~ substrate concentration.
- When all enzymes are occupied by the substrate, rate of reaction is constant.

Michaelis Menten constant Km

Also known as Brigs & Haldane constant

Definition:- It is a substrate concentration to produce half of the maximum velocity

Significance:- It denotes the strength of association betn enzyme & substrate. [Affinity of enzymes towards substrate]

More Km → Less affinity

Less Km → More affinity

Km value is constant for particular enzyme. It is a signature for enzyme-substrate pair.

Michaelis Menten equation.

$$V_0 = \frac{V_{max}[S]}{K_m + [S]}$$

V_0 : Initial velocity
V_{max} : Maximum velocity
[S] : Substrate concentration
K_m :- Michaelis Menten constant

Hexokinase (HK)	Glucokinase
High Affinity	Low affinity
Low Km	High Km
Can act at glucose 10 mg/dl	Act when Glu >100 mg/dl
Act on All Hexoses	In well fed State
⊖ by Glu 6 Ⓟ	Only on Glucose
Not inducible	Inducible
	HK IV

Line weaver Burk plot / Double reciprocal plot

$$\frac{1}{V} = \frac{K_m}{V_{max}} \times \frac{1}{[S]} + \frac{1}{V_{max}}$$

Significance :- It is not possible to achieve the higher substrate concentration every time, so we can calculate Km & Vmax at initial velocity by using double reciprocal plot.

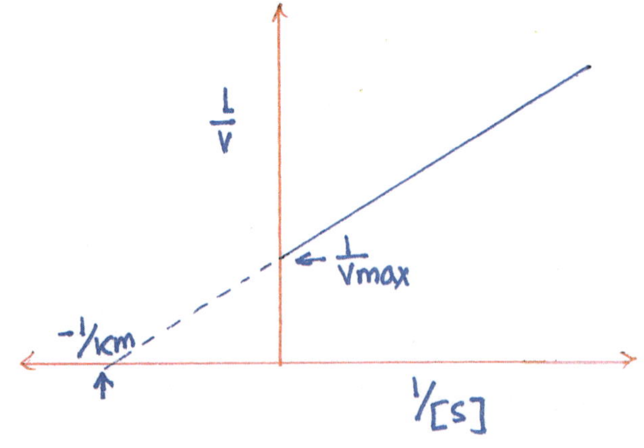

It is useful in calculating the Vmax & Km value of enzyme catalysed reaction with and without Inhibitor (Drugs).

3. Product concentration

Increase in product concentration will decreases the velocity of the reaction

S + E ⟶ ES ⟶ (P)

4. Temperature

V ∝ Temp

Optimum temp :- temp at which there is maximum activity of an enzyme.

For most of the enzymes in the body

- 37°C → Optimum temp
- 50°C → Denaturation occurs
- 70°C → Inactivation of enzymes.
  [exception- Muscle adenylate kinase]

37°C

V

0 10 20 30 40 50 60 70

Temp °C

- Bell shaped curved
- Mechanism :- Collision theory

↑↑ Temp ⟶ ↑↑ collision betn Enz & substrate ⟶ ES complex formation

- Q_{10} :- Increase in enzyme activity by increasing temp by 10°C

5. PH

- pH decide the charge on AA molecules at the active site of an enzyme.
- change in pH causes conformational change in active site that affect the substrate binding

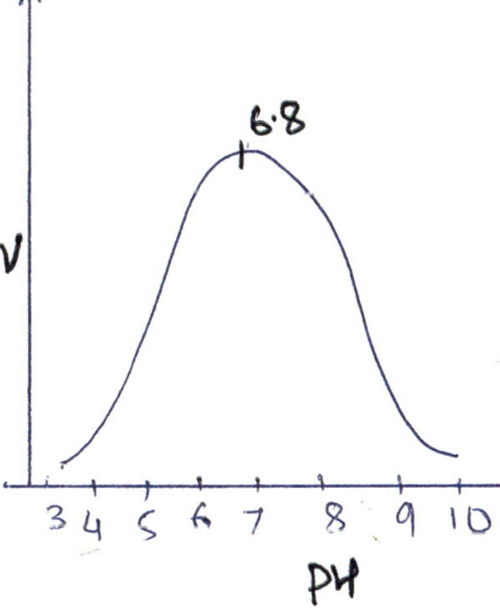

Optimum pH : 6.8

Exception : Pepsin pH 1-2
Acid phosphatase : 3-5
Alk. phosphatase : 9-10

6. Effect of Activators [Metal ions]

Some metal ions increases the reaction velocity

Examples:- Cl^- → Amylase

Mg^{2+}, Ca^{2+} → ATPase

Cu & Fe → Cytochrome oxidase

Zn → SOD, Carbonic anhydrase

7. Effect of Inhibitors

Inhibitors decreases the velocity of the reaction. This property is useful in the treatment of various disorders.

- Inhibition
  - Reversible
    - Competitive
    - Noncompetitive
    - Uncompetitive
  - Irreversible

ENZYME INHIBITION

Reversible	Irreversible
1. competitive	suicide
2. Noncompetitive	feedback
3. Uncompetitive	

Inhibitor :- substance or compound which decreases enzyme velocity.

Reversible Inhibition	Irreversible Inhibition
① Inhibitor binds to the enzyme by non covalent interactions. (weak forces)	① Inhibitor binds to the enzyme by covalent interactions.
② can be reversed by – removal of inhibitor – ↑↑ substrate concentration	② cannot be reversed
③ Examples - Therapeutic drugs	③ Inhibitor binds with an enzyme, causes destruction of active site. Example :- poisonous gases.

Competitive Inhibition

- Inhibitor is a structural analogue of substrate
- Inhibitor competes with the substrate to bind with active site of an enzyme.
- Reversible by increasing the concentration of substrate
- Km increases, Vmax remains unchanged * mcq

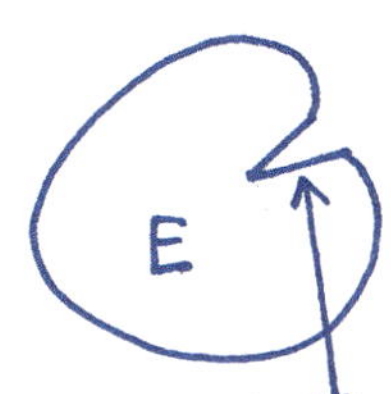

Substrate

Inhibitor

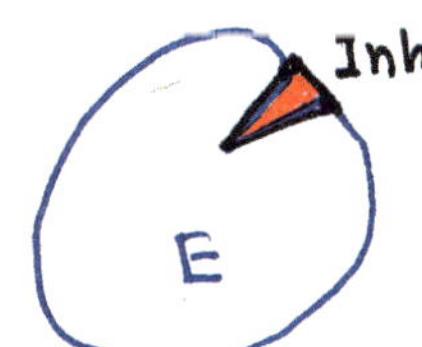

Competitive Inhibition

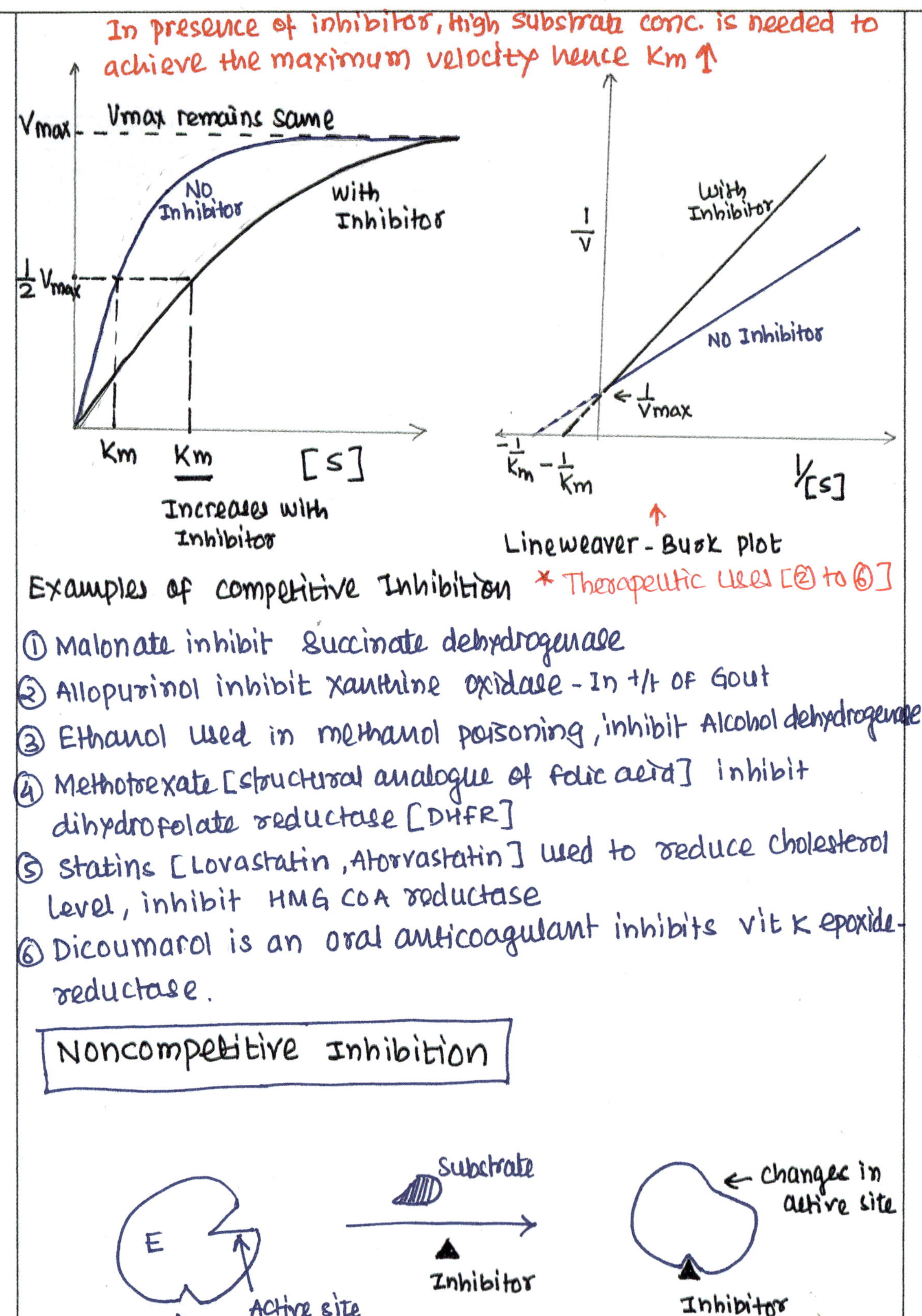

In presence of inhibitor, High substrate conc. is needed to achieve the maximum velocity hence Km ↑

↑ Lineweaver - Burk plot

Examples of competitive Inhibition * Therapeutic Uses [② to ⑥]

① Malonate inhibit succinate dehydrogenase
② Allopurinol inhibit xanthine oxidase - In t/t of Gout
③ Ethanol used in methanol poisoning, inhibit Alcohol dehydrogenase.
④ Methotrexate [structural analogue of folic acid] inhibit dihydrofolate reductase [DHFR]
⑤ Statins [Lovastatin, Atorvastatin] used to reduce cholesterol level, inhibit HMG COA reductase
⑥ Dicoumarol is an oral anticoagulant inhibits vit K epoxide-reductase.

Noncompetitive Inhibition

Noncompetitive Inhibition

① No structural similarity between substrate & inhibitor
② No competition for binding with active site
③ Bind to other site and causes conformational changes in the active site of an enzyme
④ Enzyme-inhibitor complex or Enzyme-substrate-Inhibitor complex is formed
⑤ Mostly irreversible, can't be overcome by ↑↑ [S]
⑥ Km - unchanged, Vmax → lowered.

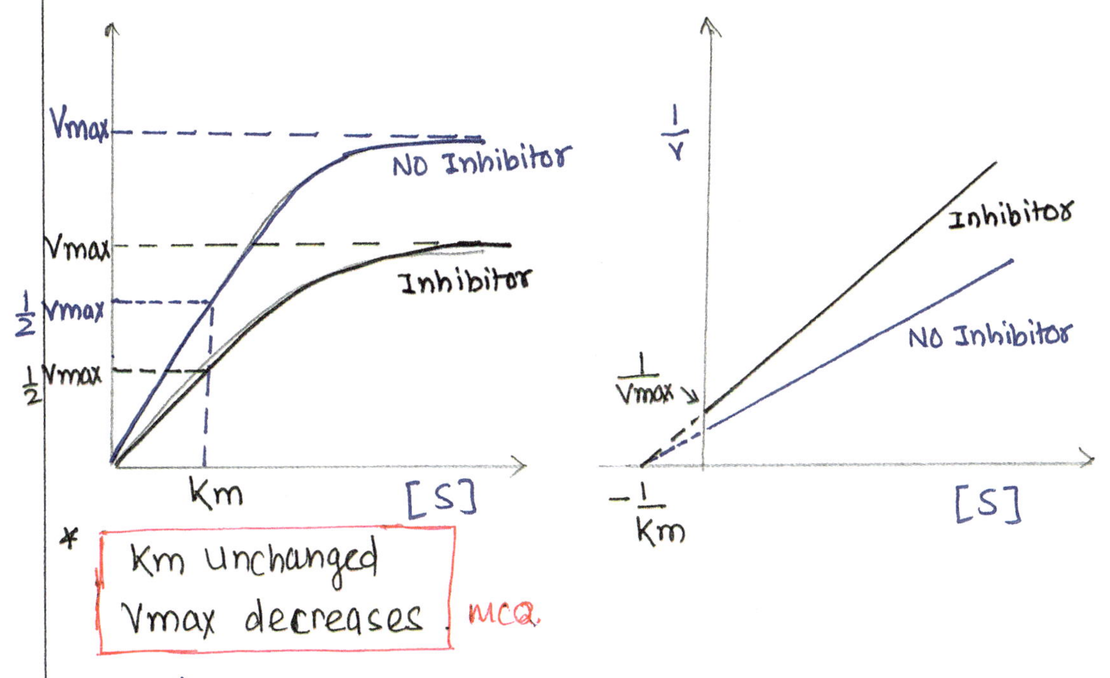

* Km unchanged
Vmax decreases. MCQ.

Examples

① Fluoride inhibit enolase

Fluoride bulb is used to collect blood sample for plasma glucose estimation. Fluoride ions inhibit enolase enzyme, hence glycolysis. Glucose level in the plasma remains constant. If, Not collected in fluoride bulb, Glucose level will decrease by 5-10 mg/dl/hr due to continuous glycolysis.

② Complex IV of ETC inhibited by cyanide, CO, H_2S
③ Arsenite inhibit α ketoglutarate DH
④ Pb, Ag, Hg inhibit serine proteases.
⑤ Iodoacetate inhibit Glyceraldehyde 3 Ⓟ DH

Uncompetitive Inhibition

Inhibitor binds with E-S complex to form E-S-I complex

$I + ES \longrightarrow ESI$

Km : decreases

Vmax : decreases

Example : ① placental ALP inhibited by phenylalanine
② Mycophenolate ⊖ IMP DH in purine synthesis used in cancer therapy.

Irreversible Inhibition

covalent binding of inhibitor to enzyme molecule

causes distortion of active site & inactivates the enzyme

can not be reversed

Examples

① DFP nerve gas → Inhibit serine proteases & acetyl-cholinesterase
[Diisopropyl Fluorophosphate]

② Disulfiram → Inhibit Acetaldehyde DH

ethanol ⟶ Acetaldehyde —X→ (Acetaldehyde DH) Acetic acid

↑ Accumulation of acetaldehyde

causes - vomiting, diarrhea, Abd. Pain

Disulfiram is an antabuse used in t/t of Alcohol addiction.

Suicide Inhibition OR Mechanism based Inhibition

- Inhibitor itself is converted into more potent form by the enzyme that is inhibited

Allopurinol ⟶ (Xanthine oxidase) Alloxanthine

HX →(XO) X →(XO) Uric acid

competitive inhibitor

Allopurinol is used in t/t of Gout

Aspirin inhibit cyclooxygenase - Analgesic, anti platelet drug

Difluoro methyl ornithine (DFMO) inhibits ornithine decarboxylase.

- Used in trypanosomiasis (sleeping sickness)

Feedback Inhibition

The final product of the pathway inhibit the 1st rate limiting step of the reaction.

Example:

Succinyl CoA + Glycine

↓ ALA synthase ← ⊖ feedback inhibition (from Heme)

δ ALA

↓

↓

↓

Heme

Feedback inhibition at genetic level, is called as Feedback regulation.

Example :- Dietary cholesterol decreases cholesterol synthesis.

BAQ :- Enumerate four important drugs that act as a competitive inhibitors.

SAQ :- Competitive inhibitors c̄ therapeutic uses.

LAQ :- What is enzyme inhibition? Explain it with suitable examples.

SAQ :- Differentiate between competitive and noncompetitive inhibition.

ENZYME PATTERN IN LIVER DISEASES

SAQ | Clinical Case

① Alanine transaminase (ALT) OR
Serum Glutamate Pyruvate Transaminase (SGPT)

EC NO : 2.6.1.2
Transferase – transaminase with PLP as coenzyme

Alanine + αKG ⇌ (SGPT/ALT, PLP) Pyruvate + Glutamate

Plasma nonspecific, intracellular enzyme
↓ damage, Proliferation, necrosis, neoplasm of hepatocytes
SGPT leaksout in plasma
Conc. of SGPT ↑ } Viral Hepatitis, Cirrhosis of liver, liver cancer, Alc. liver ds

More specific for liver diseases
Normal range :- 5 to 35 IU/L

② Aspartate transaminase (AST) OR
Serum Glutamate Oxaloacetate Transaminase (SGOT)
EC NO : 2.6.1.1
Transferase – transaminase with PLP as coenzyme

Aspartate + α KG ⇌ (SGOT/AST, PLP) Oxaloacetate + Glutamate

Distributed in cytosol + mitochondria of liver & Heart

Normal range : 5 to 40 IU/L

Concentration of SGOT increases in :- Heart diseases. (Ac. Myocardial Infarction)
Liver diseases
- Viral hepatitis
- Alc. liver disease
- Cirrhosis of liver
- liver cancer

In Alcoholic liver disease

↓

Acetaldehyde

↓

Damages mitochondria

↓

AST ↑↑ as compared to ALT

Normal AST : ALT ratio (De Ritis ratio) is <1

AST : ALT ratio > 1 indicates Alc liver ds.

> 2 strongly suggestive of Alc Liver ds.

③ Alkaline phosphatase (ALP)

Important enzyme to differentiate between hepatic disease and obstructive pathology

It is an ectoenzyme - Present on surface of biliary epithelium

6 Isoenzymes → Biliary (α_1)
- Placental (α_2HS)
- Liver (α_2HL)
- Bone (pre-β)
- Intestine (γ)
- Leukocyte ALP

In Hepatic diseases → ALP ↑ 2-3 times

In obstructive diseases → 10-12 times

obstruction to bile flow ⟶ Irritates bile duct epithelium

↓

release of ALP from bile epithelium

↓

↑ ALP Level in serum

ds - Gallstone
- cancer of head of pancreas

In Bone diseases → upto 25 or more time increase
- Pagets ds
- Osteomalacia
- Rickets
- Metastasis in bone

④ 5' Nucleotidase :- Normal in hepatitis
↑↑ in Obstructive diseases

⑤ γ-Glutamyl transferase :- ↑↑ in Alcoholic liver disease
obstructive liver disease

CARDIAC MARKERS

SAQ / Clinical case

SAQ:- Enumerate different markers used for the diagnosis of MI. Write about their course in MI

Clinical case

A 40 years old man, chronic smoker, had sudden onset of chest pain and perspiration. ECG shows ST elevation and T wave inversion. Serum CK-MB levels were increased.

What is your probable diagnosis → Acute Myocardial infarction

Other enzymatic markers → LDH1, SGOT

Other markers → Serum Troponin I & T

Earliest enzymatic marker to rise → CK-MB *MCQ

Most specific → cardiac troponins

Cardiac Marker: Clinical laboratory tests which are used to detect a cardiac dysfunction.

Ideal marker
1. should able to detect the disease early.
2. specific & sensitive
3. should tell about prognosis
4. should help in monitoring

Cardiac markers are mainly used for diagnosis of
- Acute Myocardial Infarction (MI) → coronary vasospasm, Atherosclerosis → ↓↓ Blood supply to Heart → Distal to block → Damage, Necrosis, Ischaemia of Heart muscle → Chest Pain, ECG changes
- Congestive cardiac failure → Inability of Heart to pump out the blood

Indications → chest pain, perspiration
ECG changes
History of Heart disease.

Important Marker's in Myocardial infarction

Marker	Release	Peak	Return	Remark
Myoglobin	2 hrs	4-6 hrs	20-24 hrs	Earliest, not specific
CT-I	4 hrs	12-24 hrs	3-5 days	Earliest, specific
CT-T	6 hrs	72 hrs	10-14 days	specific
CK-MB	6-18 hrs	24-30 hrs	2-3 days	specific & enzyme
LDH 1	2 days	3-4 days	10-15 days	Late, specific
AST/SGOT		30-48 hrs	4-6 days	Not specific

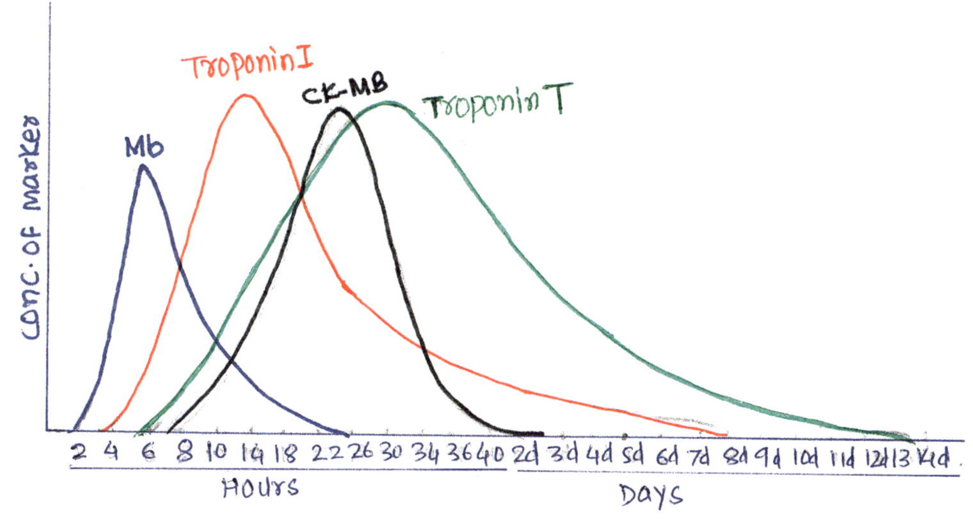

Creatine Kinase (CK)

There are 3 isoenzymes of CK, made up of two types of subunits M & B.

CK 1 → BB → Brain
CK 2 → MB → Heart → earliest enzyme to rise in MI
CK 3 → MM → Muscle

Cardiac Troponins

- For bed side testing / point of care testing / immediate result.
- Specific markers for MI, Non enzymatic markers.
- Troponin I & Troponin T
- Rises in 4-6 hrs, reaches peak in 24-72 hrs and return to normal in 3 to 10 days.

New cardiac Biomarkers

1. Ischaemia Modified Albumin (IMA)
2. Glycogen Phosphorylase BB
3. PAPP-A [Pregnancy Associated Plasma protein A] → Assessment of future MI
4. Brain Natriuretic peptide (BNP) → Assessment of ventricular failure
5. hs-CRP (Highly selective-C-reactive protein)
6. Myeloperoxidase (MPO)
7. Cholinesterase

Clinical case

A 40 year old man was brought to the hospital with chief complaints of chest pain and perspiration. He was chronic cigarette smoker. On investigation ECG [Electrocardiogram] showed ST elevation & T wave inversion. Troponin I & Troponin T levels were high on point of care testing.

What is your probable Diagnosis?

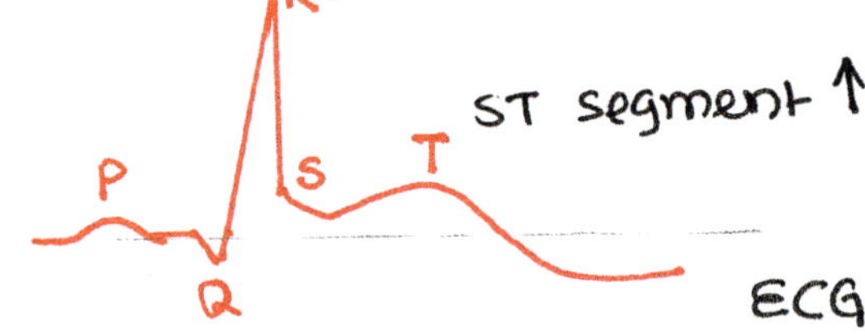

Troponin I & T → Elevated

Diagnosis → Acute Myocardial Infarction

ISOENZYMES

SAQ: What are the Isoenzymes? Give isoenzymes of LDH and ALP.

Definition:- Different forms of same enzyme catalysing the same reaction are called as Isoenzymes/ Isozymes.

Example:-

$$\text{Lactate} \underset{\text{LDH1, LDH2, LDH3, LDH4, LDH5}}{\overset{NAD^+ \rightarrow NADH+H^+}{\rightleftharpoons}} \text{Pyruvate}$$

LDH1, LDH2, LDH3, LDH4, LDH5 } All isoenzymes catalyses same reaction

How they are formed?

Mnemonic - True Lady Is Allowed

True Isoenzymes - Originate from different genes, may be present on different chromosomes.

Example:- Amylase < Pancreatic, Salivary

Lactate dehydrogenase (LDH) - all Isoenzymes (LDH 1,2,3,4,5) are present in same individual & all individuals of population.

Isoforms - Formed due to post-translational modification of protein. Example - ALP

Alloenzymes - Originate from different alleles of same locus of the gene.

Example - Glucose 6 (P) dehydrogenase - 400 isoenzymes only one variant is present in one individual.

Characteristics [How they differ]

1. Electrophoretic mobility :- $LDH_1 > LDH_2 > LDH_3 > LDH_4 > LDH_5$
   CK BB > CK-MB > CK-MM
2. Sensitivity to Heat : ALP α 2HL (Liver) - Sensitive to Heat
   ALP α 2HS (placenta) - resistant to Heat.
3. Susceptability to Inhibitors :- Tartarate labile acid phosphatase
4. Requirement of cofactors :- Cytosolic Isocitrate DH → $NADP^+$
   mitochondrial Isocitrate DH → NAD^+
5. Antibodies produced - Different antibodies are produced in response to isoenzymes

⑥ Tissue of origin: ALP – Bone, placenta, liver, Intestine, Biliary tract

⑦ Km value & affinity: Glucokinase & Hexokinase

Diagnostic uses of Isoenzymes

① LDH [Lactate dehydrogenase]

Tetramer composed of H & M units, 5 isoenzymes.

Isoenzyme	Subunit	Tissue of origin	Serum concentration
LDH 1	H_4	Heart	17-27%
LDH 2	H_3M	RBC, Brain	27-37%
LDH 3	H_2M_2	Brain, Leukocytes	18-25%
LDH 4	HM_3	Liver	3-8%
LDH 5	M_4	skeletal muscle	2-5%

Electrophoresis LDH 1 > LDH 2 > LDH 3 > LDH 4 > LDH 5

Flipped pattern :- Normally LDH 2 > LDH 1

In MI concentration of LDH 1 increases.

It is specific and late marker in MI

② Creatine kinase (CK)

Dimer, composed of M & B units.

Creatine ⇌ phosphocreatine (ATP → ADP; CK 1, CK 2, CK 3)

Isoenzyme	Subunit	Tissue	Diseases
CK 1	BB	Brain	Brain diseases.
CK 2	MB	Heart	Myocardial infarction
CK 3	MM	Muscle	Muscular dystrophies, Myopathies

CK-MB is earliest enzymatic marker to rise in MI Started rising in 6 hrs, reaches peak in 24 hrs and return to normal in 2-3 days.

Electrophoretic mobility CK 1 > CK 2 > CK 3

Normal level :- 5 to 25 IU/L

OR 3 to 5% of total CK activity.

③ Alkaline phosphatase (ALP)

- Important enzyme to differentiate between hepatic diseases and obstructive pathology
- It is an ectoenzyme, present on cell surface

Isoenzyme	Tissue of origin	Diseases
α_1 ALP	Biliary canaliculi [10%]	Obstructive jaundice
α_2 HL-ALP	Hepatic cells [25%]	Hepatitis
H_2 HS-ALP	Placenta	
pre β-ALP	Bone - osteoblast [50%]	Pagets ds, Rickets, osteomalacia, tumor
γ-ALP	Intestine	
Leukocyte ALP	By Leukocytes	Lymphoma

Regan Isoenzyme :- Carcinoplacental varient of ALP - found in carcinoma of lung, liver and G.I.T.

Kasahara Isoenzyme : Fetal intestinal varient - Inappropriate expression in malignancy.

Nagao Isoenzyme : Germ cell varient - Inappropriate expression in malignancy.

ALP in Obstructive Jaundice

It is an ectoenzyme

↓

present on surface of biliary epithelium

↓

obstruction to flow of bile [common bile duct stone, ca. of head of pancreas]

↓

Irritation of biliary epithelium & release of ALP into the blood.

↓

10-12 times of the normal

In Hepatic diseases (Viral Hepatitis) → 2 to 3 times of Ⓝ ALP

In obstructive Jaundice (Gall stone) → 10-12 times of Ⓝ

In Bone disorders & tumor of Bones → 25 times of Ⓝ level

USES OF ENZYMES

1. Diagnostic uses
2. Therapeutic uses
3. Laboratory uses
4. Industrial uses

Important LAQ :- Describe the uses of enzymes in diagnosis of various diseases.

Enzymes

Plasma specific
- Functional in plasma
- concentration is more in plasma

example - clotting factors, Lipoprotein lipase

Plasma non specific
- present in cells
- Nonfunctional in plasma
- Low concentration in plasma

example - SGOT, SGPT

Whenever Proliferation, Necrosis, Damage, Neoplasm of cell
↓
concentration increases in cell and leak out in plasma
↓
↑↑ concentration in plasma
↓
used in diagnosis of diseases.

1. Liver diseases

Hemolytic Jaundice	Hepatic Jaundice	Obstructive Jaundice
Normal SGOT	SGPT ↑↑↑	ALP ↑↑↑
SGPT	SGOT ↑↑↑	GGT ↑↑↑
ALP	ALP ↑	5'NT ↑↑↑
GGT	GGT ↑	SGPT ↑
5'NT	5'NT ↑	SGOT ↑

Enzyme	Normal level	
SGPT (ALT)	5-35 IU/L	More specific for liver ds.
SGOT (AST)	5-40 IU/L	Increases in Liver & Heart ds.
ALP	40-125 IU/L	Marker for obstructive liver ds.
AST : ALT ratio	< 1	Normal
	> 2	Alcoholic liver ds.

MCQ
GGT γ-Glutamyl transferase – Marker for Alcoholic liver ds. and obstructive liver ds.

② Cardiac markers

CK-MB - earliest enzymatic marker in MI
Also important in diagnosis of reinfact

start rising in 4-6 Hrs
peak 24 Hrs
return to normal in 2-3 days.

LDH 1 → Specific & Late marker
SGOT → Nonspecific, Late marker
Glycogen phosphorylase
Cholinesterase
Nonenzymatic cardiac Markers – Troponin I & T
Myoglobin
HS-CRP
Ischaemia modified Albumin
PAPP-A

③ Pancreatic diseases.

Serum Amylase : ↑ in Acute and chronic pancreatitis
also in salivary gland disorders.

MCQ *
Serum lipase :- specific for diagnosis of Acute pancreatitis.

④ Bone diseases

ALP - Pre β isoenzyme → Increases in Rickets, osteomalacia Pagets ds, Bone tumors.

⑤ Prostate cancer

Prostate specific antigen (PSA) → Increases MCQ
Acid phosphatase (ACP) → Increases in prostate cancer

⑥ Muscle disorders

CK-MM → Muscular dystrophies, myopathies.
LDH 4,5 → Muscle disorders.
Aldolase →

Decrease in enzyme level

Wilsons disease → ↓↓ Ceruloplasmin
G6PD deficiency → ↓↓ Glucose 6 Phosphate dehydrogenase
Severe liver ds → ↓↓ Amylase

Que:- Enumerate any four therapeutic uses of enzymes.

Mnemonic - ASPARginase.

① Asparginase - In Acute Leukemias.
② Streptokinase / Urokinase - thrombolytic ← Ac. MI, Deep venous thrombosis, Cerebrovascular episodes.
③ Papain - Antiinflammatory
④ α_1 Antitrypsin - Emphysema
⑤ Recombinant tPA (Tissue plasminogen activator) - Ac. MI
⑥ Pancreatin - Digestive disorders

Que:- Enumerate four important laboratory uses of enzymes.

used in estimation of

① Glucose → Hexokinase, GOD-POD
② Urea → Urease
③ Uric acid → Uricase
④ Cholesterol → Cholesterol oxidase - Peroxidase
⑤ Triacylglycerol → lipase

Also in ELISA :- Horseradish peroxidase [HRP]
Alkaline phosphatase
Glucose oxidase

r DNA technology :- Taq polymerase
Restriction endonuclease
DNA ligase

Regulation of enzyme activity

LAQ: Describe the mechanisms involved in the regulation of enzyme activity

Two major mechanisms

Regulation of intrinsic catalytic efficiency → Regulation of change in activity of existing enzyme

1. Allosteric regulation
2. Covalent modification
3. Proteolysis
4. Compartmentalization
5. Cascade mechanism
6. Protein-protein interaction

Quantity → Regulation of change in enzyme concentration

1. Induction
2. Repression
3. Degradation

SAQ: Define allosteric regulation and give some examples.

1. Allosteric regulation [SAQ]

Allo – other] Small regulatory molecule binds to the
Steric – site] allosteric site of an enzyme

Enzymes having site other than active site of an enzyme are known as allosteric enzymes

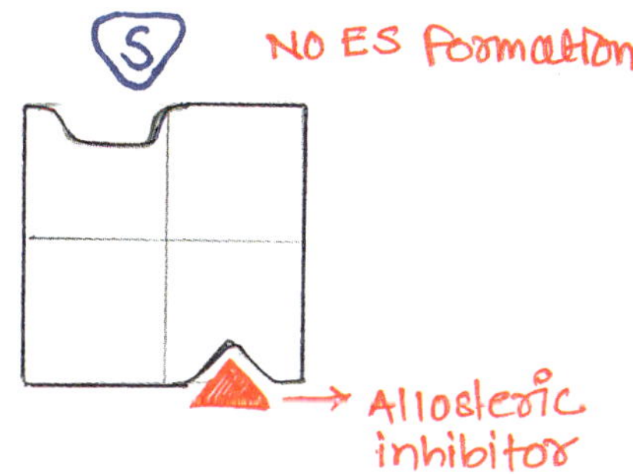

Tense form of an Enzyme

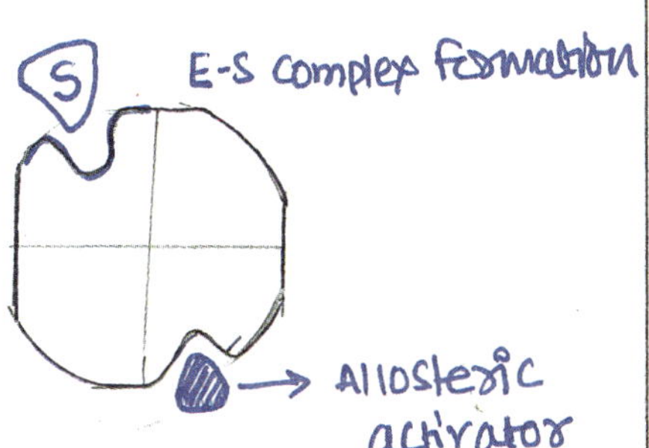

Relaxed form of an Enzyme

Allosteric inhibitor / Negative modulator] favors the formation of tense state, decreases the enzyme activity, Increases the Km value.

Allosteric activator / Positive modulator] favors the formation of relaxed state, Increases the enzyme activity, Decreases the Km value

Positive modulator shift the saturation curve to the left, towards the lower substrate concentration hence decreases Km value while negative modulator increases the Km value.

Examples

Allosteric enzyme	Allosteric activator	Allosteric Inhibitor
Acetyl CoA carboxylase [Fatty acid synthesis]	citrate	palmitoyl CoA
Aspartate transcarbamoylase [pyrimidine synthesis]	ATP	CTP
Phosphofructokinase [Glycolysis]	AMP	citrate

2. Covalent Modification

SAQ

Rapid in action and short term regulation

Phosphorylation

Addition of Ⓟ
Donor is ATP
Enzyme- Protein kinases

Dephosphorylation

Removal of Ⓟ
Enzyme- Phosphatase

Enz-Ser-OH —(ATP → AMP, Kinase)→ Enz-Ser-O-Ⓟ

Some enzymes are active in phosphorylated state
example :- Glycogen phosphorylase, Phosphorylase b kinase, Citrate lyase, HMG CoA Reductase kinase, Gluconeogenesis

MCQ * Some enzymes are active in dephosphorylated state
example :- Glycogen synthase, Acetyl CoA carboxylase, HMG CoA reductase, Pyruvate dehydrogenase, Glycolytic enzymes.

3. Proteolysis

Inactive precursor of enzymes are converted into active form by cleavage of peptide.

Proenzymes / Inactive zymogens

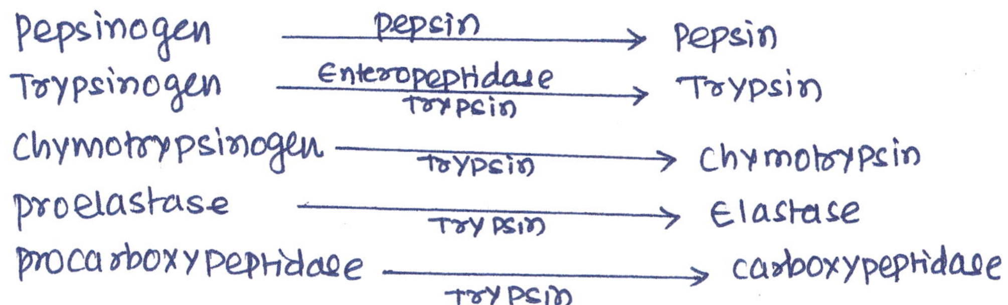

4. Compartmentalization

Controlling the translocation of enzymes to the subcellular compartment

Glycine + Succinyl CoA —ALA synthase (mitochondria)→ δ ALA →→→→ Heme

Heme inhibit the transport of ALA synthase to mitochondria.

5. Cascade Mechanism

Amplification of small signal into Larger response.

Example :- ① Glycogen metabolism regulation through Hormones
② Activation of clotting factors

6. Protein-Protein interaction

Regulation occurs through a small protein molecule

Example :- Ca^{2+} - calmodulin complex

↓

Activates various kinases

7. INDUCTION

by regulation of gene expression

Increase in enzyme synthesis in response to specific inducer occurs at genetic level.

① Glucagon induces pyruvate carboxylase [Gluconeogenesis]

Glucagon
↓
G-protein coupled receptor
↓
GTP
↓
ATP —(Adenyl cyclase)→ cAMP
↓
Active protein kinase A
↓
transcription factors
↓
DNA
↓
↑ mRNA
↓
↑ protein [pyruvate carboxylase]

② Insulin induces Glucokinase [Glycolysis]

8. REPRESSION

by regulation of gene expression

Decreases enzyme synthesis by acting at genetic level

① Pyruvate kinase [glycolysis] is repressed by glucagon

② Pyruvate carboxylase [Gluconeogenesis] is repressed by insulin

Glucagon → G-protein coupled receptors → GTP
↓
↑ cAMP
↓
Activate protein kinase A
↓
transcription factor → DNA → ↓ mRNA → ↓ Pyruvate kinase

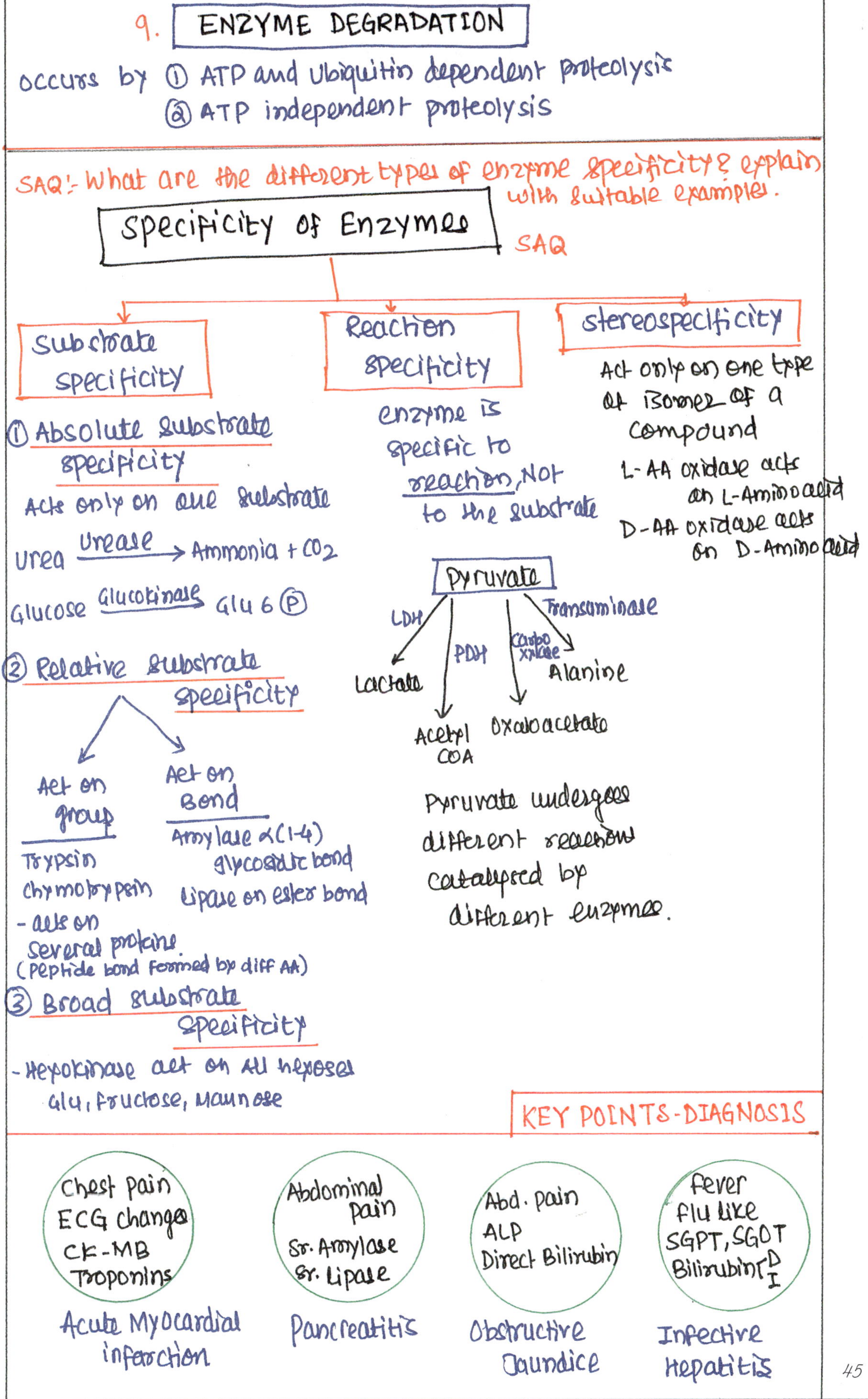

9. ENZYME DEGRADATION

occurs by ① ATP and Ubiquitin dependent proteolysis
② ATP independent proteolysis

SAQ:- What are the different types of enzyme specificity & explain with suitable examples.

Specificity of Enzymes

SAQ

Substrate specificity

① Absolute substrate specificity

Acts only on one substrate

Urea $\xrightarrow{\text{Urease}}$ Ammonia + CO_2

Glucose $\xrightarrow{\text{Glucokinase}}$ Glu 6 (P)

② Relative substrate specificity

- Act on group
  - Trypsin
  - Chymotrypsin
  - acts on several proteins (Peptide bond formed by diff AA)
- Act on Bond
  - Amylase α(1-4) glycosidic bond
  - Lipase on ester bond

③ Broad substrate specificity

- Hexokinase act on All hexoses Glu, fructose, Mannose

Reaction specificity

enzyme is specific to reaction, Not to the substrate

Pyruvate:
- LDH → Lactate
- PDH → Acetyl CoA
- Carboxylase → Oxaloacetate
- Transaminase → Alanine

Pyruvate undergoes different reactions catalysed by different enzymes.

Stereospecificity

Act only on one type of isomer of a compound

L-AA oxidase acts on L-Amino acid

D-AA oxidase acts on D-Amino acid

KEY POINTS-DIAGNOSIS

Findings	Diagnosis
Chest pain, ECG changes, CK-MB, Troponins	Acute Myocardial infarction
Abdominal pain, Sr. Amylase, Sr. Lipase	Pancreatitis
Abd. pain, ALP, Direct Bilirubin	Obstructive Jaundice
Fever, Flu like, SGPT, SGOT, Bilirubin (D, I)	Infective Hepatitis

Clinical cases

1. A 45 year old man, chronic smoker, had sudden onset of chest pain and perspiration. ECG examination shows ST- elevation and T wave inversion. Serum CK- MB levels were raised.

What is your probable diagnosis?
What are the other markers for above mentioned condition?

Δ Acute Myocardial infarction

Other markers – Troponin I & T, LDH-1, Mb, SGOT etc.

2. A 56 year old male patient was brought to emergency department with chief complaints of nausea, vomiting and pain in epigastric & rt. upper quadrant of abdomen. pain was severe, disturbing and relieved by crouching. Serum amylase level was - 790 IU/L.

What is your probable diagnosis?

Δ Acute pancreatitis

Enzyme specific to pancreatitis is serum lipase > Amylase *MCQ
More specific

3. An old man came to Hospital with complaints of abdominal pain, pruritus and yellow colored urine.
Lab investigation: SGOT- 72 IU/L, SGPT- 89 IU/L, ALP- 926 IU/L
Direct Bilirubin - 16 mg/dl, Indirect- 1.2 mg/dl

What is your probable diagnosis?
Enumerate the causes of above condition.

Δ obstructive Jaundice

Causes :- Common bile duct stones
Gallstone
Cholangiocarcinoma
Cancer of head of pancreas
stricture of CBD

4. A 19 yrs old boy having flu like symptoms came to OPD with nausea, vomiting and yellow discoloration of skin & sclera.
Lab investigation shows - SGOT- 923 IU/L, SGPT- 897 IU/L
ALP- 190 IU/L, Direct bilirubin → 9.8 mg/dl
Indirect bilirubin → 8.7 mg/dl

What is your probable diagnosis?

Δ Hepatic Jaundice / Infective Hepatitis

CHEMISTRY OF CARBOHYDRATES

MCI COMPETENCY BI 3.1

SAQ

1. Define and classify carbohydrates with suitable examples.
2. Biologically important Disaccharides.
3. Homopolysaccharides
4. Heteropolysaccharides and GAG's (Most Imp)
5. Dietary fibres and it's role in diet.
6. Glycemic index
7. Functions of Dextran
8. Isomers of Monosaccharides.

Important clinical cases.

Mucopolysaccharidosis — ① Hurler syndrome
② Hunter Syndrome

① A 13 yr old child presented with distention of abdomen, edema feet and stunted growth

On examination :- Short stature
coarse facies
Short neck
Corneal clouding
Hepatosplenomegaly

X-ray – Skeletal abnormalities

Laboratory test – ↑ Heparan sulfate & ↑ Dermatan sulfate
Decreased Iduronidase activity in WBC's

Diagnosis :- Mucopolysaccharidosis – Hurler syndrome.

② A child brought to the Hospital with distended abdomen and stunted growth

On examination :- Short stature, Coarse facies, short neck
No corneal clouding
Hepatosplenomegaly

X-ray – Skeletal abnormalities

Laboratory test – ↑ HS & ↑ DS
Decreased activity of Iduronate sulfatase

Diagnosis :- Mucopolysaccharidosis – Hunter syndrome

CARBOHYDRATES

Definition:- Polyhydroxy aldehydes or ketones or compounds which produce them on hydrolysis are known as carbohydrates

old definition - Hydrates of carbon c̄ formula $C_nH_{2n}O_n$
but Lactic acid, acetic acid are not carbohydrates
and deoxyribose is a carbohydrate but Not Fits into $C_nH_{2n}O_n$.

SAQ:- Define and classify carbohydrates with suitable examples.

Classification

- **Monosaccharides** [Simple sugar, single unit]
- **Disaccharides** [two sugar units, connected by glycoside bond]
- **Oligosaccharides** [3-10 sugar units] → Example ① Stachyose ② Raffinose
- **Polysaccharides** [More than 10 units]

Monosaccharides — On the basis of: No. of C-atoms / functional group

No. of C-atoms		functional group: Aldehyde [CHO]	functional group: Ketone [C=O]
Triose	3C	Glyceraldehyde	DHA
Tetrose	4C	Erythrose	Erythrulose
Pentose	5C	Ribose	Ribulose
Hexose	6C	Glucose	Fructose
Heptose	7C	Glucoheptose	Sedoheptulose

Tests for Monosaccharides (Mnemonic - MBBS)

Molisch test :- It is a group test for all carbohydrates.

Benedict's test : for reducing sugars : Bedside test for detection of sugar in urine. Semiquantitative test, gives idea about approximate amount of reducing sugar in urine (useful in DM). Green → 0.5 gm%, Yellow - 1 gm%, orange - 1.5 gm%, Red - 2 gm%, Brick Red → > 2 gm%.

Barfoed's test :- for monosaccharides only.

Seliwanoff's test : for Ketose sugar e.g. Fructose

Disaccharides

Reducing

1. Lactose : Glucose + Galactose
   β(1-4) glycosidic bond

2. Maltose : Glucose + Glucose
   α(1-4) glycosidic bond

Osazone

carbohydrate derivatives formed when reducing sugars are reacted with phenyl hydrazine at boiling temp.

Lactosazone - Powder puff-shaped crystals.

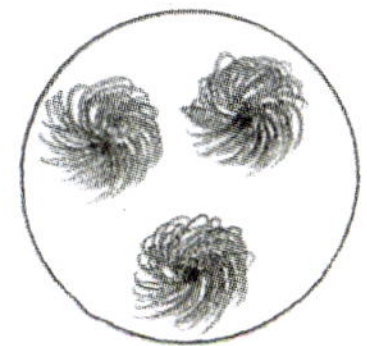

Maltosazone - petal shaped sunflower shaped crystals.

Non-reducing

1. Sucrose
   [Glucose + Fructose]
   $\alpha_1-\beta_2$- glycosidic linkage

C_1 and C_2 both are involved in the linkage formation, Not available for osazone formation.

* Does Not form Osazone

2. Trehalose
   [Glucose + Glucose]
   $[\alpha_1-\alpha_1]$
   widely used to prolong food shelf life.

Glucose, Fructose, Mannose] Differ around C_1 & C_2 ⇒ This difference is masked when Phenylhydrazine get attached at C_1 & C_2 ⇒ Needle shaped, Broom shaped

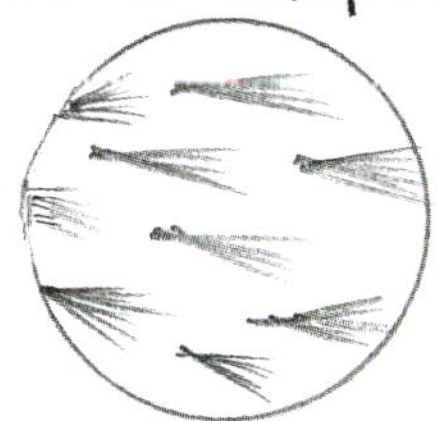

Polysaccharide

Homopolysaccharides
[composed of same sugar units]

Heteropolysaccharides
[constituent monosaccharides are different]

HOMOPOLYSACCHARIDES

1. Starch
2. Glycogen
3. Cellulose
4. Dextrin
5. Dextran
6. Inulin
7. Chitin

>10 Monosaccharides
Constituent monosaccharides are same
Example :- G-G-G-G-G-G-G-G-G-G

1. **Starch** :- Storage form of carbohydrates in plant
Composed of **Amylose** + **Amylopectin**

Amylose:
① Straight chain polymer of D-glucose
② 20% of starch
③ Water soluble
④ α(1-4) glycosidic linkage betn glucose molecules.

Amylopectin:
① Branched chain polymer of D-glucose
② 80% of starch
③ Insoluble in water
④ α(1-4) in straight chain & α(1-6) in branch point

2. **Glycogen** :- Multibranched polysaccharide of glucose.
Main storage form of glucose in liver and skeletal muscle in animals.
Homopolysaccharide of glucose with α(1-4) glycosidic linkage in linear chain and α(1-6) at branching point.
Branching makes molecule more globular, compact and less space consuming.

3. **Cellulose** :- Polymer of glucose with β(1-4) glycosidic linkage
can't digested by humans due to lack of enzyme
Cellulose acts as a dietary fibre
Increases bulk of stool, relieves constipation
Decreases absorption of glucose and cholesterol from intestine.
Important in management of DM and obesity

4. **Dextrin**

Partial degradation product of starch. Homopolymer of glucose.

5. **Dextran**

It is a complex, branched, homopolymer of glucose.

It is a plasma volume expander and used in the treatment of hypovolumic shock (MW > 60000 Da)

Also has antithrombotic activity (MW 40000 Da)

It is used in eyedrops as a lubricant

It provides parenteral nutrition.

6. **Inulin** - Polymer of fructose

Used in assessment of glomerular filtration rate

Better marker for measurement of GFR over urea and creatinine. *MCQ

↓

Filtered through glomerulus

Neither reabsorbed, Nor secreted by the renal tubules.

Also can act as a dietary fibre.

7. **Chitin** - Present in exoskeleton of insects, cell walls in fungi

Polymer of N-acetylglucosamine [amine derivative of glucose]

SAQ:- Enumerate important homopolysaccharides. Write about their functions.

HETEROPOLYSACCHARIDES

Composed of More than 10 sugar units and constituent units (monosaccharides) are different.

1. Glycosaminoglycans (GAG)/Mucopolysaccharides - Most common.
2. Agar :- Made up of Glucose & Galactose, used as a bacterial culture media.
3. Agarose :- Made up of Galactose + anhydro galactopyranose used for protein and Nucleic acid electrophoresis.
4. Pectin :- Main content is galacturonic acid, used in food as gelling agent, Also it is a dietary fibre.

GLYCOSAMINOGLYCANS [GAG]

Mucopolysaccharides

Heteropolysaccharides composed of repeating units of disaccharides

↓ composed of

Amino sugar + Uronic acid

Amino sugar	Uronic acid
Glucosamine	D-Glucuronic acid
Galactosamine	L-Iduronic acid

GAGs contains Negatively charged groups [COO^-, SO_4^-, Acetyl]

↓

Repulsion from each other

- Slippery action [Hyaluronic acid, Mucous]
- Shock absorbent in [Synovial fluid, vitreous humor, extracellular matrix]

Examples :- Hyaluronic acid
Chondroitin sulfate
MCQ * Dermatan sulfate - N-Acetyl galactosamine + L-Iduronic acid
Heparin
Heparan sulfate
MCQ * Keratan sulfate → No uronic acid
N-Acetyl glucosamine + Galactose

MCQ :- HS & CS - Have role in compressibility of cartilage in weight bearing

	GAGs	Composition	Present in	Functions.
1.	Hyaluronic acid	N-Acetyl glucosamine + D-glucuronic acid No sulfate group* MCQ	Synovial fluid* vitreous humor MCQ	Lubricant Shock absorbent cell migration wound repair.
2.	Chondroitin sulfate	N-Acetyl galactosamine + D-glucuronic acid	Cartilage Bone Tendon	Shape and structure
3.	Dermatan sulfate	N-Acetyl galactosamine + L Iduronic acid	skin valves	Shape and structure
4	Heparin	Glucosamine + Glucuronic acid/ Iduronic acid	Lungs mast cells Blood Skin	Anticoagulant * MCQ * causes release of LPL from capillary endothelium
5	Heparan sulfate	Glucosamine + Iduronic acid/ Glucuronic acid	Skin Kidney Basement membrane Synaptic vesicles	-Negative charge on basement membrane * MCQ - Attachment of LPL to capillary endothelium *MCQ - Receptor on cell
6	Keratan sulfate	N-Acetyl glucosamine + Galactose * MCQ	cornea cartilage	Transparency of cornea shape & structure

SAQ:- Enumerate different glycosaminoglycans. Add a note on their role in our body.

Proteoglycan :- Mainly contains carbohydrates >95%. (GAGs)

Glycoproteins :- >95% proteins
<5% carbohydrates

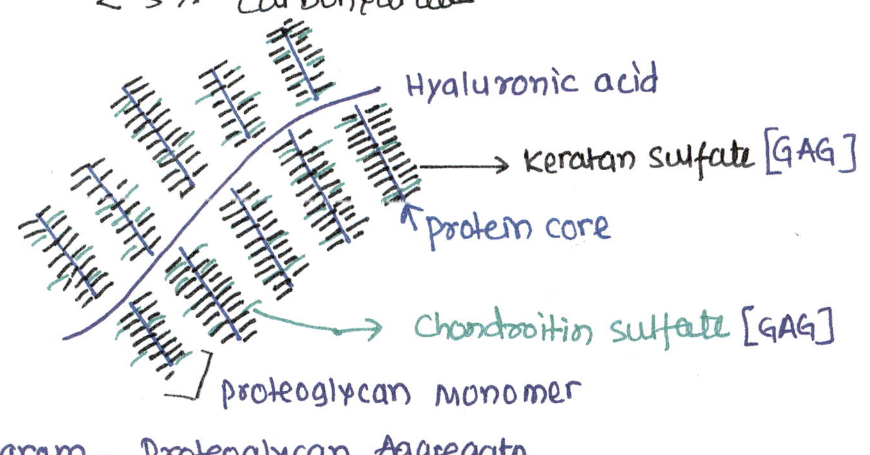

Diagram - Proteoglycan Aggregate

GLYCEMIC INDEX [GI]

Definition:- It is a value used to measure how much specific foods increases your blood sugar level. [0 to 100]

It is a system of assigning a number to carbohydrates containing foods according to how much food increases blood sugar.

Importance:-

To lose weight or to maintain a healthy weight

In the Management of chronic diseases related to obesity such as DM & coronary artery diseases.

Examples

1. Low glycemic index foods :- green vegetables, Eggs
2. Medium glycemic index foods : sweet corn, Bananas, oat
3. High glycemic index foods : white rice, potatoes, breads, sugary foods & drinks

GI of Glucose → 100
Rice → 87
Ice cream → 51
Kidney beans → 24
Soyabeans → 16

Glucose, Galactose, Lactose, Trehalose, Maltose, Isomaltose } GI = 100

SAQ :- Define glycemic index. Mention glycemic index of some important foodstuffs.

SAQ :- Describe the importance of glycemic index.

$$\text{Glycemic index} = \frac{\text{Area under the curve after 50 gm Test meal}}{\text{Area under the curve after 50 gm Glucose}} \times 100$$

GI < 55 → Low GI
55-69 → Medium GI
70 OR More → High GI

GI: Low, Medium, High

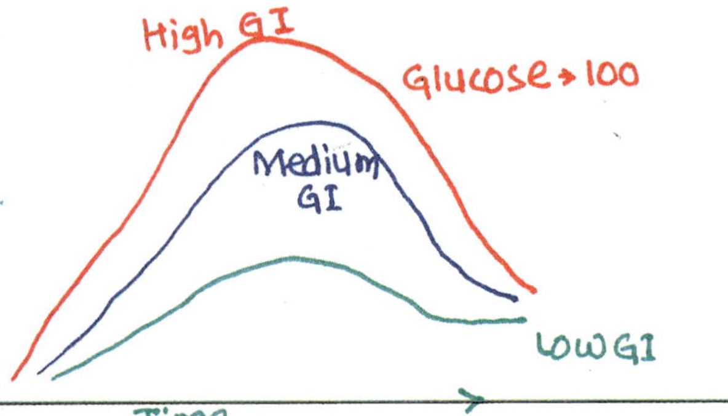

DIETARY FIBRES

BAQ :- Enumerate four important dietary fibres.

SAQ :- What is a role of dietary fibre in human health?

Definition :- plant components that are NOT broken down by human digestive enzymes.

Names : Cellulose
Hemicellulose
pectin
Lignin
Gums
Resistant starch
Oligosaccharides
Lactulose [Galactose + fructose] → synthetic disaccharide
Inulin etc.

Sources :- Fruits, dark green vegetables, whole grains, legumes, oats, rye, root tubers, almonds etc.

Mechanism of action → Bulking (Absorb water)
Viscosity (Thicken the content of GIT)
Fermentation

Important functions

① Increases bulk of stool and relieves constipation
② Helpful in weight loss
③ Decreases absorption of glucose from intestine and helps in reduction of blood sugar level in DM.
④ Decreases absorption of cholesterol from intestine and help to reduce cholesterol in patient with CAD, Atherosclerosis, DM etc.
⑤ Reduces the risk of colon cancer and diverticulosis.
⑥ Helpful in gastrointestinal disorders

ISOMERS

Stereoisomers:- compounds having same molecular formula but differ in orientation of groups in space.

Example: D-Glucose and L-Glucose

```
      CHO                 CHO
       |                   |
  H - C - OH         HO - C - H
       |                   |
 HO - C - H           H - C - OH
       |                   |
  H - C - OH         HO - C - H
       |                   |
  H - C - OH         HO - C - H
       |                   |
     CH2OH               CH2OH

  [D-Glucose]         [L-Glucose]
```

Orientation of -OH group to right/left around the carbon atom

```
      CHO       |        CHO
       |        |         |
  H - C - OH    |   OH - C - H
       |        |         |
     CH2OH      |       CH2OH
```

Functional Isomers:- compounds having same molecular formula but different functional groups.

Example:- Glucose [CHO], Fructose [C=O]

Epimers:- Compounds having same molecular formula but differ around single C-atom, other than anomeric carbon.

MCQ ✱ Examples:- Glucose and Galactose are epimers at 4th
Glucose and Mannose are epimers at 2nd

Optical isomers:- optically active compounds that differ in their direction of rotation of plane polarized light
to the right (+) or d form
to the left (-) or l form

Anomers:- Same molecular formula but differ in orientation of groups around anomeric C-atom. [Carbonyl carbon, 1st in glucose]

```
        CH2OH
         |
  H      |-------O     H
   \    H          \   |
    |  OH      H    |--1
  OH|                /
     ---------------|  [OH]
    H              OH
```

α-D Glucose

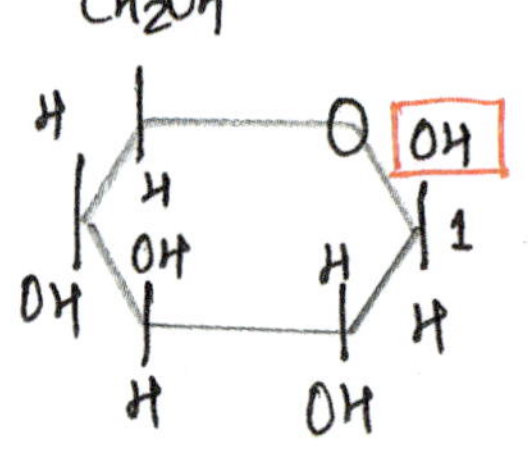

β-D Glucose

α → OH is below the plane away from Oxygen

β - OH is above the plane towards Oxygen atom

METABOLISM OF CARBOHYDRATES

MCI COMPETENCY BI 3.2- 3.10

LAQ :- Regulation of blood glucose level.
LAQ :- Glycogen metabolism, regulation & disorders.
LAQ :- Gluconeogenesis with regulation
LAQ : Glycolysis with energetics and regulation
LAQ :- TCA cycle with energetics.

SAQ - ① HMP shunt
② Galactosemia
③ R-L cycle and importance of 2,3 BPG
④ Pyruvate dehydrogenase complex
⑤ Glycogen storage disorders
⑥ Anaplerotic reaction's of TCA cycle
⑦ Cori cycle & Cahill cycle
⑧ Glucose transporter's
⑨ fate of pyruvate

Clinical cases

① Diabetic ketoacidosis — Most IMP case
② Diabetes mellitus - Polyuria, Polydipsia, Polyphagia, Benedict's test
③ Galactosemia - Oil droplet cataract, Jaundice, Hypoglycemia
④ Von Gierk's disease - Hepatorenomegaly, Hyperlipidemia, ↑ UA, ↑ LA
⑤ G6PD deficiency - hemolysis, Drug History.
⑥ Lactose intolerance
⑦ Sucrose intolerance
⑧ pompe's disease :- Hypotonia, Heart involvement, Also Lysosomal d.
⑨ Mc Ardles disease :- Exersize intolerance
⑩ Hereditary fructose intolerance

Digestion and absorption of Carbohydrates

Dietary carbohydrates : Starch
Glycogen
Cellulose
Sucrose
Lactose
Fructose
Glucose etc.

Digestion

Mouth – Partial digestion
By the action of Salivary amylase
Amylase breaks α(1-4) glycosidic bond

↓

Stomach – (Nil)
only hydrolysis of sucrose by HCl
Low pH stops action of Amylase

↓

Intestine – complete digestion
Pancreatic amylase acts on α(1-4) glycosidic linkage
Disaccharidase acts on disaccharides [Sucrase, Lactase, Maltase, Isomaltase]

Intestinal amylase act on terminal α(1-4) linkage and liberates free glucose.

Lactase acts on lactose

Lactose ⟶ Glucose + Galactose

Isomaltase act on α(1-6) linkage, breaks the branching point

Maltase

Maltose ⟶ Glucose + Glucose

Sucrase

Sucrose ⟶ Glucose + Fructose

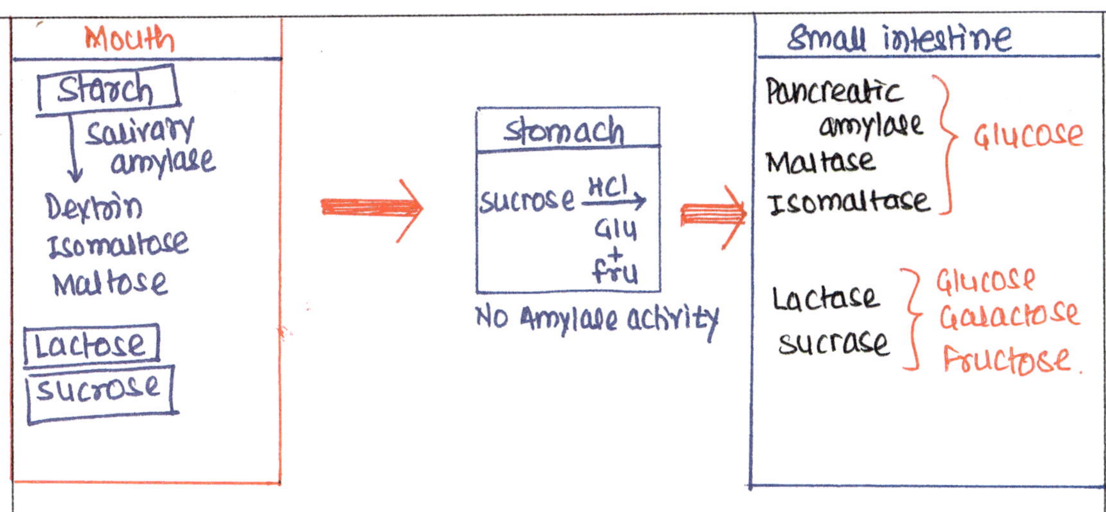

Diagram - Digestion of Carbohydrates

Absorption

MCQ * Glucose → by secondary active transport along c̄ Na^+ (symport)

Fructose → by facilitated diffusion

Pentoses → by passive diffusion

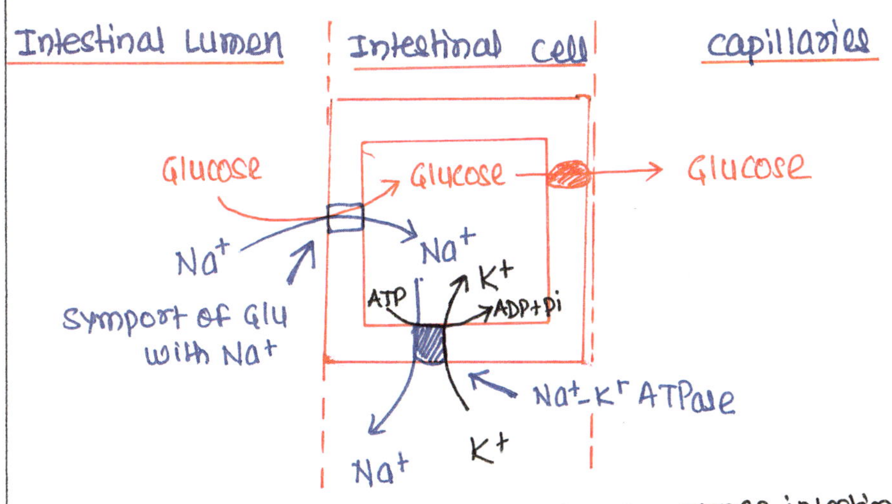

Diagram : transport of glucose across intestinal epithelium

Clinical case

A month old baby had loose stool after every breastfeed. He was crying continuously and irritable. His pediatrician put him on Lactose-free commercial preparation and miracle happened, Now baby is normal.

Probable diagnosis :- Lactose intolerance

Deficient enzyme :- Lactase (intestinal brush border enzyme)

Treatment :- Lactose free diet
Commercial preparations are available in market

Symptoms of Lactose intolerance

Abdominal pain, cramps, diarrhea, flatulence etc.

Clinical case. 2

A 30 year old Male had a loose stool after consumption of sucrose rich diet. He had abdominal pain, cramps and flatulence. Every time when he takes such type of diet, he gets simillar symptoms.

Probable diagnosis :- Sucrose intolerance

Deficient enzyme : **Sucrase**

Treatment :- Avoid sucrose in diet
sugar-free in diet.

Glucose transporters GLUTs → 1 to 12
facilitated diffusion

Transporter	Tissue location	Function	
GLUT 1	Brain, kidney, colon, placenta, RBC's, Retina	Basal glucose uptake	Widely distributed Abundant in RBC's Major in Brain & placenta
GLUT 2	Liver Pancreas B/L surface of intestine PCT [Kidneys]	Blood glucose regulation	
GLUT 3	Neuron placenta kidney		
GLUT 4	Skeletal Muscle Heart Adipose tissue	Insulin dependent	* GLUT 4 8 12 Insulin sensitive
GLUT 5	small intestine, Testis sperm	fructose transport	
GLUT 7	Liver ER	Glucose transport in ER	

GLYCOLYSIS

Embden-Meyerhof pathway

LAQ: Describe the steps in glycolysis. Add a note on site, importance, energetics and regulation of glycolysis.

Important aspects

1. It occurs in all cells of the body
2. Aerobic and anaerobic pathway
3. It is the only source of energy in RBC's
4. Anaerobic glycolysis provides energy to muscle
5. Glycolysis provides carbon skeleton for synthesis of nonessential amino acids and glycerol part of fat.
6. This pathway is less active in cardiac muscles. whenever there is ischaemic attack, it has poor prognosis.
7. Reversible steps in glycolysis are also involved in gluconeogenesis.
8. Cancer cells derive most of their energy from anaerobic glycolysis.

Glucose ⟶ Pyruvate/ Lactate

③ Irreversible steps
① Hexokinase
② Phosphofructokinase
③ Pyruvate kinase

Commited step
Rate limiting step
Key enzyme
} Phosphofructokinase

② Substrate level phosphorylation
① 1,3-bisphosphoglycerate kinase
② pyruvate kinase

Clinical laboratory Application

For estimation of plasma glucose, we collect the sample in fluoride bulb. Fluoride ions inhibit enolase enzyme by non-competitive inhibition in glycolysis & stops utilisation of Glucose.

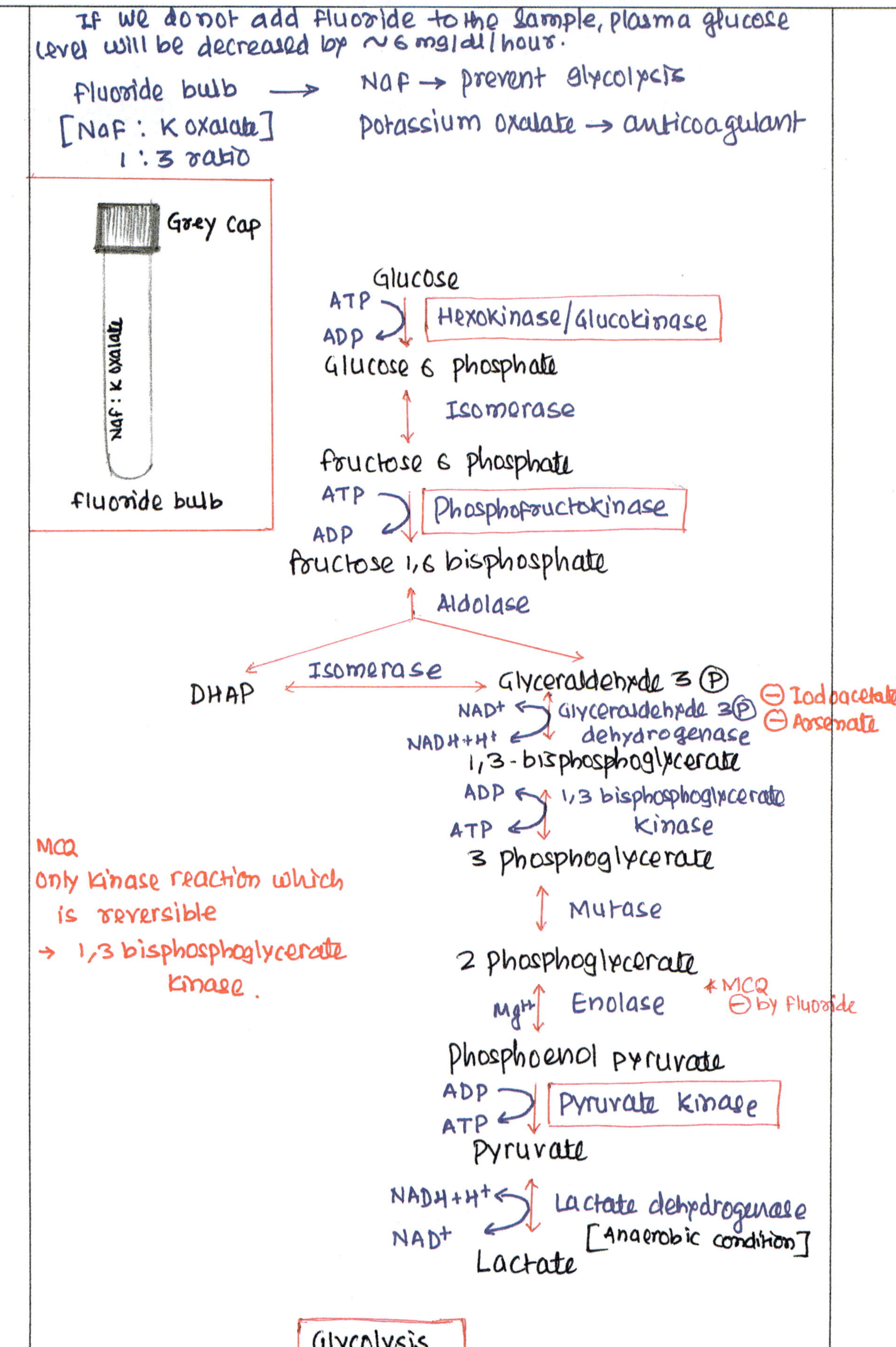
If we donot add fluoride to the sample, plasma glucose level will be decreased by ~6 mg/dl/hour.
Fluoride bulb → NaF → prevent glycolysis
[NaF : K oxalate] 1 : 3 ratio
Potassium oxalate → anticoagulant
Grey cap
NaF : K oxalate
fluoride bulb
Glucose
ATP
ADP
Hexokinase/Glucokinase
Glucose 6 phosphate
Isomerase
Fructose 6 phosphate
ATP
ADP
Phosphofructokinase
Fructose 1,6 bisphosphate
Aldolase
DHAP
Isomerase
Glyceraldehyde 3 (P)
NAD+
NADH+H+
Glyceraldehyde 3(P) dehydrogenase
⊖ Iodoacetate
⊖ Arsenate
1,3-bisphosphoglycerate
ADP
ATP
1,3 bisphosphoglycerate kinase
3 phosphoglycerate
Mutase
2 phosphoglycerate
Mg++
Enolase
* MCQ ⊖ by Fluoride
Phosphoenol pyruvate
ADP
ATP
Pyruvate kinase
Pyruvate
NADH+H+
NAD+
Lactate dehydrogenase [Anaerobic condition]
Lactate
MCQ
Only kinase reaction which is reversible
→ 1,3 bisphosphoglycerate kinase.
Glycolysis

Regulation

By ③ Irreversible steps
① Hexokinase/Glucokinase
② Phosphofructokinase
③ Pyruvate kinase

② Hormones - Insulin/Glucagon

Hexokinase	Glucokinase (Hexokinase IV)
Act on Hexose sugar	Act only on Glucose
High affinity for glucose	Low affinity
Low km value	High km value
Inhibition by Glu 6 Ⓟ	No inhibition by Glu 6 Ⓟ
Not inducible	Inducible enzyme
Act even at low glucose concentration [<10mg/dl]	Act when Glucose > 100 mg/dl
Present in All tissues	Liver, pancreas Stimulated by Insulin

Enzyme	Activation	Inhibition
Hexokinase		Glucose 6 Ⓟ
Glucokinase	Insulin	Glucagon
Phosphofructokinase (PFK)	Insulin, AMP fructose 6 Ⓟ fructose 2,6 bisphosphate PFK-2	Glucagon ATP Citrate cAMP
Pyruvate kinase	Insulin Fructose 1,6 bisphosphate	Glucagon ATP cAMP

Fructose 6 Ⓟ → (PFK 1) → Frucose 1,6 bisphosphate

Fructose 6 Ⓟ → (PFK-2) → Fructose 2,6 bisphospate → ⊕ PFK 1

Pyruvate Kinase → Active in dephosphorylated state → Favors glycolysis
[Insulin]

Energetics

Aerobic pathway Glucose ⟶ Pyruvate (2)

2 NADH = 2 × 2.5 ATP	= 5 ATP
2 substrate level Phosphorylation	= 4ATP
2 ATP utilized	− 2ATP
Net synthesis	**7 ATP**

Anaerobic pathway :-

2 substrate leve phosphorylation	4 ATP
2 ATP utilized	− 2 ATP
Net synthesis	**2ATP**

In Muscle – Absent Glucose 6 phosphatase

2 substrate level Phosphorylation	4 ATP
1 ATP utilized	−1 ATP
	3 ATP

Rapoport Leubering Cycle

Imp. SAQ

LAQ :- Describe the steps of RL cycle. Add a note on importance of 2,3 BPG.

SAQ :- Importance of 2,3 BPG in erythrocytes.

It is a supplementary pathway for generation of 2,3-bisphosphoglycerate in erythrocytes. [10% glucose diverted to RL cycle]

Hb saturation %

$\uparrow pCO_2$
2,3 BPG
Temp
$[H^+]$

pO_2

Glucose
↓
Glucose 6 Ⓟ
↓
fructose 6 Ⓟ
↓
fructose 1,6 bisphosphate
↙ ↘
DHAP ⟷ Glyceraldehyde 3 Ⓟ
↓
1,3 bisphosphoglycerate → (Bisphosphoglycerate Mutase) → 2,3 BPG
ADP → ATP
RL Cycle
↓
3 phosphoglycerate ← (Phosphatase) ← 2,3 BPG
↓
2 phosphoglycerate
↓
PEP
↓
Pyruvate

Clinical significance

2,3 BPG level increases in

1. Anemia
2. High altitude
3. Fetal Hb
4. COPD

It causes release of O2 to the tissues.

functions of 2,3-BPG

1. It occupy the central pocket of Hb molecule and responsible for unloading of oxygen to the tissues.
2. Maintenance of buffer in RBC's.

PDH complex

SAQ:- Describe the enzymes and coenzymes of Pyruvate dehydrogenase complex

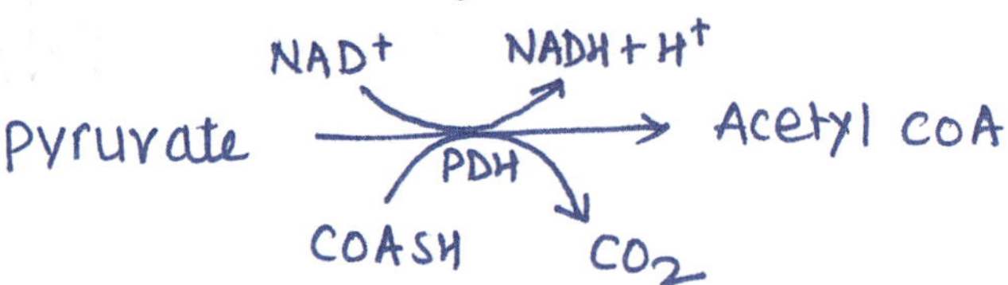

Multienzyme

Connecting link between glycolysis & TCA cycle

Irreversible and committed step in oxidation of glucose

It is composed of 3 enzymes & 5 coenzymes *MCQ

E1 - Pyruvate dehydrogenase
E2 - Dihydrolipoyl transacetylase
E3 - Dihydrolipoyl dehydrogenase

TPP
NAD^+
FAD
CoA
Lipoic acid

Pyruvate
E1 PDH
CO_2
TPP
Hydroxy ethyl TPP
E2 Dihydrolipoyl Transacetylase
Acetyl Lipoamide
Oxidised Lipoamide
COASH
Acetyl CoA
Reduced Lipoamide
E3 Dihydrolipoyl dehydrogenase
FAD
$FADH_2$
NADH $+H^+$
NAD^+
ETC
2.5 ATP

Regulation

① Allosteric regulation by products - NADH, Acetyl CoA

② Covalent Modification : Phosphorylation & Dephosphorylation
PDH is active in dephosphorylated state

Clinical significance : Wernicke - korsakoff syndrome

In chronic alcoholics deficiency of TPP occurs

because:- Defective absorption of Thiamine
Defective conversion of Thiamine → TPP

Symptoms :- Global confusion
Ophthalmoplegia
Ataxia
Memory loss
Psychosis

↑ Pyruvate level

* symptoms due to ↓ ATP synthesis.

Fate of Pyruvate

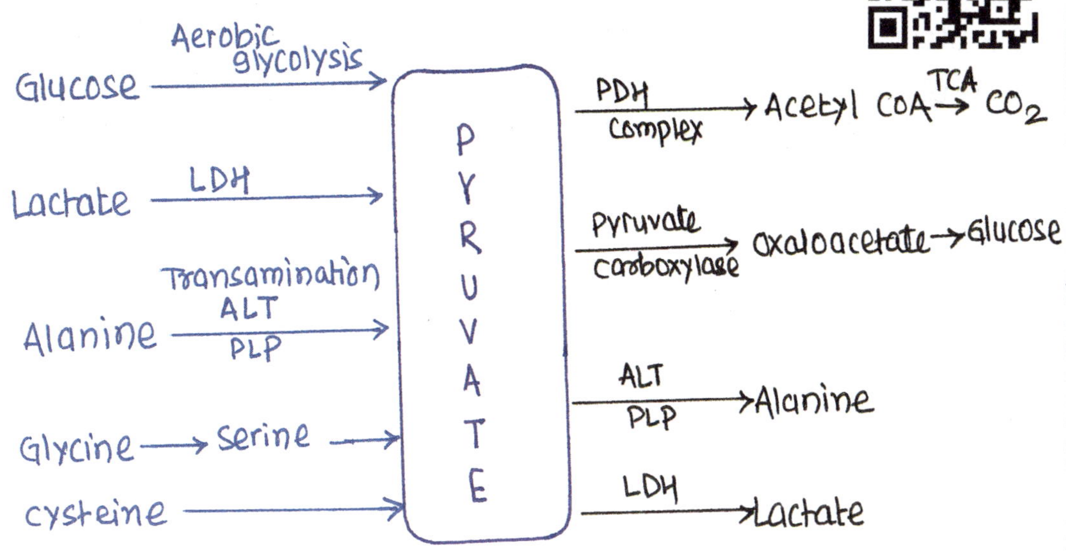

fate of Acetyl CoA

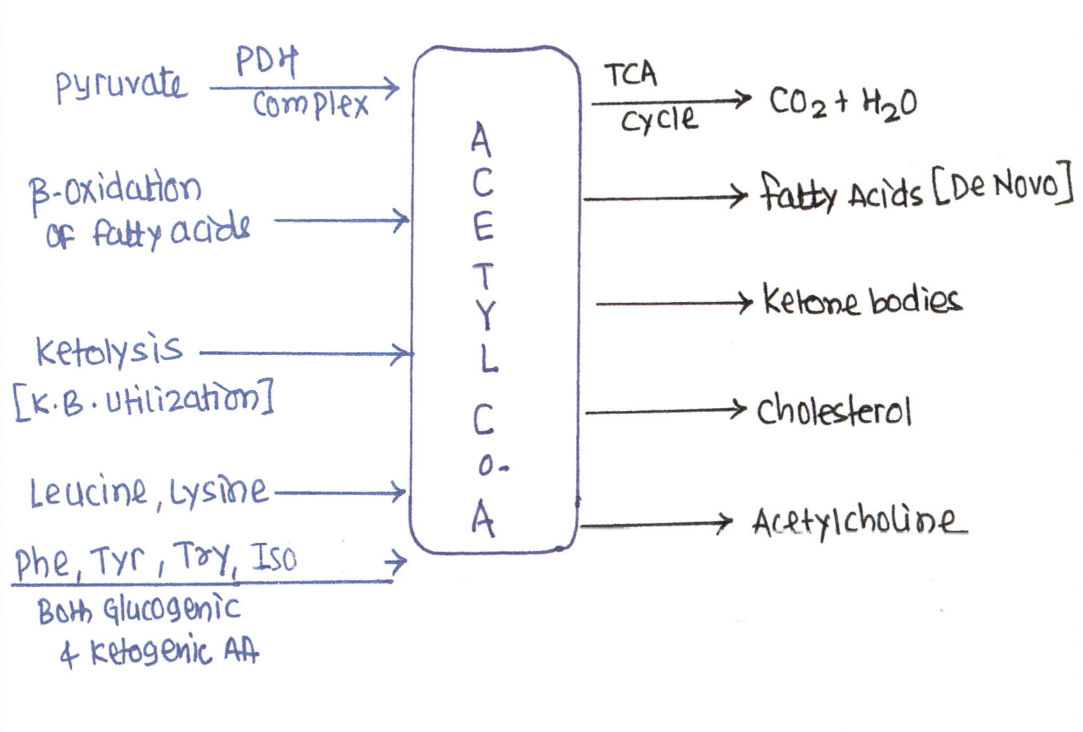

KREBS CYCLE OR TCA CYCLE

Citric acid cycle

LAQ:- Describe the steps of TCA cycle. Add a note on site, regulation, amphibolic role and anaplerotic reactions of TCA cycle.

Sir Hans Krebs – 1937

1. It is final common oxidative pathway for conversion into CO_2
2. It gives NADH & $FADH_2$
3. catalyst – oxaloacetate for conversion of fat into energy
4. Important for synthesis of nonessential AAs, Heme, Purines
5. provides intermediates for various pathways
6. provides energy (ATP)

Mnemonic to remember: Citrate Is Krebs starting Substrate for Making oxaloacetate

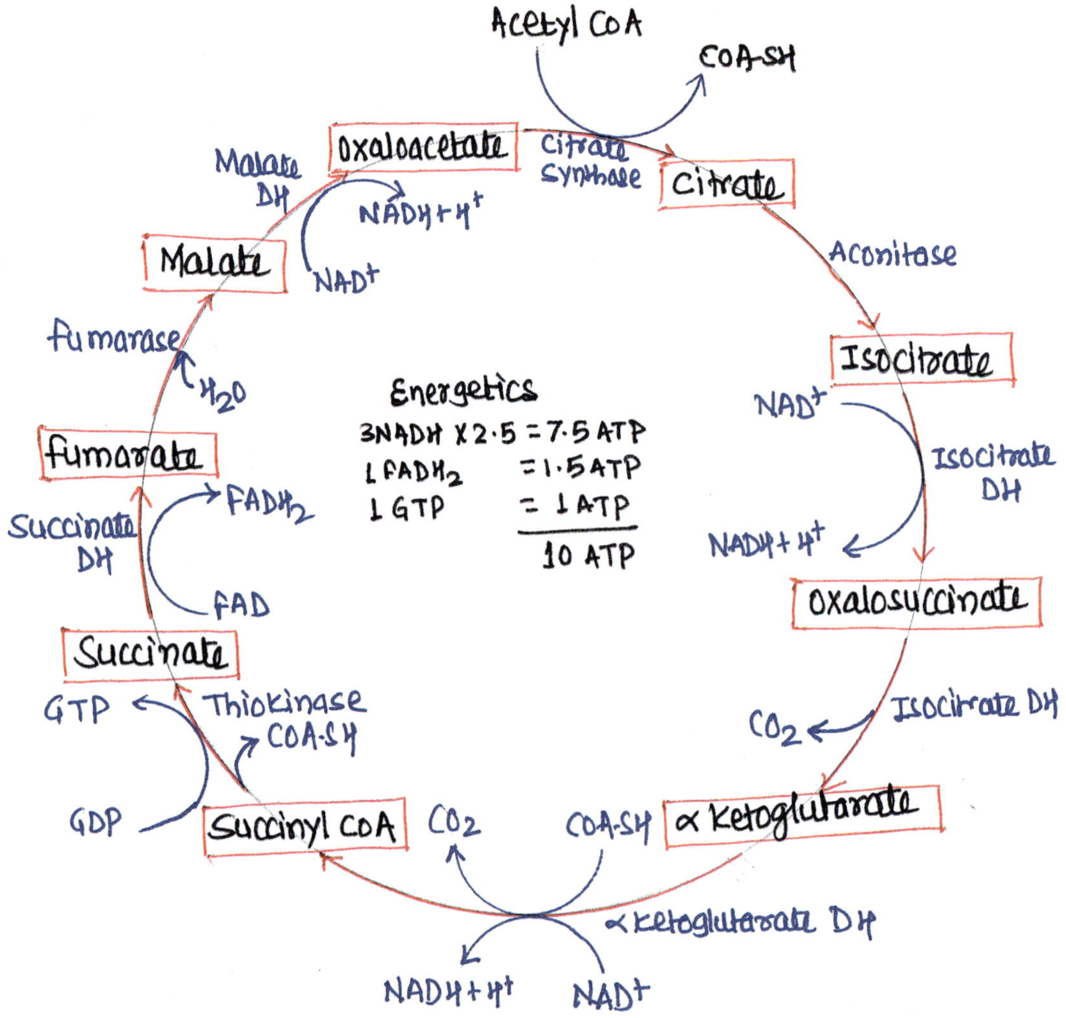

Acetyl CoA → COA-SH

Oxaloacetate → (Citrate synthase) → Citrate

Citrate → (Aconitase) → Isocitrate

Isocitrate → (Isocitrate DH; NAD^+ → $NADH + H^+$) → Oxalosuccinate

Oxalosuccinate → (Isocitrate DH; CO_2) → α Ketoglutarate

α Ketoglutarate → (α Ketoglutarate DH; COA-SH, CO_2; NAD^+ → $NADH + H^+$) → Succinyl CoA

Succinyl CoA → (Thiokinase; COA-SH; GDP → GTP) → Succinate

Succinate → (Succinate DH; FAD → $FADH_2$) → Fumarate

Fumarate → (Fumarase; H_2O) → Malate

Malate → (Malate DH; NAD^+ → $NADH + H^+$) → Oxaloacetate

Energetics

3NADH × 2.5	= 7.5 ATP
1 $FADH_2$	= 1.5 ATP
1 GTP	= 1 ATP
	10 ATP

Inhibitors : FAM - AKS [Mnemonic]

Fluoroacetate → Aconitase

Arsenite → α ketoglutarate

Malonate → Succinate DH ⊛ MCQ

Regulation ① Availability of substrate

② citrate synthase, Isocitrate DH, α-ketoglutarate DH } Important steps Allosterically regulated

ATP & NADH will inhibit the pathway

Amphibolic role

① Catabolic role

Acetyl CoA → CO_2

② Anabolic role

Oxaloacetate → Aspartate

Citrate → Acetyl CoA

α ketoglutarate → Glutamate, GABA

Succinyl CoA → Heme synthesis, Ketolysis

Anaplerotic reactions : Replenishing the intermediates of TCA cycle.

① Pyruvate → oxaloacetate

② Glutamate → α ketoglutarate

③ Arginine, proline, Histidine → α ketoglutarate

④ Propionyl CoA → succinyl CoA

Valine, Methionine, Isoleucine, Threonine → succinyl CoA

⑤ Phenylalanine, Tyrosine → Fumarate

Urea cycle → fumarate

⑥ Pyruvate → Malate

⑦ Aspartate → Oxaloacetate

Cori cycle Lactic acid cycle

Metabolic pathway in which Lactate created by anaerobic glycolysis in muscle is transported to liver and transformed to glucose. Glucose is again utilized by the muscle.

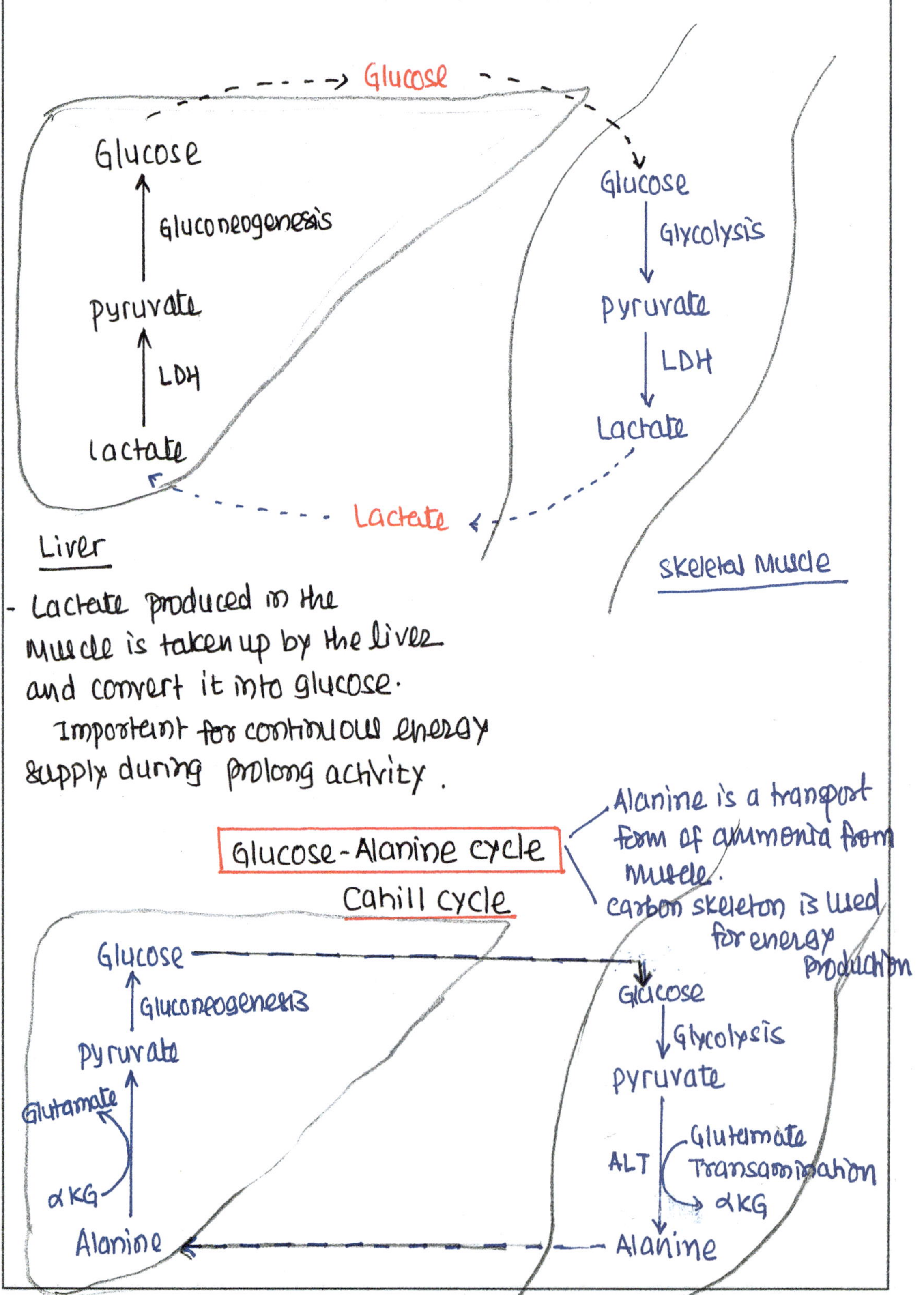

GLUCONEOGENESIS

four Important LAQs.

Describe the synthesis of Glucose from Lactate
Glycerol
Propionyl CoA
Glucogenic amino acids

Definition:- formation of glucose from non carbohydrate precursors

4 Important substrate — ① Lactate
② Glycerol
③ Propionyl CoA
④ Glucogenic aminoacids.

MCQ

① Not a complete reversal of glycolysis
② Irreversible steps in glycolysis are reversed by ④ Imp. enzymes.
③ 6 ATPs utilized per molecule of glucose
④ Important in starvation – provides glucose
Gluconeogenesis is a source of glucose after depletion of glycogen stores.
⑤ Source of glucose for brain in starvation

Site:- cytosol and mitochondria
Liver and Renal cortex

③ Irreversible steps in glycolysis, Reversed by ④ Imp enzymes.

① Pyruvate carboxylase
② Phosphoenol pyruvate carboxykinase

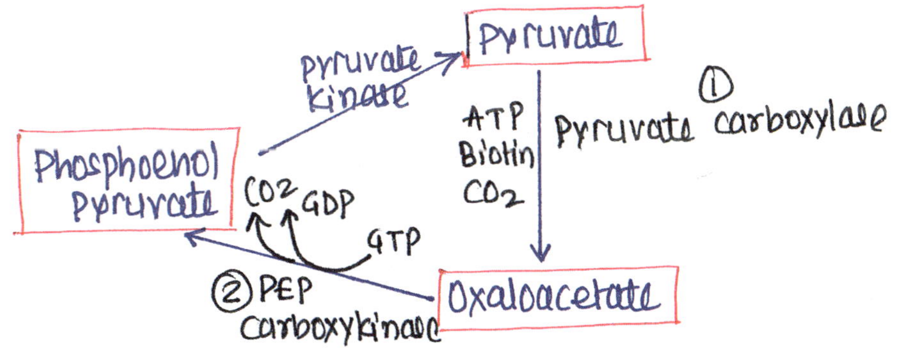

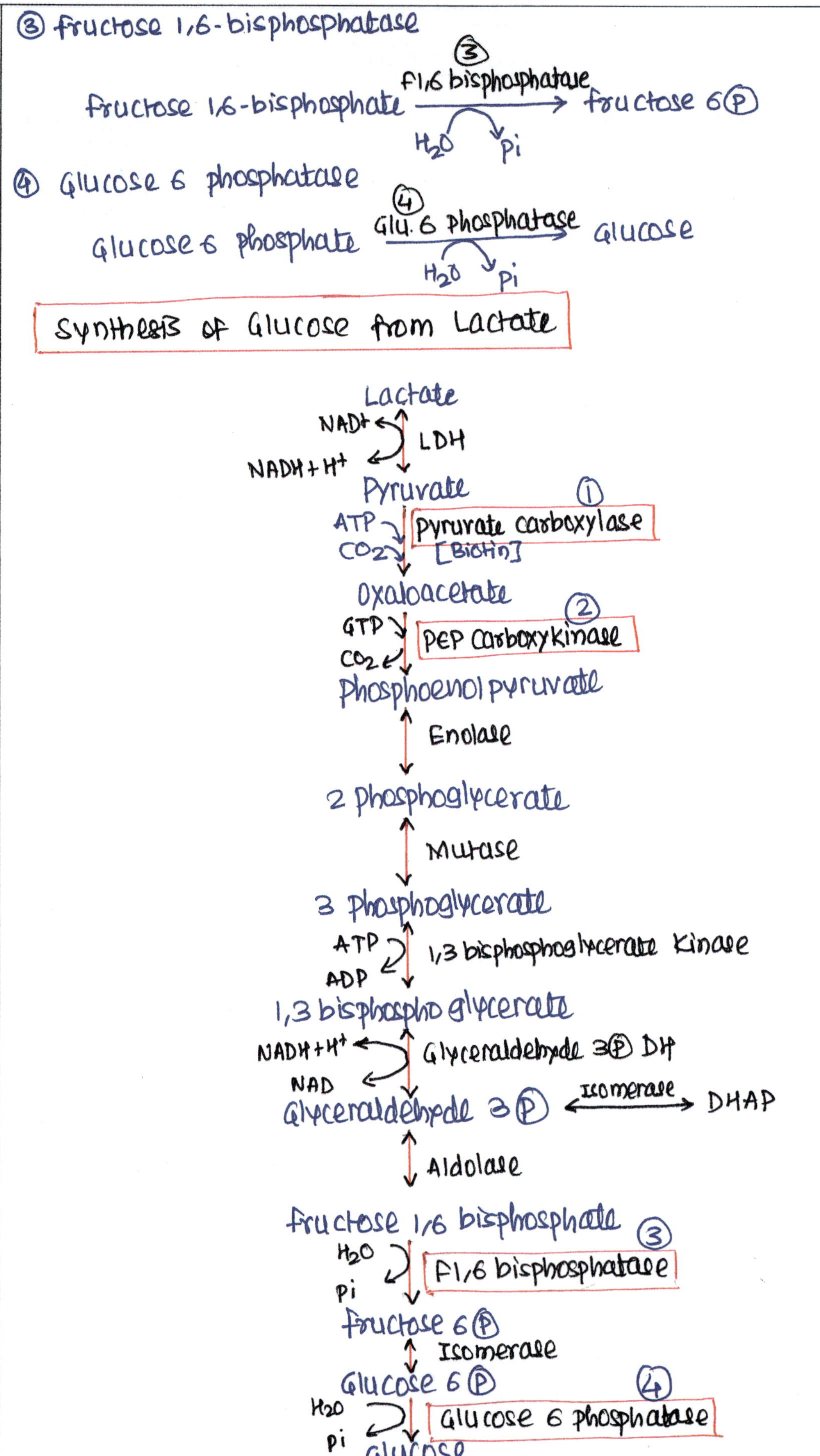

③ fructose 1,6-bisphosphatase
fructose 1,6-bisphosphate
③ F1,6 bisphosphatase
fructose 6 Ⓟ
H_2O
Pi
④ Glucose 6 phosphatase
Glucose 6 phosphate
④ Glu. 6 phosphatase
Glucose
H_2O
Pi
Synthesis of Glucose from Lactate
Lactate
NAD^+
LDH
$NADH + H^+$
Pyruvate
①
ATP
Pyruvate carboxylase
CO_2
[Biotin]
Oxaloacetate
②
GTP
PEP carboxykinase
CO_2
Phosphoenol pyruvate
Enolase
2 phosphoglycerate
Mutase
3 phosphoglycerate
ATP
1,3 bisphosphoglycerate Kinase
ADP
1,3 bisphospho glycerate
$NADH + H^+$
Glyceraldehyde 3Ⓟ DH
NAD
Glyceraldehyde 3 Ⓟ
Isomerase
DHAP
Aldolase
fructose 1,6 bisphosphate
③
H_2O
F1,6 bisphosphatase
Pi
fructose 6 Ⓟ
Isomerase
Glucose 6 Ⓟ
④
H_2O
Glucose 6 phosphatase
Pi
Glucose

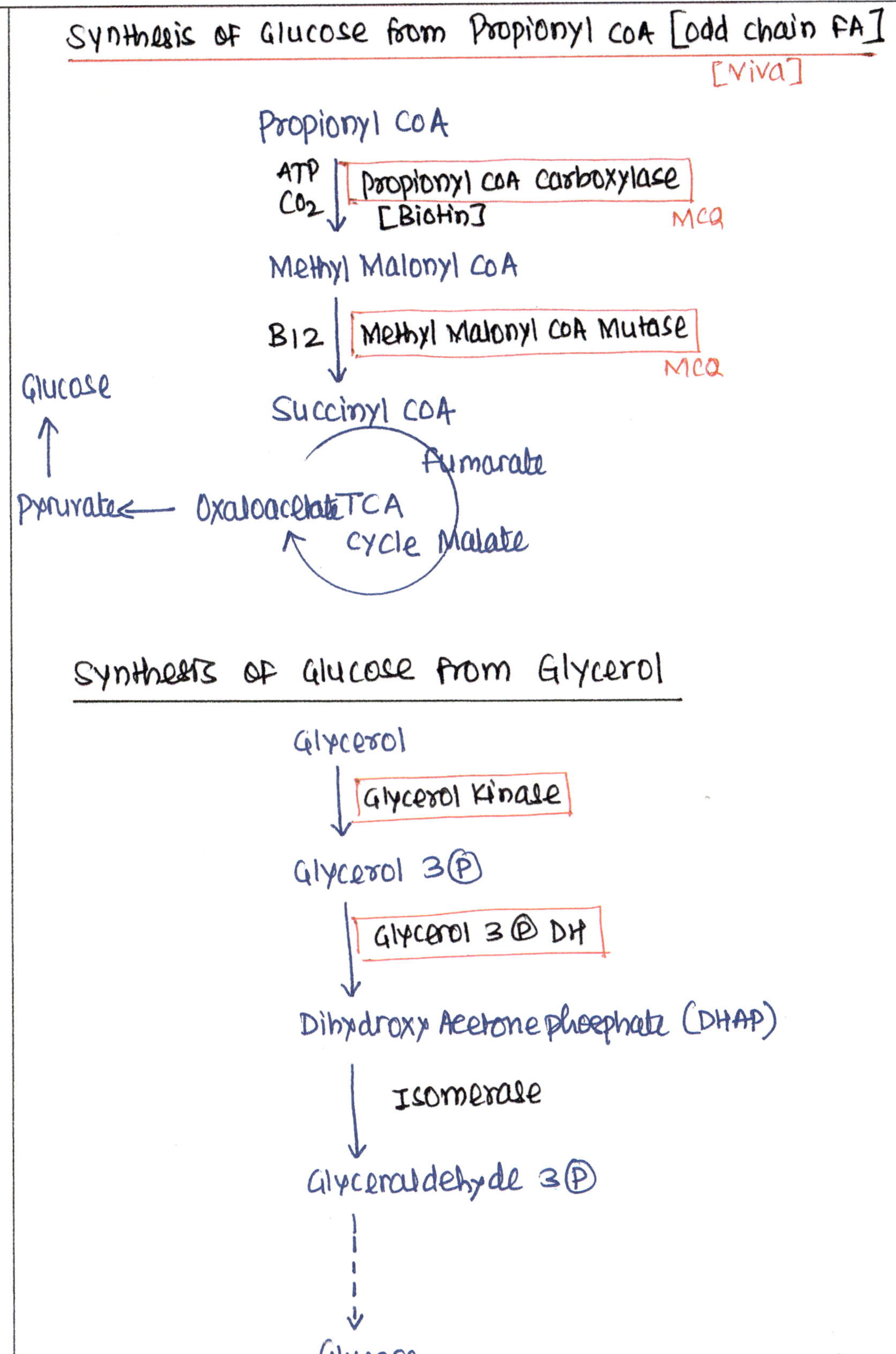
Synthesis of Glucose from Propionyl CoA [odd chain FA]
[Viva]
Propionyl CoA
ATP
CO_2
Propionyl CoA carboxylase
[Biotin]
MCQ
Methyl Malonyl CoA
B12
Methyl Malonyl CoA Mutase
MCQ
Succinyl CoA
Fumarate
TCA cycle
Malate
Oxaloacetate
Pyruvate
Glucose
Synthesis of Glucose from Glycerol
Glycerol
Glycerol Kinase
Glycerol 3 (P)
Glycerol 3 (P) DH
Dihydroxy Acetone phosphate (DHAP)
Isomerase
Glyceraldehyde 3 (P)
Glucose

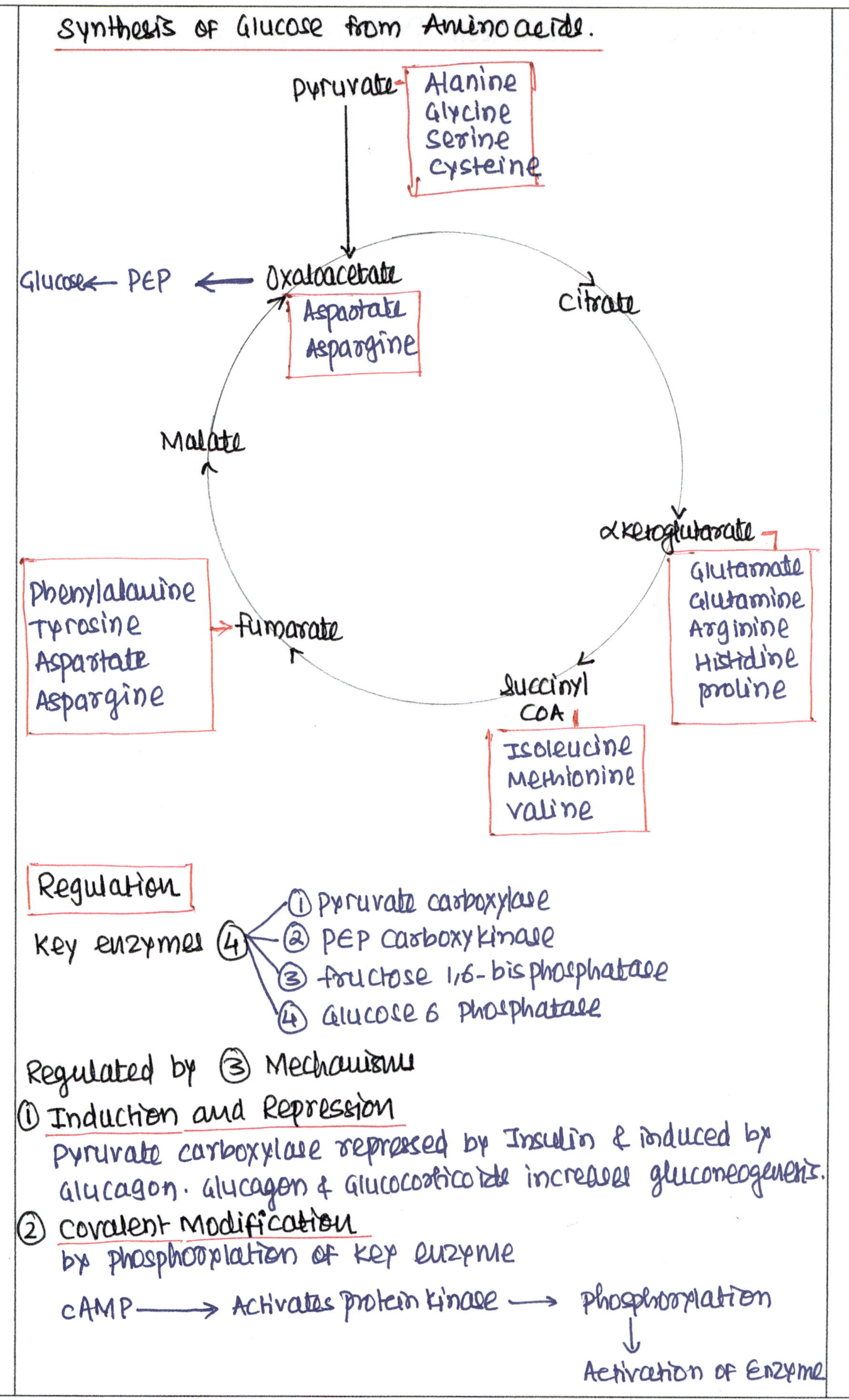
Synthesis of Glucose from Aminoacids.
Pyruvate
Alanine
Glycine
Serine
Cysteine
Glucose ← PEP ← Oxaloacetate
Aspartate
Aspargine
Citrate
Malate
α ketoglutarate
Glutamate
Glutamine
Arginine
Histidine
proline
Phenylalanine
Tyrosine
Aspartate
Aspargine
fumarate
Succinyl COA
Isoleucine
Methionine
valine
Regulation
Key enzymes (4)
① Pyruvate carboxylase
② PEP Carboxykinase
③ fructose 1,6-bisphosphatase
④ Glucose 6 Phosphatase
Regulated by ③ Mechanisms
① Induction and Repression
Pyruvate carboxylase repressed by Insulin & induced by Glucagon. Glucagon & Glucocorticoide increases gluconeogenesis.
② Covalent Modification
by phosphorylation of key enzyme
cAMP ⟶ Activates protein kinase ⟶ Phosphorylation
↓
Activation of Enzyme

③ Allosteric regulation

Acetyl CoA is an activator of pyruvate carboxylase
Citrate activates F2,6 bisphosphatase
F 2,6 bisphosphate is an inhibitor of F1,6 bisphosphatase.

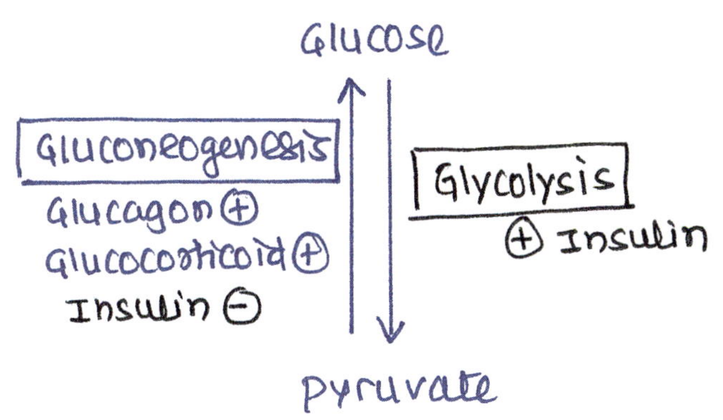

BAQ:- Why ethanol causes hypoglycemia?

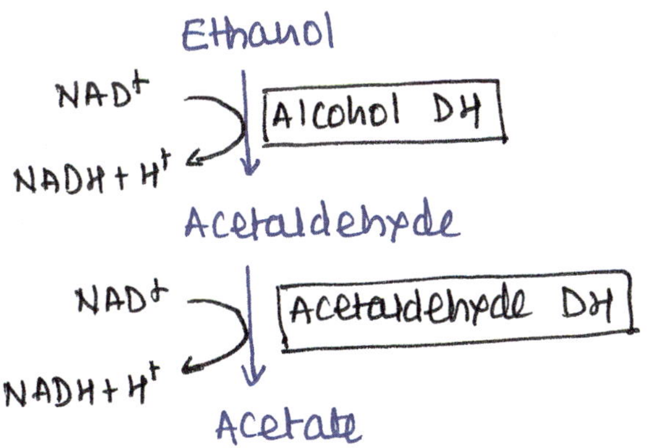

During metabolism of ethanol, cytoplasmic NADH is raised that causes reversal of Pyruvate → Malate → oxaloacetate

Also favors pyruvate to Lactate formation. This will inhibit gluconeogenesis and hypoglycemia occurs.

GLYCOGEN METABOLISM

LAQ :- Describe the glycogen metabolism with its regulation.

Glycogen

- Homopolysaccharide of glucose with α(1-4) glycosidic linkage in linear chain and α(1-6) at branching point.
- Branching makes molecule more globular, compact and less space consuming

Metabolism of glycogen

① Glycogenesis :- synthesis of glycogen

② Glycogenolysis :- degradation of glycogen

Glycogenesis

Glucose ⟶ Glycogen

It is a synthesis of glycogen from glucose

Key enzyme - Glycogen synthase

Site - Cytosol

Tissue - Liver and muscle

Steps

1. Activation of glucose to UDP-glucose
2. Addition of glucose to glycogen primer/glycogenin
3. Elongation up to 10-12 glucose units.
4. Branching by forming α(1-6) glycosidic linkage
5. Elongation
6. Branching
7. Glycogen formation

GLYCOGENESIS

Glucose

ATP ↓ Glucokinase

Glucose 6 phosphate

↓ Mutase

Glucose 1 Phosphate

UTP ↓ UDP-Glucose Pyrophosphorylase

UDP- Glucose

UDP ← ↓ Glycogen synthase

Addition of activated glucose to glycogen primer (GP) or glycogenin

G – G – G – G – G – G – G – (GP)

↑ α(1-4) Linkage

↓ Glycogen synthase

Elongation up to 10-12 glucose residues

G-G-G-G-G-G-G-G-G-G-G-G-G-G-G-G-(GP)

↓ Branching enzyme
transfer 6G to form new branch

formation of branch point

G-G-G-G-G-G-G-G-G-G-(GP)
with branch G-G-G-G-G-G attached at the fifth G ← α(1-6) Linkage

↓

Elongation and branching by glycogen synthase and branching enzyme

↓

GLYCOGEN

GLYCOGENOLYSIS

GLYCOGEN

↓ Glycogen phosphorylase

Removal of glucose in the form of glucose 1 Ⓟ

↓ Glycogen phosphorylase → Glucose 1 Ⓟ

Limit dextrin formation

G-G-G-G-G-G-G-G-G-G-G-GP
with a branch G-G-G-G

↓ Transferase α(1-4)-α(1-4) glucan transferase

Transfer of trisaccharide unit from branching point to linear chain

G-G-G-G-G-G-G-G-G-G-G-G-G-G-GP
with a single G branch ← α(1-6) linkage

↓ Debranching enzyme α(1-6) glucosidase → [free glucose] G

Breaking of α(1-6) linkage

G-G-G-G-G-G-G-G-G-G-G-G-G-G-GP

↓ Glycogen phosphorylase

Removal of glucose 1 Ⓟ one by one

↓ Glycogen phosphorylase

Glucose 1 Ⓟ

↓ Mutase

Glucose 6 Ⓟ

↓ Glucose 6 phosphatase

Glucose

GLYCOGENOLYSIS

Glycogen ⟶ Glucose

Key enzyme :- Glycogen phosphorylase

Site - cytosol

Tissue - liver and muscle

Steps

1. Step by step removal of glucose in the form of glucose 1 Ⓟ till the formation of limit dextrin.
2. Limit dextrin formation
3. Transfer of trisaccharide unit from branching point to linear chain by transferase.
4. Breaking of α(1-6) linkage and liberation of free glucose by debranching enzyme
5. Again removal of glucose in the form of glucose 1 Ⓟ
6. Glucose 1 Ⓟ is converted into glucose 6 Ⓟ
7. Glucose formation.

Regulation of Glycogen metabolism

① Hormonal :- Glucagon / Insulin

② Allosteric :- ATP, Glucose and Glucose 6 Ⓟ will inhibit glycogenolysis.

③ Calcium :- Activates glycogen phosphorylase.

Key enzymes — ① Glycogen synthase
② Glycogen phosphorylase

Two state - ① Dephosphorylated
② phosphorylated

Two Imp. Hormones - ① Glucagon
② Insulin

MCQ

Glycogen synthase active in dephosphorylated state

Glycogen phosphorylase active in phosphorylated state

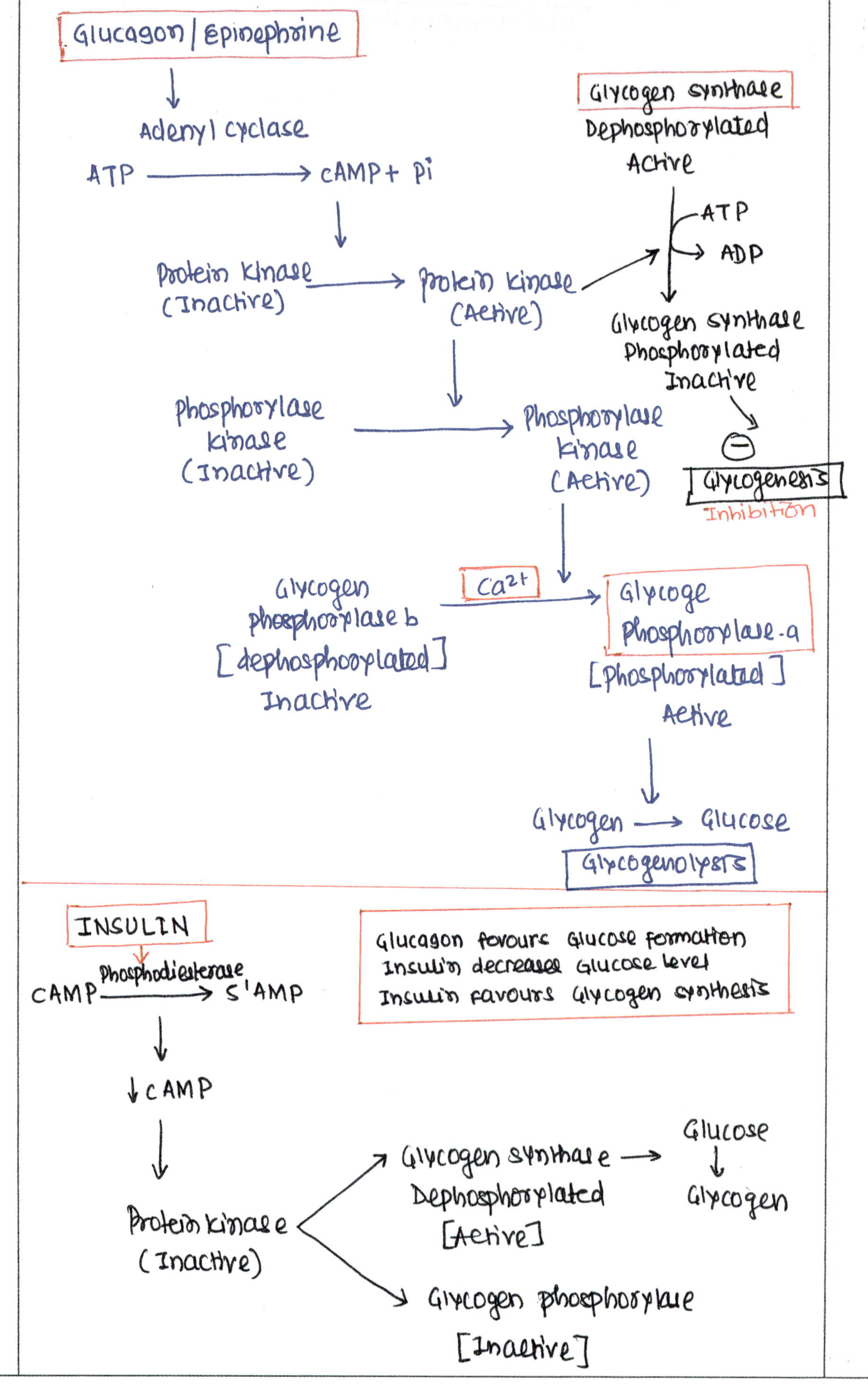
Glucagon / epinephrine
Adenyl cyclase
ATP
cAMP + Pi
Protein kinase (Inactive)
Protein kinase (Active)
Glycogen synthase Dephosphorylated Active
ATP
ADP
Glycogen synthase Phosphorylated Inactive
Glycogenesis
Inhibition
Phosphorylase kinase (Inactive)
Phosphorylase kinase (Active)
Glycogen phosphorylase b [dephosphorylated] Inactive
Ca^{2+}
Glycoge Phosphorylase-a [Phosphorylated] Active
Glycogen
Glucose
Glycogenolysis
INSULIN
CAMP
Phosphodiesterase
5'AMP
↓CAMP
Protein kinase (Inactive)
Glycogen synthase Dephosphorylated [Active]
Glucose
Glycogen
Glycogen phosphorylase [Inactive]
Glucagon favours Glucose formation
Insulin decreases Glucose level
Insulin favours Glycogen synthesis

SAQ

Glycogen storage disorders

Mnemonic - Versatile Player Corey Anderson Made His - Glycogen

Type	Name	Deficient Enzyme	Clinical features
Type I	Von Gierke's disease	Glucose 6 phosphatase MCQ	Fasting hypoglycemia Hyperlipidemia Hyperuricemia Lactic acidosis Hepatomegaly.
Type II	Pompe's disease	Lysosomal Acid Maltase [α(1-4) glucosidase] MCQ	Hypotonia, Floppy infant Poor feeding Glycogen accumulation in heart, liver, tongue breathing problem
Type III	Cori's disease [Limited dextrinosis] [Forbe's disease]	Debranching enzyme	fasting hypoglycemia Hepatomegaly, myopathy Dextrin accumulates
Type IV	Anderson's disease [Amylopectinosis]	Branching enzyme	Hepatosplenomegaly Cirrhosis of liver death at early age
Type V	McArdle's disease	Muscle phosphorylase MCQ	Exercise intolerance second wind myoglobinuria Accumulation of glycogen in muscles.
Type VI	Her's disease	Liver Phosphorylase	Hepatomegaly fasting hypoglycemia

Mnemonic to remember Type III & IV → ABCD

Anderson's disease - Branching enzyme

Cori's disease - Debranching enzyme.

Von Gierke's disease - Type Ia GSD

Deficient enzyme :- Glucose 6 Phosphatase

Key points - Fasting hypoglycemia
- Uric acid ↑↑
- TG, Cholesterol ↑↑
- Lactic acid ↑↑

Glu 6 Ⓟ
↓ ✗ Glucose 6 Phosphatase
Glucose

C/F :- Onset 3-4 months of life
- Doll like face c̄ fat cheeks
- Thin extremities
- Protuberant abdomen
- Short stature
- Hepatomegaly, kidney ↑

No splenomegaly
Muscle involvement

Reason for **Hyperuricemia**

Glu 6 Ⓟ —HMP shunt→ Ribose 5 Ⓟ
Glu 6 Ⓟ ✗→ Glucose
Ribose 5 Ⓟ ↓ IMP, AMP, GMP purine nucleotides
↓ degradation
Uric acid ↑

Reason for **Hyperlipidemia**

Glucose 6 Ⓟ —✗→ Glucose
↓
Acetyl CoA
→ ↑ Cholesterol
→ Fatty acids ↓ Triacylglycerol ↑

Treatment :- Protein rich diet, symptomatic t/t
Allopurinol for Hyperuricemia
Statin's for Hypercholesterolemia

Pompe's disease - GSD Type II

Deficient enzyme :- Lysosomal acid maltase
α(1-4) glucosidase

Also belongs to- Lysosomal storage disorder

C/F :- Age of onset → 1st few months of life
- Hypotonia
- Myopathy (floppy infant)
- Involvement of Heart, liver and tongue
- poor feeding, failure to thrive
- breathing problem, SN deafness

Lab Δ :- leukocyte, liver, muscle - biochemical assay for enzyme (acid maltase)
Biopsy - Accumulation of glycogen.

T/t :- Recombinant enzyme - α glucosidase
[Myozyme, Lumizyme]

McArdle's disease - GSD V

clinical case

Deficient enzyme :- Muscle glycogen phosphorylase MCQ

C/F :- Exercise intolerance [Fatigue & Cramp in early few minutes of exercise]
- Second wind
  [If He/she takes rest for few minute, He/she can resume exercise with little discomfort]
- Myoglobinuria [due to rhabdomyolysis] on prolonged activity

Lab Δ :- Muscle biopsy - Accumulation of glycogen
↓ Activity of glycogen phosphorylase - Enzyme assay

T/t :- Dietary modification
Training - walking, cycling [Aerobic exercises]

Clinical case - 1

An infant brought to the hospital with chief complaints of distended abdomen and short stature. On examination, liver and kidneys were enlarged. Laboratory finding shows Low plasma glucose, Uric acid - 10 mg%, serum cholesterol - 278 mg%.

What is your probable diagnosis?

Clinical case 2

A neonate was brought to the hospital with hepatomegaly. Cardiac ultrasound indicates cardiac hypertrophy. The neonate was floppy with diminished muscle tone. The activity of α-Glucosidase was low in leukocyte, muscle and liver.

What is your probable diagnosis?

Clinical case 3

A patient came to OPD with chief complaints of fatigability that was worsened on exertion. He lagged behind when he went out with his friends. A muscle biopsy showed subsarcolemmal accumulation of lakes of glycogen.

What is your probable diagnosis?

Clinical case 4

A 6-Year-old child was brought to the hospital with chief complaints of excessive fatigue and muscle pain after playing in the school. Blood investigations were done immediately after the exercise & the result showed low plasma Glucose & Lactate levels. Muscle glycogen was abnormally high.

What is the most probable diagnosis in this child?

Answer → McArdle's disease [GSD type V]

Defective enzyme – Muscle glycogen phosphorylase

Hexose Monophosphate Shunt [HMP Shunt]

SAQ:- Write a note on HMP shunt with its significance

Other names

Pentose phosphate pathway - Pentose sugars are intermediate in this Pathway

Phosphogluconate pathway - phosphogluconate is an intermediate

Monophosphate ; because of generation of monophosphates
In Glycolysis - bisphosphates

Shunt pathway :- It is an alternative pathway for oxidation of glucose into CO_2

Site

Oxidative phase :- liver, adipose tissue, RBC's

Non oxidative phase ;- All cells of the body

Significance

provides **NADPH** and **Pentose sugar** — synthesis of DNA, RNA NAD, NADP, FAD etc.

Functions of NADPH

1. Required for the synthesis of fatty acids, cholesterol, steroids etc
2. Free radical scavenger :- required for the action of glutathione reductase in production of reduced glutathione.
3. Maintenance of RBC membrane integrity

NADPH → Glutathione Reductase → Glutathione ↓ Reduced Glutathione

prevents MetHb formation.

4. Transparency of lens
5. Detoxification by NADPH dependent cyt P450 system

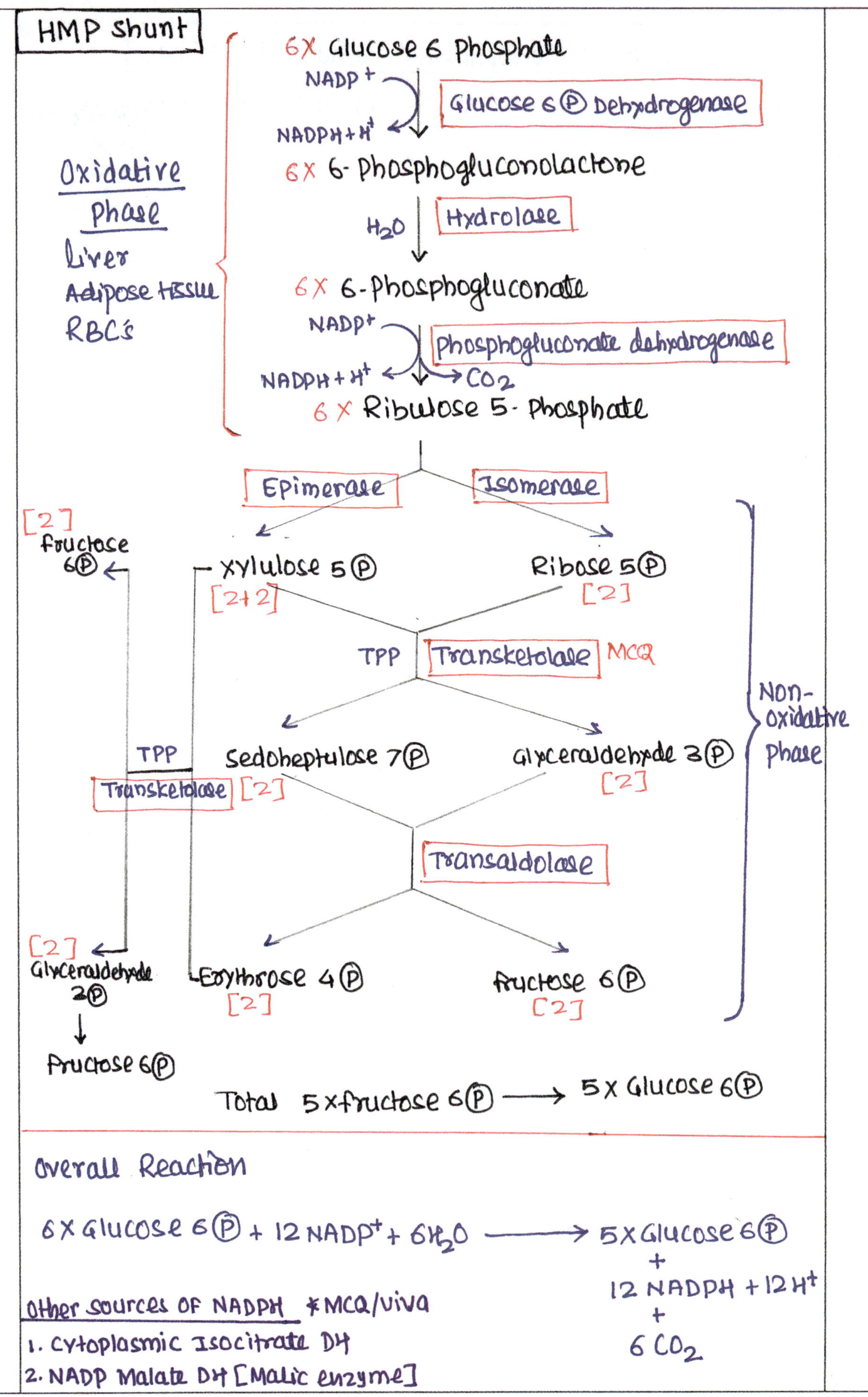

HMP shunt
6X Glucose 6 Phosphate
NADP+
NADPH+H+
Glucose 6 Ⓟ Dehydrogenase
Oxidative Phase
Liver
Adipose tissue
RBC's
6X 6-Phosphogluconolactone
H2O
Hydrolase
6X 6-Phosphogluconate
NADP+
NADPH+H+
Phosphogluconate dehydrogenase
CO2
6 X Ribulose 5-Phosphate
Epimerase
Isomerase
[2]
fructose 6Ⓟ
Xylulose 5Ⓟ
[2+2]
Ribose 5Ⓟ
[2]
TPP
Transketolase
MCQ
Non-oxidative phase
Sedoheptulose 7Ⓟ
[2]
Glyceraldehyde 3Ⓟ
[2]
TPP
Transketolase
Transaldolase
[2]
Glyceraldehyde 3Ⓟ
Erythrose 4Ⓟ
[2]
Fructose 6Ⓟ
[2]
Fructose 6Ⓟ
Total 5xfructose 6Ⓟ → 5x Glucose 6Ⓟ
Overall Reaction
6X Glucose 6Ⓟ + 12 NADP+ + 6H2O → 5X Glucose 6Ⓟ + 12 NADPH + 12H+ + 6 CO2
Other sources of NADPH *MCQ/viva
1. Cytoplasmic Isocitrate DH
2. NADP Malate DH [Malic enzyme]

Clinical Importance

* Glucose 6 ⓟ dehydrogenase deficiency [G6PD deficiency]

oxidant drugs
[primaquine, sulfonamides, Acetanilide]
↓
↑ free Radicals

G6PD deficiency → ↓ NADPH → ↑ Met Hb formation
↓ Glutathione reductase activity

Damaged RBCs

Hemolytic Jaundice
(↑ Heme breakdown)

- X-linked recessive disease
- Drug induced hemolytic anemia
- As such asymptomatic but when exposed to oxidant drugs, leads to hemolytic anemia
- Provides resistance against Malaria.

Clinical case

A 34 year old Man suffering from fever c̄ chills was diagnosed as malaria patient. He was given antimalarial drug primaquine. After few days, he complained of passing red colored urine and weakness. Laboratory data showed Hb level 7 gm% and unconjugated bilirubin 6.1 mg%.

What is your probable diagnosis? Hemolytic Jaundice due to G6PD deficiency.

Why rise in unconjugated bilirubin?

G6PD ↓ → ↓ NADPH → ↑ MetHb, ↓ Glutathione reductase → Rapid destruction of RBCs → Heme breakdown → ↑ unconjugated bilirubin

Why Anemia in this disease?

↓ NADPH → ↓ Reduced Glutathione → RBC Loss of membrane integrity → Hemolysis (Anemia)

GALACTOSEMIA — Galactose Metabolism

SAQ :- Write a note on galactosemia.
Write in short about galactose metabolism and disorder related c̄ it.

GALACTOSE —(Aldose reductase)→ Galactitol (Dulcitol) → Oil droplet cataract

GALACTOSE ↓ Galactokinase

Galactose 1 phosphate

Galactosemia ✗ ↓ Galactose 1 Ⓟ uridyl transferase

UDP-galactose

↓ Epimerase

UDP-glucose

Galactose source
→ Milk & fruits
→ Absorb c̄ Na^+
Na^+ dependent symport

- Autosomal recessive disorder
- Galactosemia refers to a group of disorders that impair the bodys ability to process and produce energy from galactose.
- It can be due to deficiency of any of the following three enzymes caused by mutation in genes.

Type 1 - Classical Galactosemia

- Defective enzyme - Galactose 1 Ⓟ uridyl transferase
- Galactose 1 Ⓟ accumulates in blood and liver, inhibit galactokinase and glycogen phosphorylase - causes hypoglycemia
- Galactose is reduced to dulcitol (galactitol). Accumulation of dulcitol in lens causes cataract due to osmotic effect. *MCQ
- Due to accumulation of galactose 1 Ⓟ, bilirubin uptake is less and conjugation is reduced, leads to increase in unconjugated bilirubin (Jaundice).

C/F :- fasting hypoglycemia
vomiting, diarrhea
failure to thrive
congenital cataract
Hepatomegaly, Jaundice
Mental retardation

Treatment :- Lactose free diet
commercial galactose free preparations

Type 2 :- Galactokinase deficiency
causes development of cataract only.

Type 3 :- Galactose epimerase deficiency.
Asymptomatic condition.

Clinical case

one year old child was brought to the Hospital with chief complaints of vomiting, jaundice, lethargy. Examination revealed Hepatosplenomegaly and cataract. Mucic acid test was positive. plasma glucose was 45mg%

What is your probable diagnosis?
Which enzyme is deficient?
What is the cause for cataract?
What is the cause for jaundice?

Clinical case-2

A newborn refuses breast milk & vomits on forceful feeding but accept glucose-water, develops diarrhea on 3rd day, by 5th day she is jaundiced with liver enlargement and eyes showed cataract. Benedict's test was positive on urine sample but plasma glucose was low by GOD-POD method.

What is your probable diagnosis? Classical Galactosemia

- Benedict's test positive → Reducing sugar → Galactose
  GOD-POD method → specific for β-D-Glucose
  Not for Galactose

- Hypoglycemia because Galactose 1 Ⓟ Inhibit galactokinase and Glycogen Ⓟrylase enzyme.

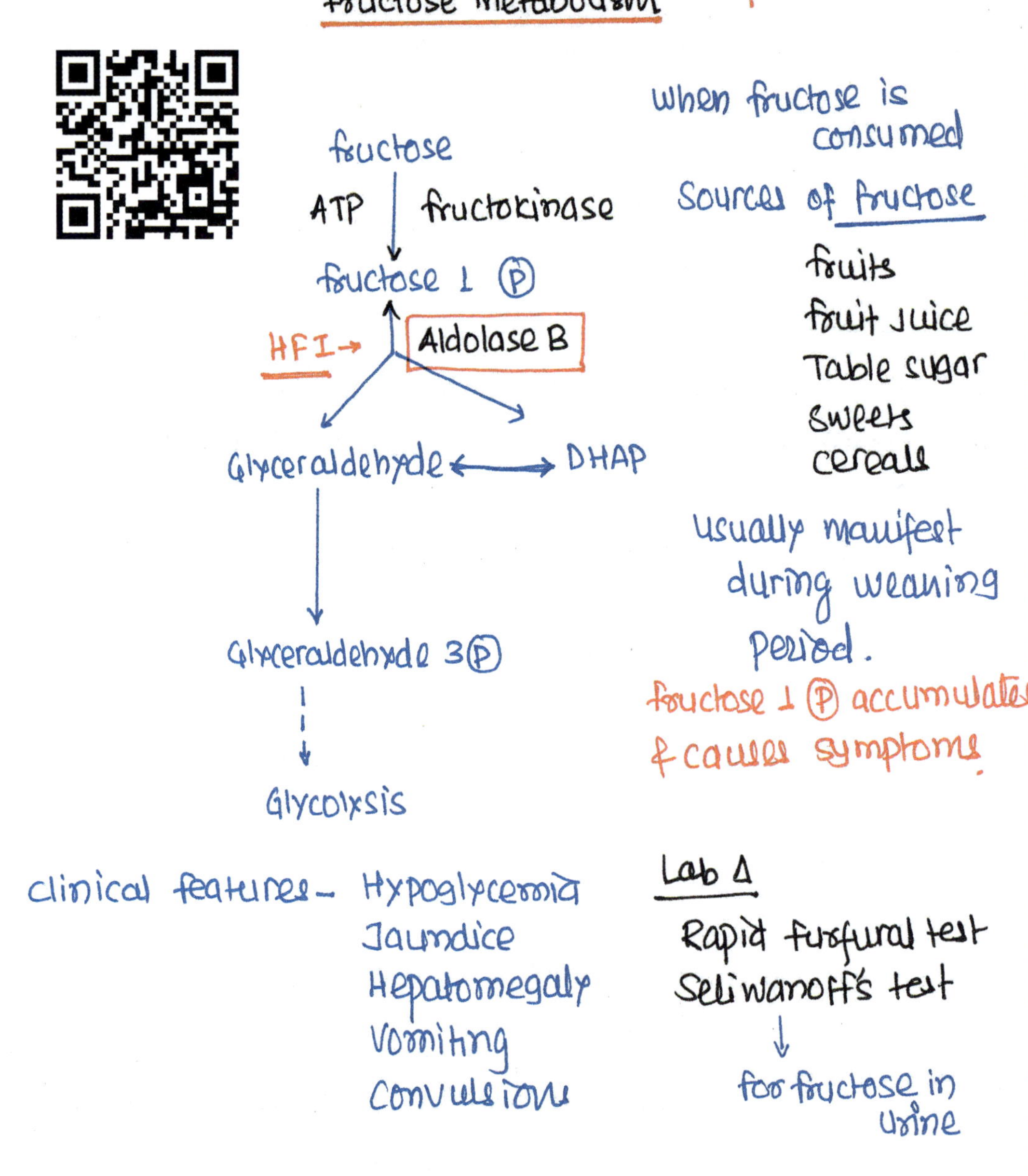

t/t : Avoid fructose & sucrose in the diet

Regulation of blood glucose

LAQ: Describe the regulation of blood glucose level in our body. [very Important LAQ]

Normal level

fasting plasma glucose :- 70-110 mg/dl
Postmeal plasma glucose :- 110-140 mg/dl
Random plasma glucose :- 70-140 mg/dl

Sources of blood glucose and utilization

Sources

1. Dietary by absorption
2. Glycogenolysis
3. Gluconeogenesis

Utilization

Glycolysis/TCA/HMP shunt
Glycogenesis
Lipogenesis

Effect of Hormones

1. Insulin → lowers blood glucose
   Increases glucose uptake by cells
   Stimulates glycolysis
   Stimulates Glycogenesis
   Inhibit gluconeogenesis
   Inhibit glycogenolysis
2. Glucagon → Increases blood glucose
   Stimulates glycogenolysis
   Stimulates gluconeogenesis
3. Epinephrine → Increases blood glucose
   Stimulates glycogenolysis and gluconeogenesis
4. Growth hormone, Cortisol, ACTH, Thyroid hormones
   Increases blood sugar level

Pathways Involved in regulation → Glycogen metabolism, Gluconeogenesis, Glycolysis

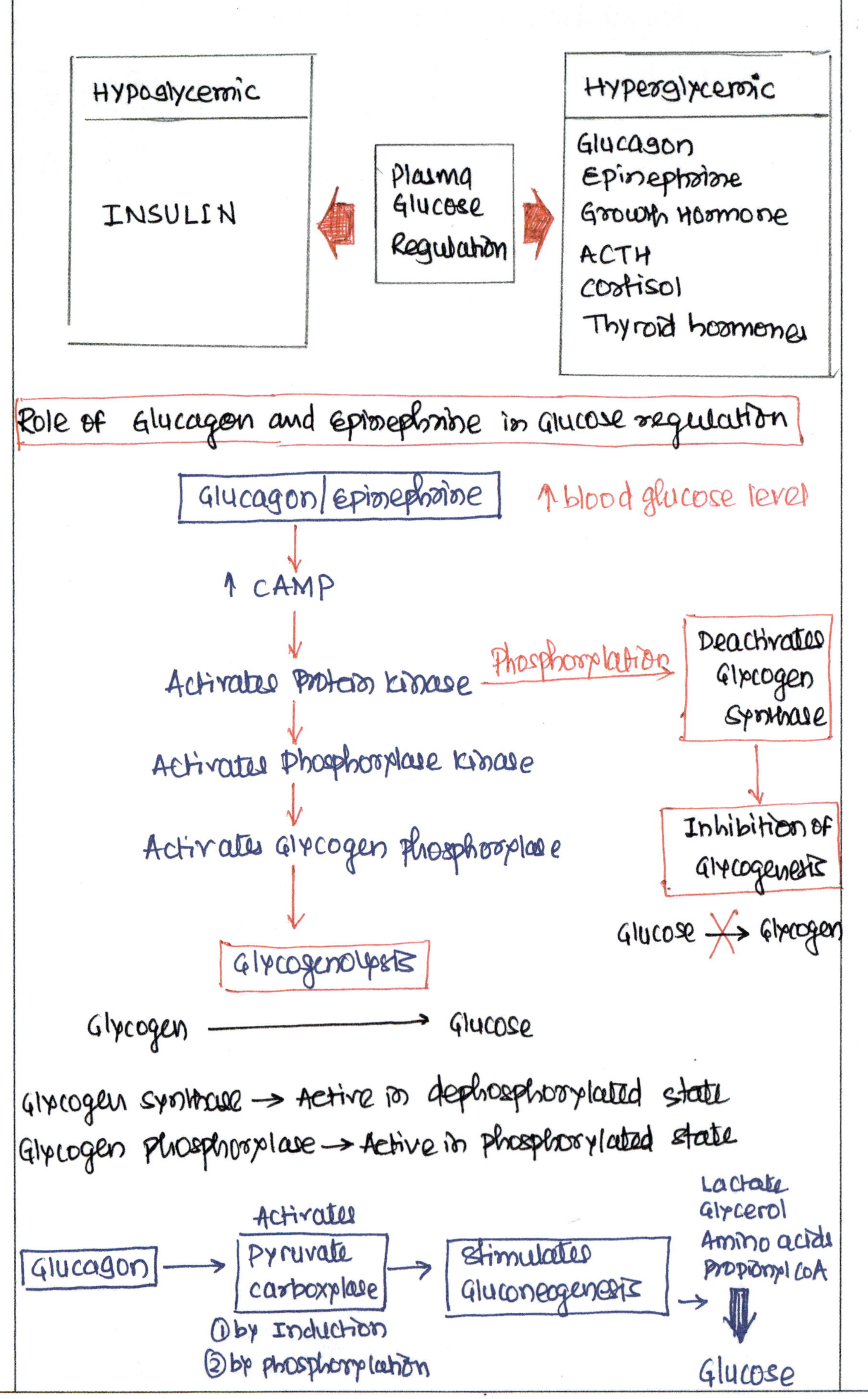

Hypoglycemic
INSULIN
Plasma Glucose Regulation
Hyperglycemic
Glucagon
Epinephrine
Growth Hormone
ACTH
Cortisol
Thyroid hormones
Role of Glucagon and epinephrine in Glucose regulation
Glucagon/epinephrine
↑ blood glucose level
↑ CAMP
Activates Protein kinase
Phosphorylation
Deactivates Glycogen Synthase
Inhibition of Glycogenesis
Glucose → Glycogen
Activates Phosphorylase kinase
Activates Glycogen phosphorylase
Glycogenolysis
Glycogen → Glucose
Glycogen synthase → Active in dephosphorylated state
Glycogen phosphorylase → Active in phosphorylated state
Glucagon
Activates
Pyruvate carboxylase
① by Induction
② by phosphorylation
Stimulates Gluconeogenesis
Lactate
Glycerol
Amino acids
Propionyl CoA
Glucose

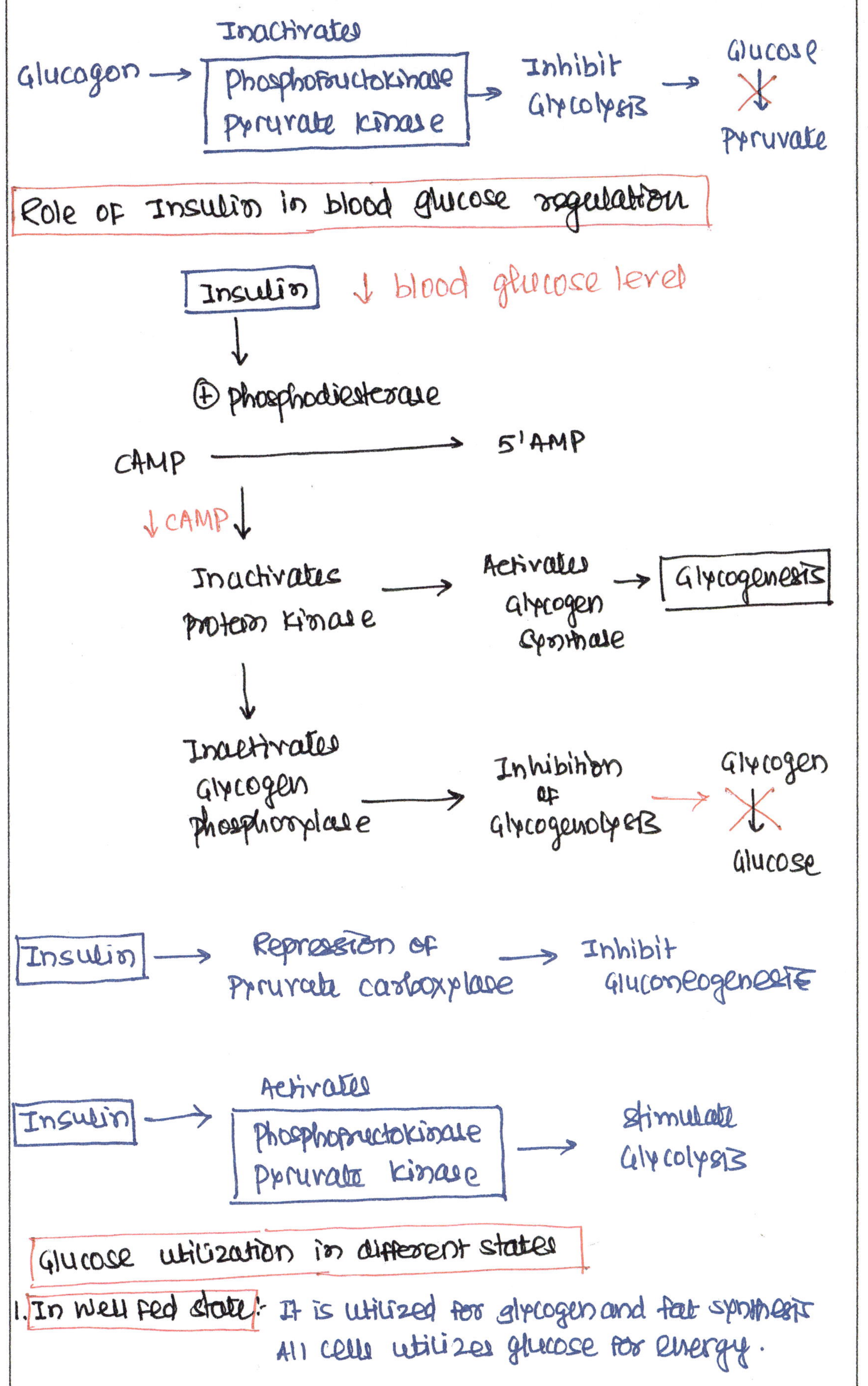

Glucagon → Inactivates [Phosphofructokinase, Pyruvate kinase] → Inhibit Glycolysis → Glucose ✗→ Pyruvate

Role of Insulin in blood glucose regulation

Insulin ↓ blood glucose level

↓

⊕ phosphodiesterase

CAMP → 5'AMP

↓ CAMP ↓

Inactivates protein kinase → Activates glycogen synthase → **Glycogenesis**

↓

Inactivates glycogen phosphorylase → Inhibition of Glycogenolysis → Glycogen ✗→ Glucose

Insulin → Repression of Pyruvate carboxylase → Inhibit Gluconeogenesis

Insulin → Activates [Phosphofructokinase, Pyruvate kinase] → Stimulate Glycolysis

Glucose utilization in different states

1. **In well fed state**: It is utilized for glycogen and fat synthesis. All cells utilizes glucose for energy.

2. **Post absorptive state**: Source of glucose is liver glycogen. Glycogenolysis will provide glucose for 4-6 hrs after meal.

3. **Starvation**:- After 16-18 hrs of fast, glycogen stores get completely depleted.
Gluconeogenesis in liver and kidneys is a main source of glucose after 16-18 hrs.
Brain utilizes glucose and ketone bodies

4. **Prolonged starvation**:- After several days, main source of glucose is by gluconeogenesis through muscle proteins. Brain completely depends on ketone bodies as a fuel.

Clinical significance

Hyperglycemia
[↑ blood sugar]

1. Diabetes mellitus
2. Hyperadrenalism
3. Hyperpituitarism
4. Hyperthyroidism

Hypoglycemia
[↓ blood sugar]

1. Improper dose of Insulin
2. Insulinoma
3. Hypoadrenalism
4. Hypopituitarism
5. Hypothyroidism.

Renal threshold ⟶ 180 mg/dl
Level of plasma glucose beyond which it is excreted in urine.

Diabetes Mellitus (DM)

Metabolic disease due to absolute or relative deficiency of Insulin.

Important clinical symptoms

1. **Polyuria** — frequent urination
   ↓
   As blood sugar level crosses the renal threshold (180 mg%), sugar (Glucose) starts appearing in urine and drags water c̄ it due to osmotic effect.

2. **Polyphagia**
   ↓
   Excessive eating
   - ↓ uptake of glucose by cells
   - cells can't convert food into energy which stimulates hunger centres
   - Causes polyphagia

3. **Polydipsia**
   ↓
   Undue thirst
   Loss of water in urine stimulates thirst centre which leads to the Polydipsia. [because of dehydration]

Diagnostic criteria

fasting plasma glucose	> 126 mg/dl (7.0 mmol/L)
2 Hr plasma glucose	> 200 mg/dl (11.1 mmol/L)
Random plasma glucose	> 200 mg/dl c̄ symptoms of DM
HbA1c (Glycated Hb) (*) MCQ	> 6.5 % [Gives idea about average blood glucose level for preceding 6-8 wks]

	Type 1 DM (IDDM)	Type 2 DM (NIDDM)
Age of onset	Childhood (up to 20 yr)	> 30 yrs age group
Prevalence	10-20%	80-90%
Defect	Insulin deficiency due to pancreatic destruction	Decreased production OR Resistance to insulin
Weight	Normal or Low	Obese
Genetic predisposition	Less	very strong
Autoantibodies	commonly found	Rare
Complication	Ketoacidosis	Hyperosmolar state
Treatment	Only insulin is effective	Oral hypoglycemic drugs. Insulin

Metabolic changes in DM

↓ INSULIN

- ↑ Gluconeogenesis → ↑ blood glucose
- ↑ Glycogenolysis → ↑ blood glucose
- ↑ Lipolysis → ↑ fatty acid & glycerol → ↑ Ketone bodies, ↑ TG

Complications

Acute complication

① Hyperosmolar state
common c̄ type 2 DM
severe dehydration and very high plasma glucose.

② Diabetic ketoacidosis
ketone bodies in urine
In type 1 DM.

③ very high or very low glucose level.

Chronic complications

① Retinopathy (involvement of retina of eye)

② Diabetic foot :- Nerve damage can affect the feeling in feet and raised blood sugar can damage the circulation, causes delayed wound healing.

③ Heart attack & stroke

④ Nephropathy (kidney problem)

⑤ Nerve damage (Neropathy)

⑥ Cataract :- Snow flakes cataract, due to deposition of <u>Sorbitol</u> in lens. MCQ

⑦ Glaucoma, Gum infection, sexual problems etc.

Glucose Tolerance Curve GTT Graphs

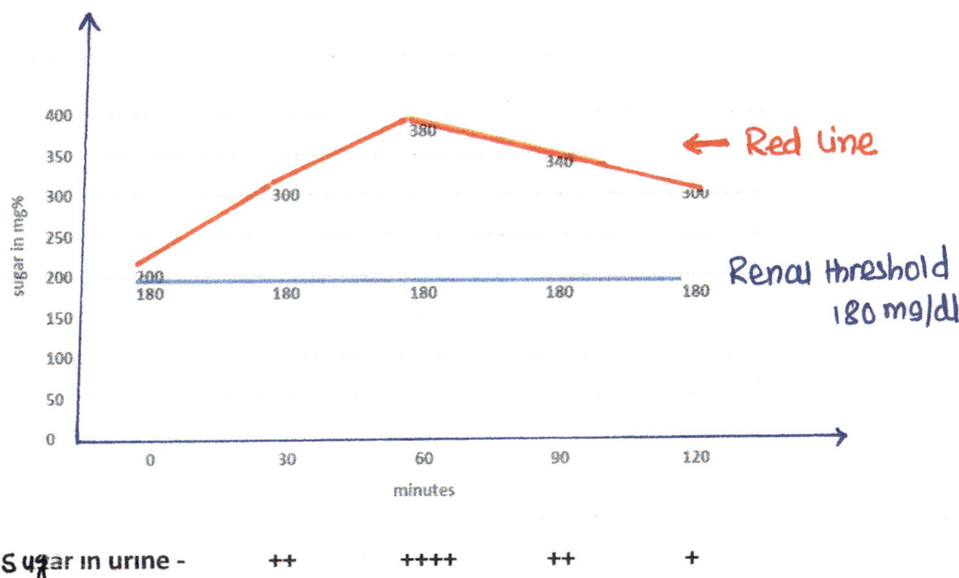

Sugar in urine - ++ ++++ ++ +

It is a glucose tolerance curve of diabetes mellitus

1. Fasting blood sugar level is above normal limit
2. The peak is above renal threshold level
3. Decline curve does not reach to its fasting within 2hrs.
4. Presence of glucose in urine.

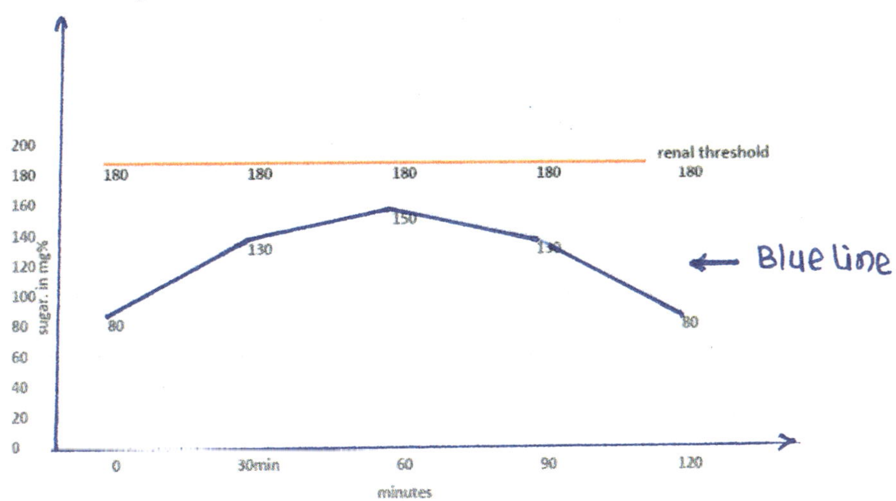

Sugar in urine - - - - -

It is Normal glucose tolerance curve because

1. Fasting blood sugar level is within the normal limits.
2. The peak is below the renal threshold level.
3. Decline curve reaches to its fasting level within 2 hrs.(120min)
4. Absence of glucose in urine.

Diabetic Ketoacidosis [DKA]

It is an acute complication of type 1 DM

Lab Δ: plasma glucose > 250 mg/dl

Low blood pH level < 7.3

Low HCO_3^- level < 15 mmol/L

presence of ketone bodies in blood & urine

Hyperkalemia (↑ serum K^+)

Clinical Manifestations :-

1. Confusion and loss of consciousness – because of dehydration and metabolic acidosis
2. Kussmaul breathing - deep and rapid breathing - due to compensatory hyperventilation to blow CO_2 out of lung in response to metabolic acidosis.
3. Dehydration :- Glycosuria (osmotic diuresis), loss of water & electrolytes
4. low blood pressure
5. Vomiting, excessive urination, Abdominal pain

↓ Insulin

↓

Elevated Glucagon, Epinephrine, cortisol, GH

Left branch:

Decreased Glucose utilization

Increased Glycogenolysis

Increased Gluconeogenesis

↑ blood glucose

Right branch:

↑ Lipolysis

↓

↑ Free FA and Glycerol

↓

↑ Ketogenesis

↓

↑ Acetone, Acetoacetate, β-OH Butyrate

↓

HCO_3^- is utilized in buffering these acids

↓

Metabolic Acidosis

In DKA pH < 7.3

In DKA pH decreases, means $[H^+]$ ion concentration increases in extracellular fluid. $[H^+]$ ions moves inside the cell in exchange of K^+ ions. That causes ↑ K^+ ion concentration in ECF hence Hyperkalemia.

When Insulin is given in the treatment, Insulin causes rapid opening of K^+ channels, that causes entry of K^+ inside the cell and causes Hypokalemia.

Hypokalemia can cause abnormal heart rythms and death hence strict K^+ monitoring is needed in t/t of DKA.

Treatment of DKA

- Insulin
- IV Fluids
- strict K^+ monitoring

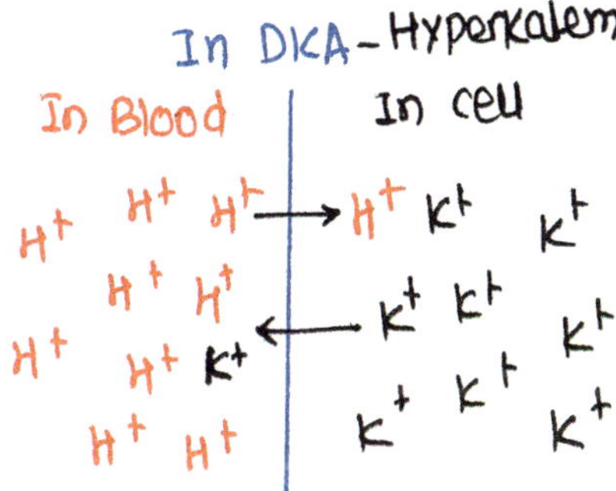

on Insulin therapy

Rapid opening of K^+ channels.

In Blood

entry of K^+ inside

In cell

Insulin therapy causes Hypokalemia

CHEMISTRY OF LIPIDS

MCI COMPETENCY BI 4.1

Short answer questions

1. Define and classify lipids c̄ suitable examples.
2. Define and classify fatty acids
3. Essential fatty acids
4. Phospholipids c̄ their clinical significance
5. Eicosanoids c̄ their clinical significance
6. Lipoproteins c̄ their functions. 7. Lipid profile tests

Important clinical cases.

Respiratory distress syndrome

- occurs in premature babies whose lungs are not fully developed.
- due to deficiency of surfactant - Dipalmitoyl Lecithin

Symptoms :- Rapid breathing c̄ grunting sound
Cyanosis (blue colored lips, fingers & toes)
Flaring of nostril
Chest retractions

Lab Δ :- Chest X-ray
Prenatal ⇒ Amniotic fluid analysis shows L/S ratio < 1

Dipalmitoyl Lecithin acts as a respiratory surfactant. It decreases surface tension of the fluid and keeps the alveoli open, prevent the collapse of lung during breathing.

L/S Ratio for full term infant > 2
[Lecithin : Sphingomyelin]

T/t :- Continuous Positive Airway Pressure
Surfactant can be given
Steroids and Antibiotic

Clinical case

A premature baby brought to the pediatric department with chief complaints of rapid breathing, blue colored lips & fingers, flaring of alae nasi and retraction of chest.

What is your probable diagnosis?

Name the deficient compound.

LIPIDS

SAQ :- Define and classify Lipids with suitable examples.

Definition :- organic compounds, relatively insoluble in water and soluble in organic solvents [Ether, benzene, chloroform, hot Alcohol, Acetone etc] and related c̄ fatty acids.

Classification

Simple Lipids [FA + Alcohol]
1. Triacylglycerol [fatty acid + glycerol]
2. Waxes [Fatty acid + Cetyl alcohol]

Complex lipid [FA + Alc + group]
1. Phospholipid [FA + Alc + N-Base + Ⓟ]
   - Lecithin
   - Cephalin
2. Glycolipid [FA + Alc + carbohydrate]
   - Cerebroside
   - Ganglioside
3. Lipoprotein [Lipid + protein]
   - Chylomicron
   - LDL
   - HDL
   - VLDL
4. Other - Sulfolipids, Amino lipids

Derived
- Fatty acids
- Glycerol
- Steroids
- Cholesterol
- Ketone bodies
- Prostaglandins

Miscellaneous
Posses characters of lipids
- Carotenoids
- Terpens
- Squalene
- Pentacepone

$$\begin{array}{l} CH_2-O-\overset{O}{\overset{\|}{C}}-R_1 \\ | \\ R_2-\overset{O}{\overset{\|}{C}}-O-CH \\ | \\ CH_2-O-\underset{O}{\underset{\|}{C}}-R_3 \end{array}$$

Triacylglycerol

GLYCEROL — FA1, FA2, FA3

Phospholipids

SAQ :- Enumerate different types of phospholipids. Add a note on their role in human body.

complex / compound lipid

composed of :- FA + Alcohol + Ⓟ

Examples

1. Phosphatidic acid [simplest precursor]
2. Lecithin
3. Cephalin
4. Phosphatidyl serine
5. Phosphatidyl threonine

] Nitrogen Base containing

6. Phosphatidyl inositol
7. Phosphatidyl glycerol
8. Cardiolipin

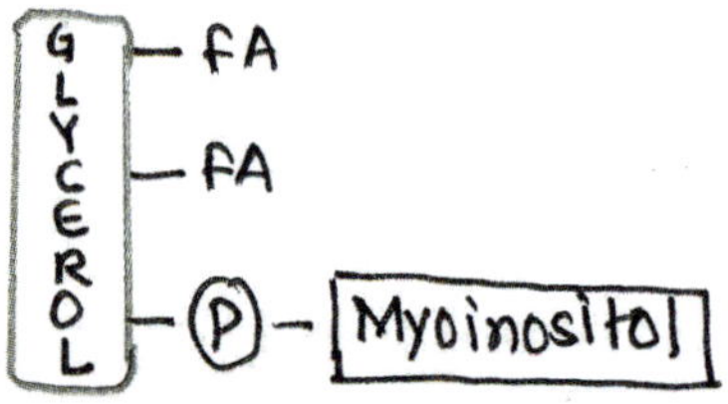

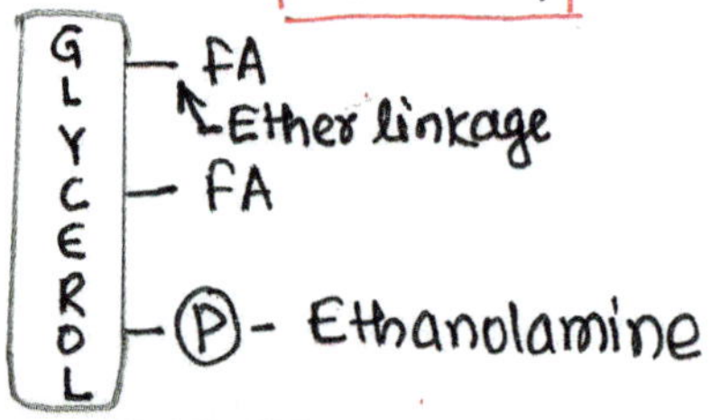

] No Nitrogen base

9. Plasmalogen
10. Platelet activating factor (PAF)

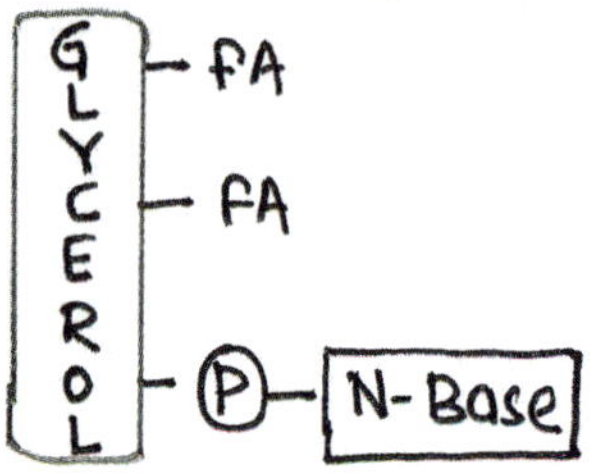

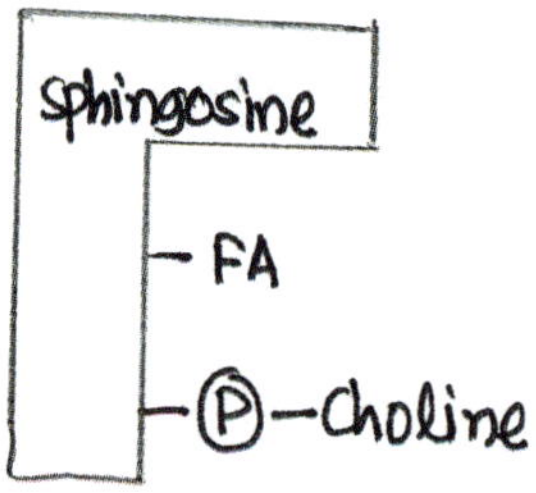

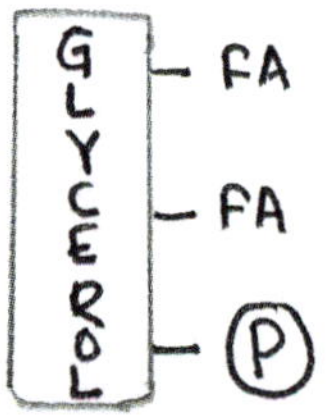

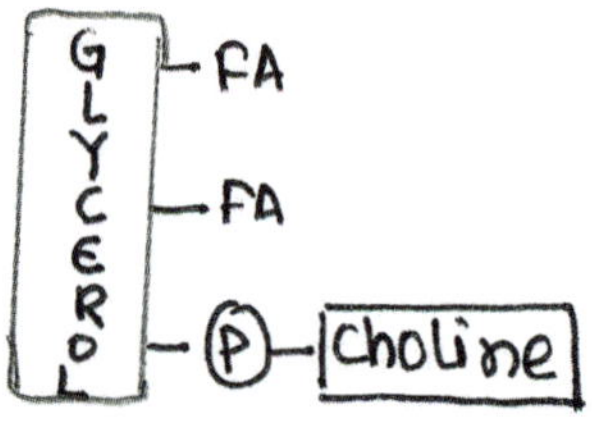

Example : sphingomyelin

Functions of Phospholipids

1. Structural component of plasma membrane and responsible for transport across the plasma membrane.

2. **Lecithin** [phosphatidyl choline] **Deficiency - ARDS**
   - Acts as a respiratory surfactant, prevents the collapse of lung. Deficiency of Dipalmitoyl lecithin causes respiratory distress syndrome.
   - It is depot for choline – help in Nerve transmission
     - Imp. in transmethylation reaction.
   - Imp. in Reverse transport of cholesterol c̄ LCAT & HDL

3. **Cephalin**
   - Helps in coagulation
   - Converts prothrombin to thrombin

4. **Phosphatidyl serine**
   - Component of cell membrane
   - Mediator of programmed cell death

5. **Phosphatidyl Inositol**
   - Phosphatidyl Inositol → Diacylglycerol, Inositol triphosphate] second messenger in hormone action.

6. **Cardiolipin**
   - First isolated from cardiac muscle
   - Present in inner mitochondrial membrane
   - Defect causes - Mitochondrial dysfunction
   - Responsible for antigenicity

7. **Platelet Activating factor**
   - Platelet aggregation
   - Mediator in – Inflammation, chemotaxis, Allergic responses.

8. **Sphingomyelin**
   - Formation of Myelin – Nerve conduction
   - Formation of lipid rafts in cell membrane

9. Helps to dissolve cholesterol in bile acids.

   Phospholipids also help in digestion and absorption of lipids.

Clinical Significance < Acute Respiratory Distress Syndrome; Antiphospholipid syndrome

Antiphospholipid syndrome is an autoimmune disease, in which antibodies are produced against phospholipids on plasma membrane. Causes formation of blood clots that causes venous thrombosis and pregnancy related complications [recurrent miscarriage, IUGR, Preterm birth] and stroke.

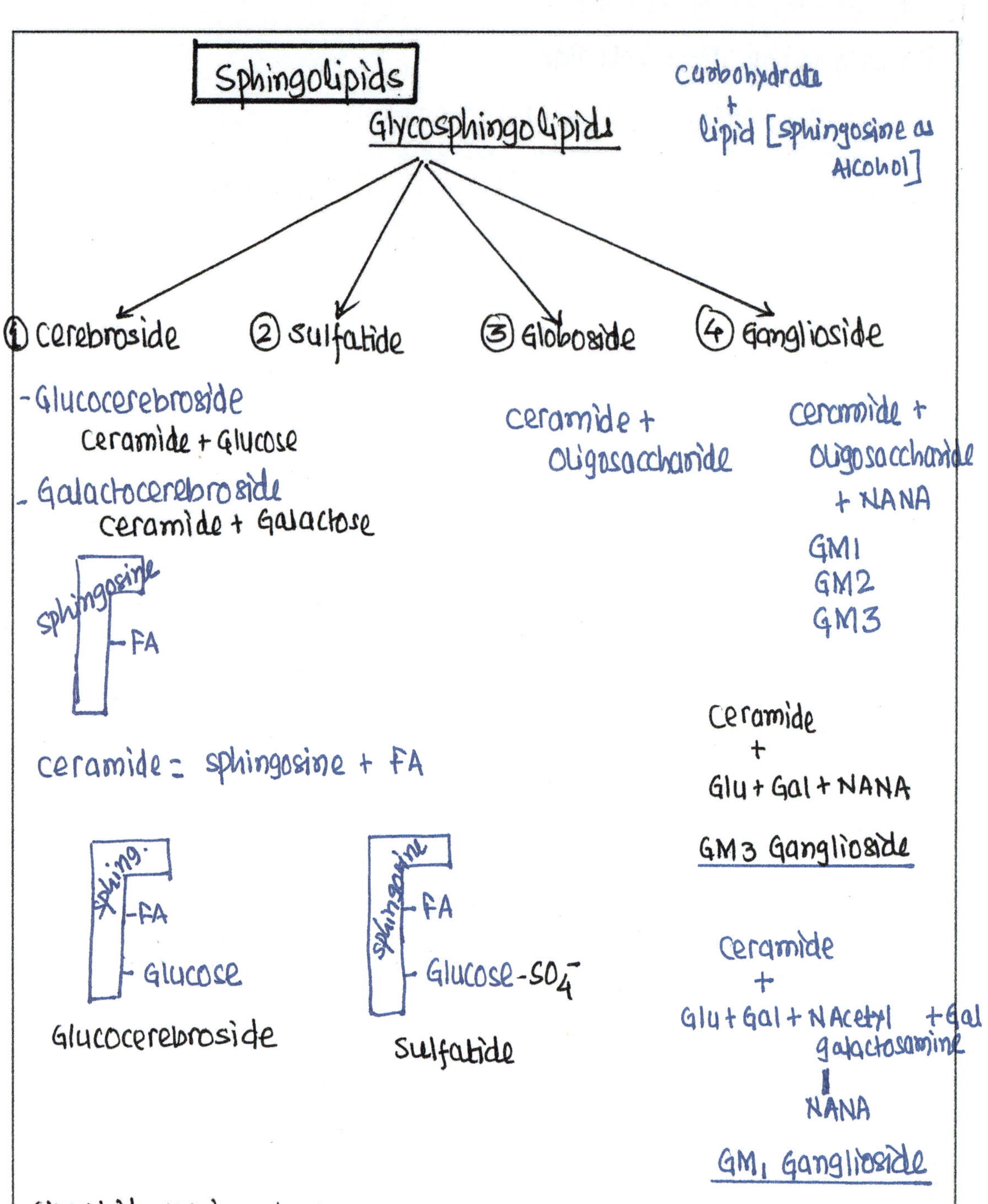

- Glycolipids are important constituent of nervous tissues & cell membrane
- They are cell surface receptors for many hormones & drugs.
- They play important role in cellular interaction, growth & development.
- GM1 ganglioside - are receptor for cholera toxin MCQ
- Defect in lysosomal hydrolases causes accumulation of glycolipids and diseases known as sphingolipidosis.

Clinical case

Lipid storage disorders / sphingolipidosis

A 23-Year-old man with H/o sudden rectal hemorrhage was referred to Medical college. On examination hepatosplenomegaly was noted. Liver biopsy demonstrated – Tissue paper macrophages. Investigations revealed anemia, leukopenia and thrombocytopenia. X-ray showed Erlenmeyer flask deformity in leg bones and eye examination revealed – No cherry red spot on retina.

What is your probable diagnosis?

Ans. Gaucher's disease

Enzyme deficient :- Glucocerebrosidase OR β-Glucosidase

AR disease – Lysosomal storage disorder

Involves Bone marrow, liver, spleen

Bone marrow → ↓ synthesis
- Anemia (↓RBC)
- leukopenia (↓WBC)
- Thrombocytopenia (↓ platelets) → cause hemorrhage & bruising

liver, spleen → Hepatosplenomegaly.

Glucocerebroside → (β-Glucosidase) → ceramide + Glucose

MCQ Accumulation of glucocerebroside in macrophages gives
* wrinkle paper OR Tissue paper appearance in Microscopy

MCQ Avascular necrosis of bones causes frequent fractures.

KGF → No cherry Red spot on Retina :- Gaucher's / Krabbe's / Fabry's ds

Disease	Deficient Enzyme	Substance Accumulate
Gaucher's disease	β-Glucosidase	Glucocerebroside
* Niemann-Pick disease	Sphingomyelinase	Sphingomyelin * MCQ
Krabbe's disease	β-Galactosidase	Galactocerebroside
Fabry's disease	α-Galactosidase	GM1 ganglioside
Tay-Sach's disease	Hexosaminidase A	GM2 ganglioside
Sandhoff disease	Hexosaminidase A & B	Globoside
Metachromatic leukodystrophy	Arylsulfatase	Sulfogalactocerebroside

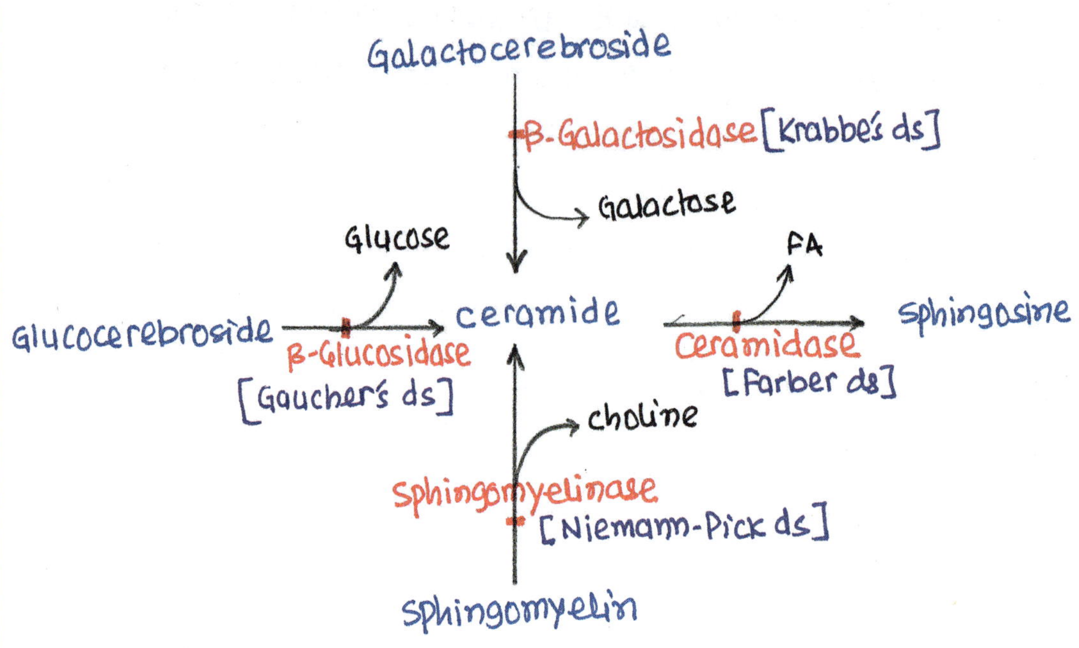

Clinical case - 2

A child was brought to the clinic with difficulty in vision. On examination - cherry Red spot on macula. There was No organomegaly. Identify the disease.

Diagnosis :- Tay-Sach's disease
AR, GM_2 gangliosidosis
Defective enzyme - Hexosaminidase A
No Hepatosplenomegaly
Microcephaly, loss of motor skills, seizures
cherry red spot on macula.

Clinical case - 3

An infant was brought to the pediatric department with chief complaint of loss of muscle control, muscle weakness, floppiness and vision problem. Eye examination showed cherry Red spot on macula. Abdomen - Hepatosplenomegaly was severe. Biopsy showed accumulation of sphingomyelin in macrophages. What is your probable diagnosis?

Ans :- Niemann-Pick disease
Enzyme defect :- Sphingomyelinase

* Cherry Red spot
* Hepatosplenomegaly

FATTY ACIDS

Aliphatic carboxylic acid c̄ formula

R – COOH

Hydrocarbon chain (R) — functional group (COOH)

Classification of fatty Acids.

On the basis of Total number of C-atoms

1. Even chain FA
   - Palmitic acid (16C)
   - Stearic acid (18C)
2. Odd chain FA
   - Propionic acid (3C)

On the basis of Length of HC-chain

a. Short chain FA (2-6)
b. Medium chain FA (8-14)
c. Long chain FA (14-24)
d. Very long chain FA (>24)

On the basis of Nature of HC chain

a. Saturated FA [No double bond]
b. Unsaturated FA [double bond]
   1. MUFA [single = bond]
   2. PUFA [More than one = bond]

Synthesis in body

a. Essential FA
   - Linoleic acid
   - Linolenic acid
b. Nonessential FA.

Nomenclature

$$\overset{8}{CH_3} - \overset{7}{CH_2} - \overset{6}{CH_2} - \overset{5}{CH_2} - \overset{4}{CH_2} - \overset{3}{CH_2} - \overset{2}{CH_2} - \overset{1}{COOH}$$

γ (C4), β (C3), α (C2) ← Carboxyl group

ω1 (C8), ω2 (C7), ω3 (C6), ω4 (C5), ω5 (C4), ω6 (C3), ω7 (C2)

→ from metyl end

Short hand representation

Palmitic acid [16 carbon c̄ No double bond]
16:0

Linoleic acid [18C, 2 double bond, at 9th & 12 c from carboxyl end]
18:2;9,12

Linolenic acid 18:3;9,12,15 | Arachidonic acid 20:4;5,8,11,14

"anoic" is added if saturated FA

"enoic" is added if unsaturated FA

Essential fatty Acid

SAQ:- Define essential fatty acids. write about their functions and related disorders.

Definition:- fatty acids that are Not synthesized in human body and need to be supplied in the diet.

Names - ① Linoleic acid → $CH_3-(CH_2)_4-CH=CH-CH_2-CH=CH-(CH_2)_7-COOH$ (18, 4, ω6, 12, 9, 7)
[C18, Δ9,12, ω6 family]

② linolenic acid

③ Arachidonic acid (semiessential)

Omega classification — MCQ

ω3 series:- ① α-linolenic acid [safflower oil, sunflower oil] (Plant oils)

② Timnodonic acid [Ecosapentaenoic acid]

③ Cervonic acid OR DHA [Docosa Hexaenoic acid]

DHA → Important for the development of brain & retina
Sources:- fish oil, Breast milk

ω6 series:- γ-linolenic acid [C18]
linoleic acid [C18] Δ9,12 [two double bond]
Arachidonic acid [C20] Δ5,8,11,14 [four = bond]

ω9 series:- Oleic acid [cis-form] [Most predominant in human body]
Elaidic acid [trans form of oleic acid]

Trans fatty acids:- present in partially hydrogenated oil & dairy products
It ↑ shelf life of food
use in processed food and bakery product
↑ LDL, ↓ HDL
↑ Risk of atherosclerosis and Heart diseases.

Dietary sources

MUFA [oleic acid]	PUFA	Saturated FA
Sesame oil	Safflower oil	Coconut oil
Mustard oil	Sunflower oil	Butter
Groundnut oil	Cottonseed oil	palm oil
Palm oil	Fish oil	
	Breast Milk	

Phrynoderma OR **Toad skin** – Multinutrient deficiency

- Due to deficiency of essential fatty acid. Also due to vit A ↓
- Seen in Infants on formula feed
  Patient on IV nutrition for prolonged time
- Involves – Skin of limbs, buttocks, back
  C/F – Abnormal thickening of the outer layer of skin [Hyperkeratosis]
  Toad like appearance of skin

T/t :- Diet c̄ essential FA
vit A supplementation

EICOSANOIDS

Includes
- Prostanoids
- Leukotriens
- Lipoxins

Precursor molecule
Arachidonic acid 20C
[20C c̄ 4 double bond]

Prostanoids → Prostaglandins (PG)
Prostacyclin (PGI_2)
Thromboxane (TXA_2)

1, 2, 3 → Indicates no. of '=' bonds

Prostaglandins – 9 classes (A to I) → Important PGs
↓
Substituent group on cyclopentane ring

Important PGs: PGI_2, PGE_2, PGD_2, $PGF_{2\alpha}$, TXA_2

COOH

Phospholipid —Phospholipase A_2 (⊖ by steroids)→ Arachidonic Acid + Lyso PAF
↓ Cyclooxygenase (⊖ by NSAID)
PGG_2
↓
PGH_2 → TXA_2 (Thromboxane) – Vasoconstriction, Aggregation of platelet

PGH_2 →
- $PGF_{2\alpha}$ – Vasoconstriction, Aggregation of platelet, Contraction of uterus
- PGD_2 – Dilation of vessels
- PGE_2 – Algesic (Pain), Uterine contraction
- PGI_2 – Prostacyclin, Inhibit Platelet Aggregation, Vasodilation

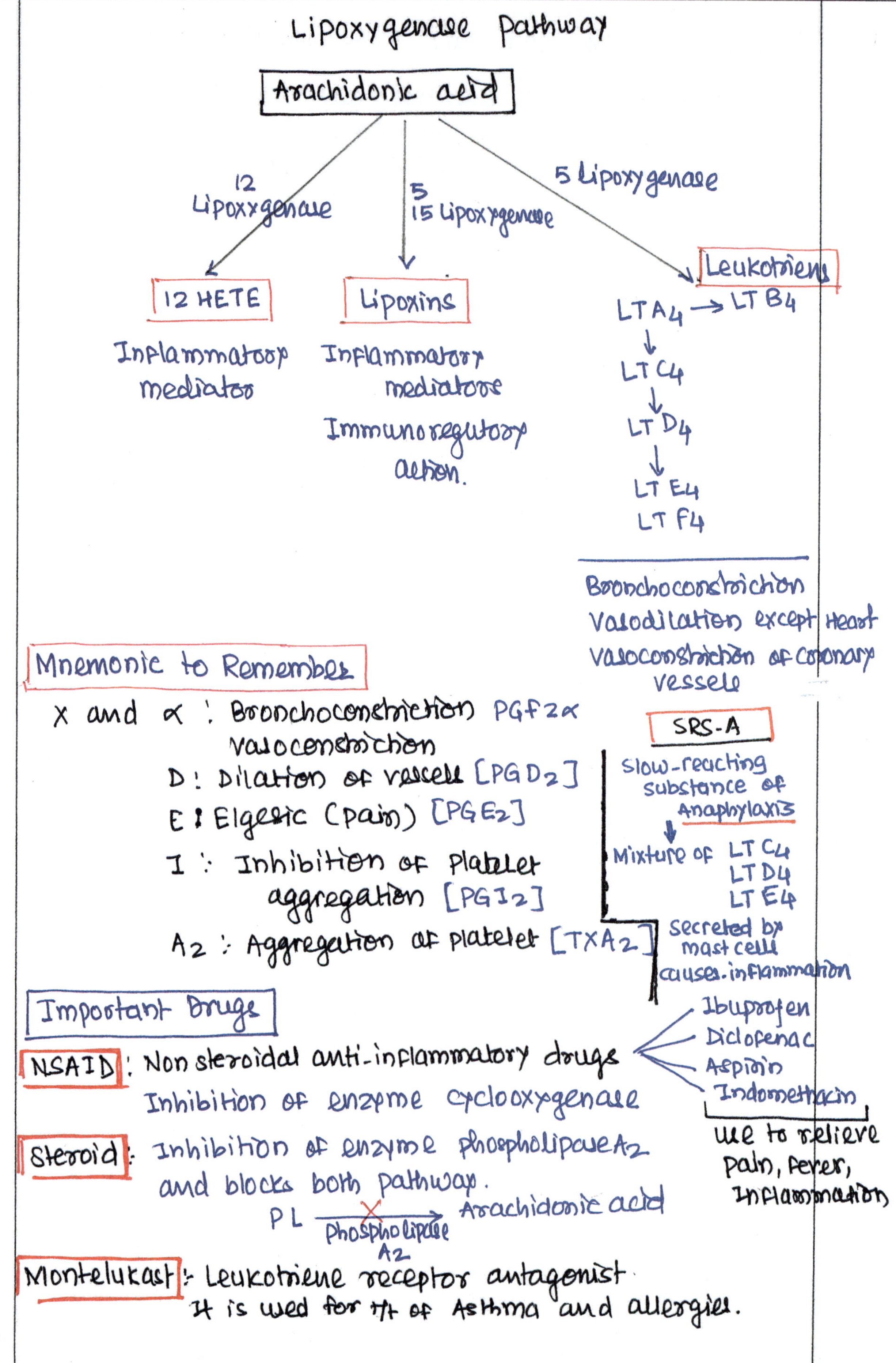
Lipoxygenase pathway
Arachidonic acid
12 Lipoxygenase
5 15 Lipoxygenase
5 Lipoxygenase
12 HETE
Inflammatory mediator
Lipoxins
Inflammatory mediators
Immunoregulatory action.
Leukotriens
LTA4 → LTB4
LTC4
LTD4
LTE4
LTF4
Bronchoconstriction
Vasodilation except Heart
Vasoconstriction of coronary vessels
SRS-A
Slow-reacting substance of Anaphylaxis
Mixture of LTC4 LTD4 LTE4
Secreted by mast cells
causes inflammation
Mnemonic to Remember
X and α : Bronchoconstriction PGF2α
Vasoconstriction
D : Dilation of vessels [PGD2]
E : Elgesic (pain) [PGE2]
I : Inhibition of Platelet aggregation [PGI2]
A2 : Aggregation of platelet [TXA2]
Important Drugs
NSAID : Non steroidal anti-inflammatory drugs
Inhibition of enzyme cyclooxygenase
Ibuprofen
Diclofenac
Aspirin
Indomethacin
use to relieve pain, fever, Inflammation
Steroid : Inhibition of enzyme phospholipase A2
and blocks both pathway.
PL → Arachidonic acid
Phospholipase A2
Montelukast : Leukotriene receptor antagonist
It is used for t/t of Asthma and allergies.

Lipoproteins

SAQ:- classify lipoproteins with their characteristics.

lipoprotein :- Lipid + protein

Lipid: Nonpolar, Amphipathic
Protein: Apoprotein

In order to transport dietary and endogenously synthesized lipid (water insoluble), it get complexed with protein to form lipoproteins.

Lipoproteins are important for transport, storage and utilization of lipid molecules.

Structure of lipoprotein

Classification

Based on density

Based on electrophoresis

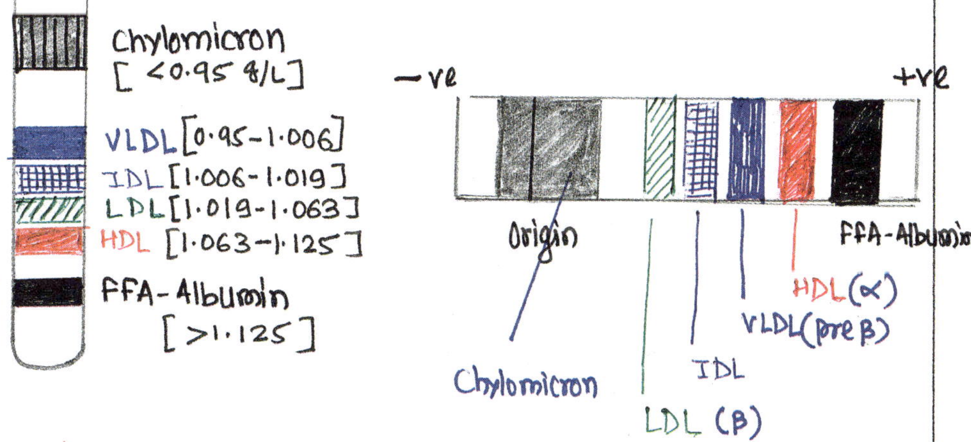

	CHYLOMICRON	VLDL	IDL	LDL	HDL
Density (g/L)	0.95	0.95–1.006	1.006–1.019	1.019–1.063	1.063–1.125
Protein	1%	7–10%	11%	21%	30–60%
Lipid	99%	90–93%	89%	79%	40–70%
Apoproteins	B-48 A, C, E	B-100 CI, II, III	B-100 E	B-100	APO AI, AII CI, II, III D, E
Major lipid	TAG PL	TAG	Cholesterol	Chol	PL cholesterol
Functions MCQ	Transport of exogenous TG from intestine to muscle & adipose tissue	Transport of endogenous TG from liver to tissues		Transport of cholesterol from liver to peripheral tissues.	Reverse transport of cholesterol from tissues to liver

Apoproteins :- protein part of lipoproteins.

Functions :- form the structure of lipoprotein

Activation of enzymes APO AI ⟶ LCAT

APO AII ⟶ lipoprotein lipase

Inhibition of enzymes APO AII, CIII } Inhibit lipoprotein lipase

Examples :- APO AI, APO AII

APO B48, APO B100

APO CI, CII, CIII

APO D, APO E

Lipoprotein a LP(a) — **Atherogenic**

Homologous to plasminogen

↓

Interferes with fibrinolysis

↓

Responsible for clot formation

Normal value :- below 30 mg/dl

> 30 mg/dl indicate increased risk of Atherosclerosis, heart attack or stroke

---

LDL [Low density lipoprotein] :- Normal range is 60–130 mg/dl

Increased level of LDL indicate risk of Atherosclerosis, Heart attack and stroke.

Also known as **Bad cholesterol**

HDL [High density lipoprotein] : Normal range 40–70 mg/dl

It is responsible for Reverse transport of cholesterol **[Good cholesterol]**

Lipid Profile

It is panel of blood tests used to find abnormalities in blood lipid concentration.

Components	Normal Range	Borderline High	High
1. Triacylglycerol →	<150 mg/dl	150-199	>200 mg/dl
2. Total cholesterol →	130-200 mg/dl	200-239	>240 mg/dl
3. HDL →	40-70 mg/dl		
4. LDL →	60-130 mg/dl	130-159	>160 mg/dl
5. VLDL →	< 30 mg/dl		
6. Cholesterol : HDL Ratio	<3.5		>5

Implications

1. To monitor and screen for your risk of cardiovascular disease.
2. for diagnosis of Hyperlipidemia or Hypolipidemia associated with genetic defects.
3. Important in management of obesity, Heart attack, Pancreatitis etc.
4. Increased levels of total cholesterol, TG, VLDL and LDL and decreased HDL level is a risk factor for cardiovascular diseases.
5. serum cholesterol level is a test for assessment of synthetic and excretory function of liver.

Laboratory Interpretation of Lipid Profile

TG →	289	mg/dl
TC →	399	mg/dl
HDL →	28	mg/dl
LDL →	313	mg/dl
VLDL →	58	mg/dl
TC/HDL →	14.2	

Interpretation

High levels of TG, Cholesterol, LDL, VLDL and TC/HDL Ratio

Low levels of HDL

→ Indicates higher risk of atherosclerosis resulting in Heart disease or stroke.

METABOLISM OF LIPIDS

MCI COMPETENCY BI 4.2- 4.7

Important LAQs

1. β-oxidation of fatty acid & energetics and disorder
2. fatty acid synthesis
3. Cholesterol synthesis and degradation
4. Ketone bodies synthesis and utilization

SAQ

1. Fatty acid synthase complex
2. Hyperlipoproteinemia
3. Metabolism of VLDL and formation of LDL
4. Reverse transport of Cholesterol
5. Metabolism of Chylomicrons
6. Role of Hormone sensitive lipase in TG metabolism.

Clinical case

Disorders of lipoprotein metabolism

Type I familial chylomicronemia → ↑ CM, ↑ TG
Milky plasma
Eruptive Xanthoma's

Type IIa familial Hypercholesterolemia
↑ Cholesterol
↑ LDL
Tendon xanthoma

Sphingolipidosis

Gaucher's disease - Tissue paper macrophages, Pancytopenia

Niemann-Pick disease - cherry Red spot on Macula
Hepatosplenomegaly

Tay sach's disease - No Hepatosplenomegaly
Loss of motor skills, seizures
Cherry red spot on macula

fatty Liver

Disorder's of fatty acid oxidation → SIDS
Refsum's disease
Jamaican Vomiting sickness
Zellweger syndrome

Digestion and Absorption of lipids

Lipids (fat) [Nill in mouth]

↓ stomach — Lingual (Acid stable lipase), Gastric lipase

Minimal digestion

↓ small intestine

Emulsification by – dispersion of lipid into smaller droplets due to ↓ surface tension

① detergent action of bile salt – Salts of Glycocholate, Taurocholate

② surfactant – DAG, MAG, Phospholipids

③ peristalsis mixing

↓

Smaller lipid c̄ ↑ surface area

TAG —Pancreatic lipase→ Monoacyl glycerol (C-2) + free fatty acid (C1 & C3)

Cholesteryl ester —cholesteryl esterase→ Cholesterol + fatty acid

Phospholipids —Phospholipase A_2→ lysophospholipid + fatty acid

↓ MAG, free FA, Glycerol, Cholesterol

Mixed micelles formation
(lipids in interior & bile salts at periphery)

FA
PL
MAG
Bile salt
[Polar groups on exterior side]

↓ Absorbed into intestinal mucosal cell by diffusion

Short & Medium chain FA directly absorbed into portal circulation

Synthesis of TG, Cholesteryl ester, Activation of long chain FA occurs in mucosal cell, incorporated into Chylomicrons

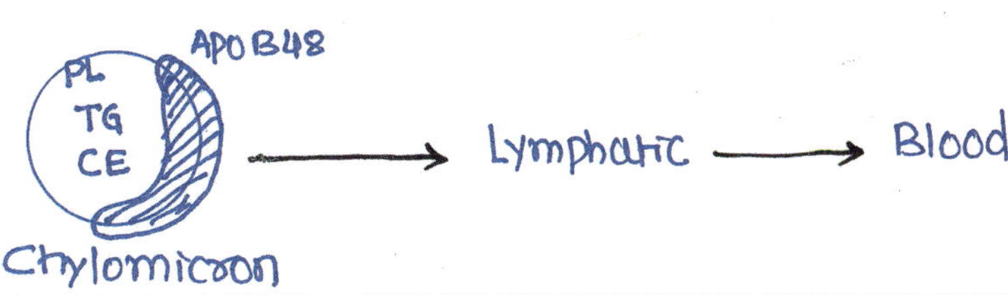

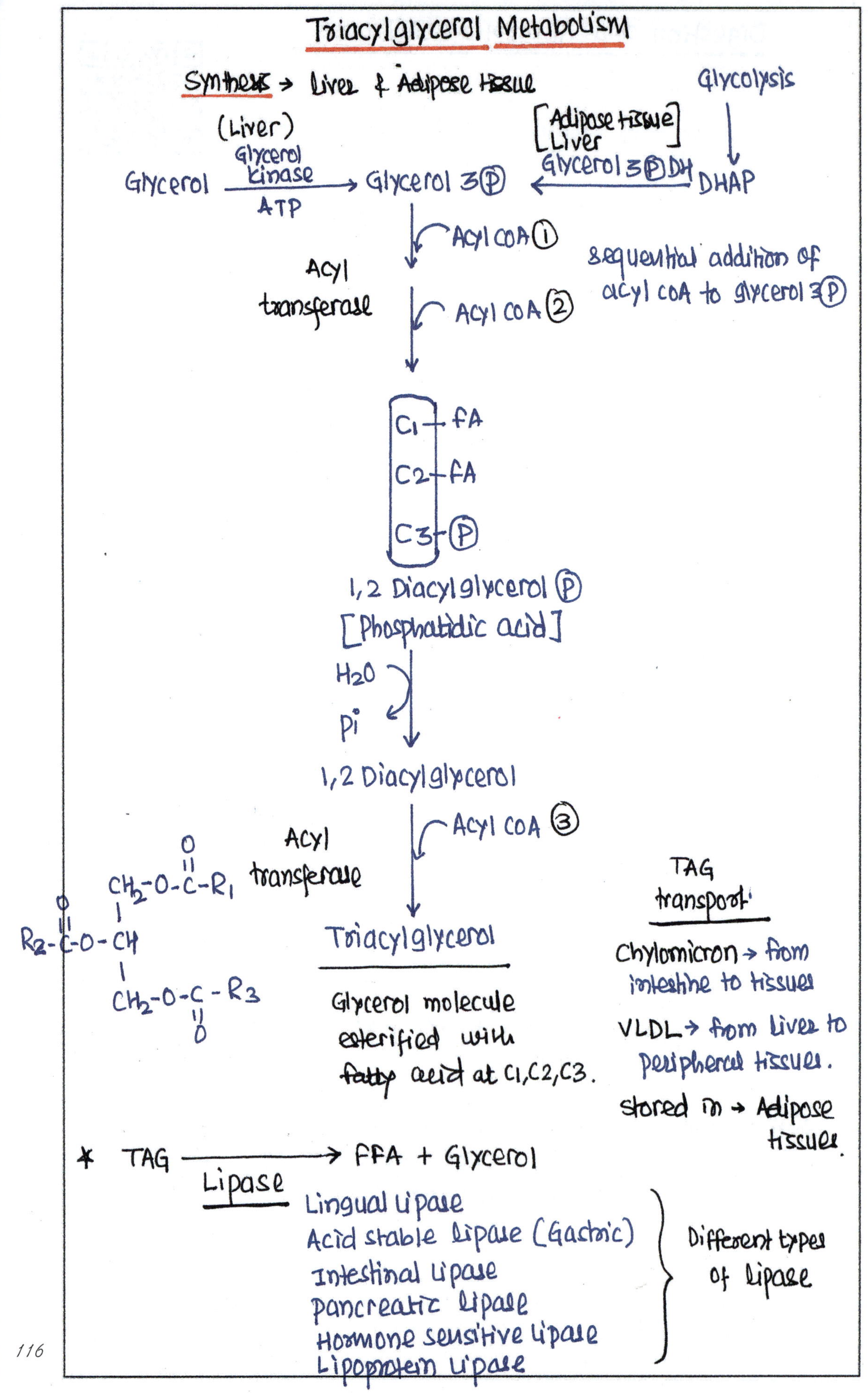

Triacylglycerol Metabolism
Synthesis → Liver & Adipose tissue
Glycolysis
(Liver)
[Adipose tissue
Liver]
Glycerol
Glycerol kinase
ATP
Glycerol 3 (P)
Glycerol 3 (P) DH
DHAP
Acyl COA (1)
Acyl transferase
Acyl COA (2)
sequential addition of acyl COA to glycerol 3 (P)
C1 - FA
C2 - FA
C3 - (P)
1,2 Diacylglycerol (P)
[Phosphatidic acid]
H2O
Pi
1,2 Diacylglycerol
Acyl COA (3)
Acyl transferase
CH2-O-C-R1
R2-C-O-CH
CH2-O-C-R3
Triacylglycerol
Glycerol molecule esterified with fatty acid at C1, C2, C3.
TAG transport
Chylomicron → from intestine to tissues
VLDL → from liver to peripheral tissues.
stored in → Adipose tissues.
* TAG → FFA + Glycerol
Lipase
Lingual lipase
Acid stable lipase (Gastric)
Intestinal lipase
Pancreatic lipase
Hormone sensitive lipase
Lipoprotein lipase
Different types of lipase

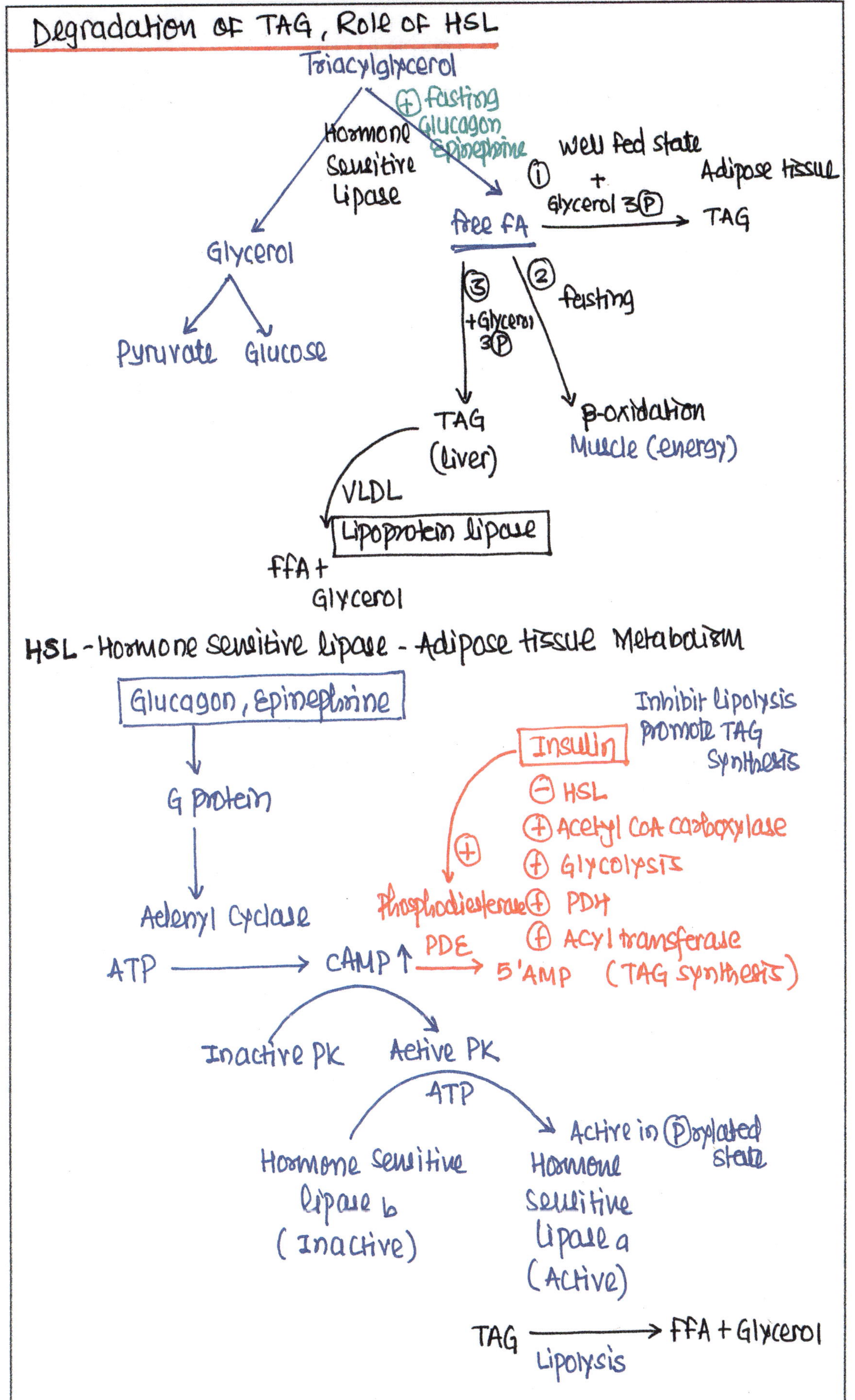

Degradation of TAG, Role of HSL
Triacylglycerol
⊕ fasting
Glucagon
Epinephrine
Hormone
Sensitive
Lipase
Glycerol
Pyruvate
Glucose
free FA
① well fed state
+
Glycerol 3Ⓟ
Adipose tissue
TAG
② fasting
β-oxidation
Muscle (energy)
③ +Glycerol 3Ⓟ
TAG
(liver)
VLDL
Lipoprotein lipase
FFA +
Glycerol
HSL - Hormone sensitive lipase - Adipose tissue Metabolism
Glucagon, Epinephrine
G protein
Adenyl Cyclase
ATP
cAMP ↑
PDE
5'AMP
Insulin
Inhibit lipolysis
promote TAG
Synthesis
⊖ HSL
⊕ Acetyl CoA carboxylase
⊕ Glycolysis
⊕ PDH
⊕ Acyl transferase
(TAG synthesis)
⊕
Phosphodiesterase
Inactive PK
Active PK
ATP
Hormone Sensitive
lipase b
(Inactive)
Active in Ⓟylated
state
Hormone
sensitive
lipase a
(Active)
TAG
Lipolysis
FFA + Glycerol

FATTY ACID OXIDATION

β-Oxidation of fatty acid

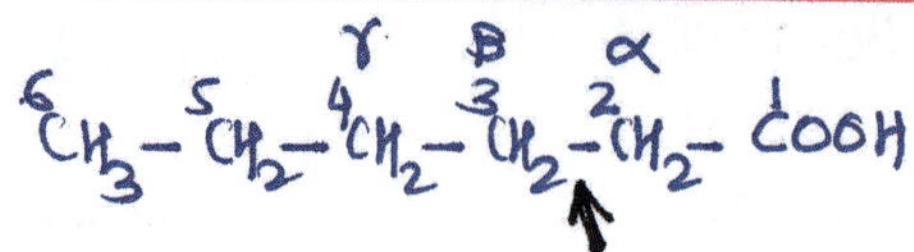

It is a most common type of oxidation of FA.
Successive removal of 2 carbon compound (Acetyl CoA)
Involved β-carbon [cleavage betn α & β carbon]
occurs in mitochondria.

Types of FA oxidation

- β-Oxidation → Mitochondria (mt)
- α-oxidation → ER/Peroxisomes
- ω-oxidation → ER/Microsomes
- Odd chain FA → mitochondria
- Unsaturated FA → mitochondria
- Very long chain FA → Peroxisomes, mt.

β-oxidation

site:- Mitochondria [liver, adipose tissue, Muscle]

3 steps
① Activation of FA
② Transport
③ β-oxidation proper

① Activation of FA

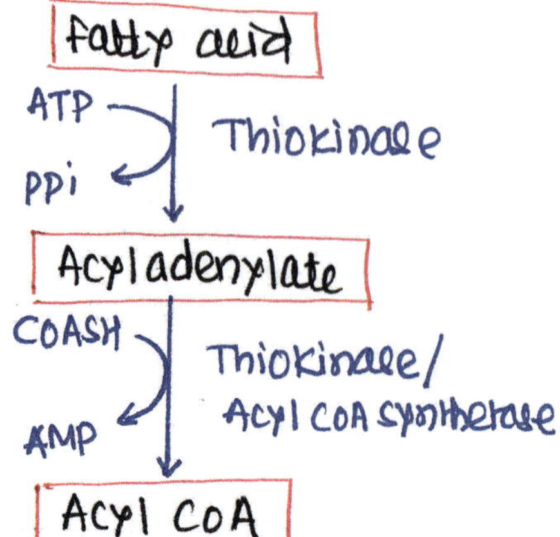

② Transport across mitochondrial membrane

ACYl COA [n]

+Carnitine ↓ Carnitine Acyl transferase I (CAT I)

ACYl carnitine

↓ Translocase

Acyl carnitine

↓ Carnitine Acyl transferase II CAT-II

Carnitine ←

ACyl COA

③ β-oxidation proper 4 Steps (In mitochondria)

① oxidation – ACYl COA DH
② Hydration – Enoyl COA hydratase
③ oxidation – β-OH ACYl COA DH
④ cleavage – Thiolase

ACYl CoA

FAD → $FADH_2$ — Acyl CoA DH — Oxidation

1.5 ATP / ETC ← $FADH_2$

↓

Transenoyl COA

H_2O — Enoyl COA Hydratase — Hydration

↓

β-OH ACYl COA

NAD^+ → $NADH+H^+$ — β-OH Acyl CoA DH — oxidation

2.5 ATP / ETC ← $NADH+H^+$

↓

Ketoacyl COA

↓ Thiolase — cleavage

ACYl COA [n-2] + ACetyl COA

↓

TCA cycle

Energetics of β-oxidation

Palmitic acid (16C)

$7\ FADH_2 \rightarrow 7 \times 1.5 = 10.5$
$7\ NADH \rightarrow 7 \times 2.5 = 17.5$] 28 ATP

8 Acetyl CoA ⟶ 80 ATP

ATP generated = 80 + 28 = 108 ATP
− 2 ATP consumed for activation of Acyl CoA

Net ATP synthesis **106 ATP**

Stearic acid (18C)

$8\ FADH_2 \rightarrow 8 \times 1.5 = 12$
$8\ NADH \rightarrow 8 \times 2.5 = 20$] 32

ATP generated 90+32−2
Net Gain = **120 ATP**

9 Acetyl CoA ⟶ 90 ATP

Regulation

- At carnitine shuttle system
- In wellfed state

↑ Insulin / ↓ Glucagon
↓ (+)
FA synthesis
↓
↑ Malonyl CoA
↓ Inhibit
Carnitine Acyl Transferase I ↓↓

In fasting

↓ Insulin / ↑ Glucagon
↓ Inhibit
FA synthesis
↓
↓ Malonyl CoA
↓
Carnitine Acyl Transferase I ↑↑

Oxidation of Odd chain FA

Odd chain FA

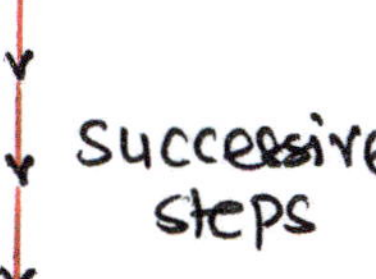

Acetyl CoA + Propionyl CoA (3C)

↓ ATP, Biotin, CO_2 — Propionyl CoA Carboxylase

Methyl Malonyl CoA

↓ B12 — Methyl Malonyl CoA Mutase

Succinyl CoA ⟶ TCA cycle

Disorders of β-Oxidation / FA oxidation

1. **SIDS : Sudden Infant Death Syndrome**

   10% cases of SIDS are due to deficiency of Medium chain Acyl CoA DH - defect in β-oxidation

   Death of infant occurs in night, due to blockade in β-oxidation pathway, if the gap between meal is more the 12 Hrs.

2. **Jamaican Vomiting Sickness**

   Due to eating of unripe ackee fruit

   Contain toxin Hypoglycin A MCQ

   Hypoglycin A inhibit Acyl CoA dehydrogenase.

3. **Refsum disease** : Neurological symptoms
   Retinitis pigmentosa
   Ataxia

   Defect in α-oxidation of fatty acid in ER & peroxisomes

   Defective enzyme :- Phytanic acid oxidase MCQ

   Accumulation of Phytanic acid in Brain causes Symptoms.

4. **Zellweger syndrome** [cerebrohepatorenal syndrome]

   Affects liver, Brain, Kidneys

   Oxidation of very Long chain FA affected due to absence of functional peroxisomes. MCQ

FATTY ACID SYNTHASE COMPLEX

It is a multi-enzyme complex that catalyzes FA synthesis

Enzyme system composed of two identical polypeptide.

1 subunit composed of – 7 Enzymes
1 protein [Acyl carrier protein]

Dimer is functionally active

All enzymes originated from single gene.

All enzymes are in equimolar concentration
well coordinated and regulated
Interference from other metabolite is less.

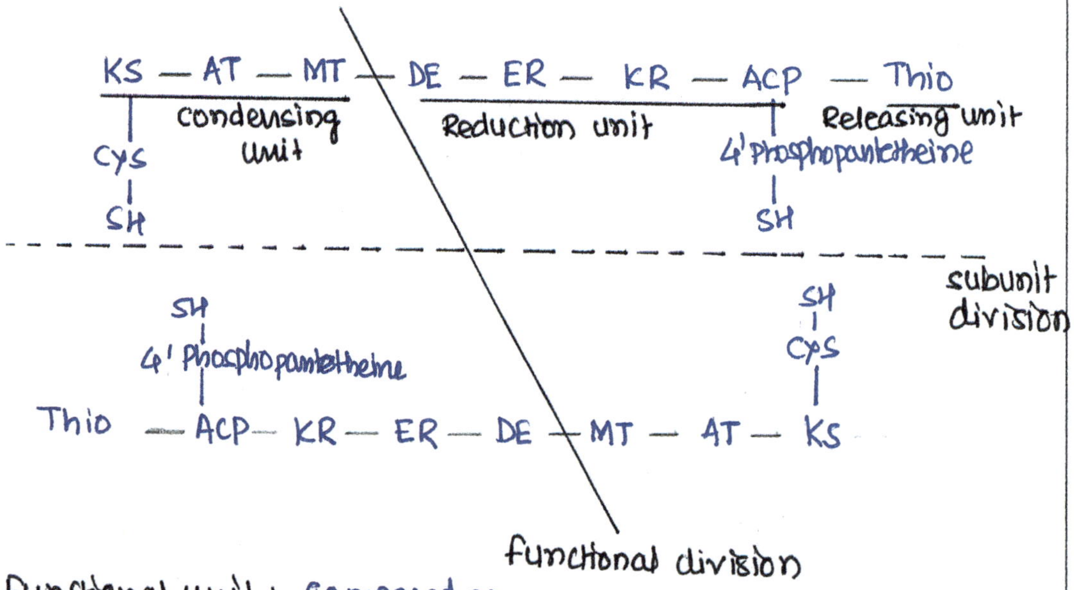

KS — AT — MT — DE — ER — KR — ACP — Thio

condensing unit | Reduction unit | Releasing unit

Cys – SH (on KS); 4' Phosphopantetheine – SH (on ACP)

- - - - - - - - - - subunit division - - - - - - - - - -

SH – 4' Phosphopantetheine (on ACP); SH – Cys (on KS)

Thio — ACP — KR — ER — DE — MT — AT — KS

functional division

Functional unit :- Composed of

Condensing unit (of one subunit) :- Ketoacyl synthase [KS]
Acetyl transacylase [AT]
Malonyl transacylase [MT]

Reduction unit (of other subunit) :- Dehydratase [DE]
Enoyl reductase [ER]
Ketoacyl reductase [KR]
Acyl carrier Protein [ACP]

Releasing unit (of other subunit) :- Thioesterase [Thio]

De Novo Synthesis of Fatty Acid

LAQ:- Describe the steps and regulation of fatty acid synthesis and Add a note on FAS complex.

Site:- Cytosol → liver, Kidney, Lung, Lactating mammary gland, Adipose tissue etc

Transport of Acetyl CoA

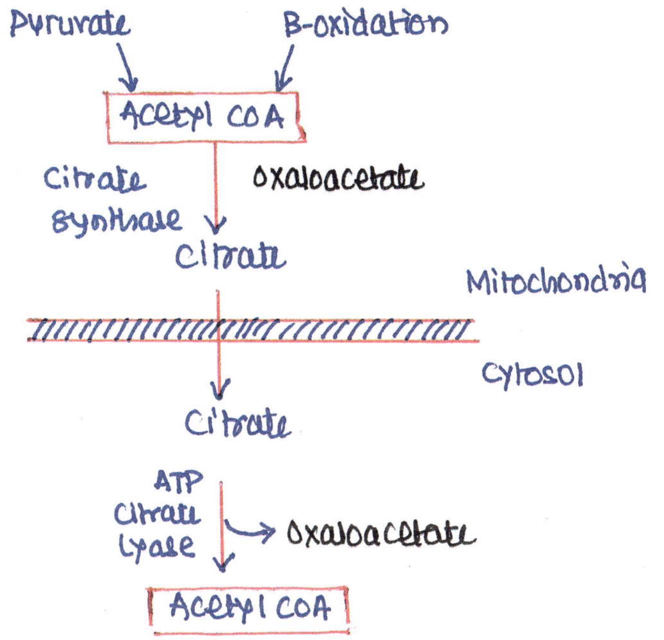

Pyruvate → Acetyl CoA ← B-oxidation

Acetyl CoA + Oxaloacetate —(Citrate synthase)→ Citrate

Mitochondria

Citrate (Mitochondria) → Citrate (Cytosol)

Cytosol

Citrate —(ATP, Citrate Lyase)→ Acetyl CoA + Oxaloacetate

Requirement : AcetylCoA
NADPH
ATP

Substrate – Acetyl CoA
Product – Palmitoyl CoA
Cofactors – NADPH, Mn^{2+}, ATP Biotin, CO_2 (HCO_3^-)

Activation of Acetyl CoA

①

$$\text{Acetyl CoA} \xrightarrow[\text{ATP, } CO_2\text{, Biotin}]{\text{Acetyl CoA carboxylase}} \text{Malonyl CoA}$$

Pathway proper :- Mnemonic to Remember

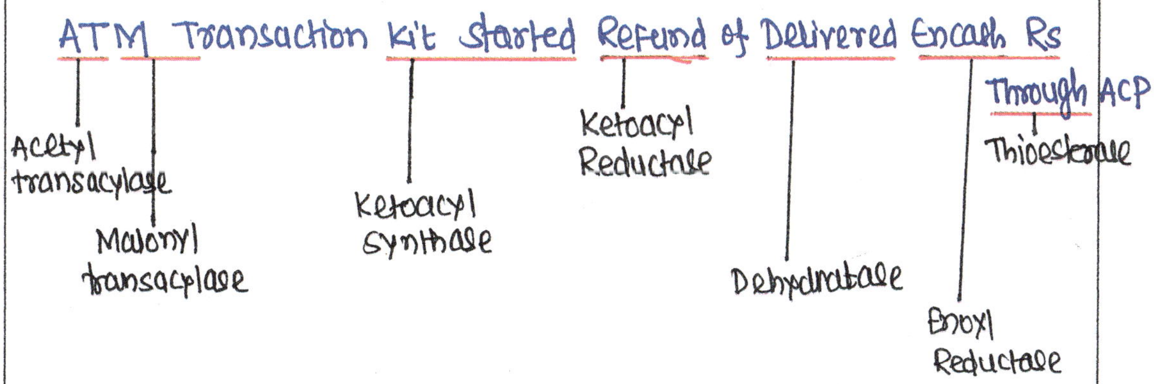

ATM Transaction Kit Started Refund of Delivered Encash Rs Through ACP

- A – Acetyl transacylase
- M – Malonyl transacylase
- Kit – Ketoacyl synthase
- Refund – Ketoacyl Reductase
- Delivered – Dehydratase
- Encash – Enoyl Reductase
- Through – Thioesterase

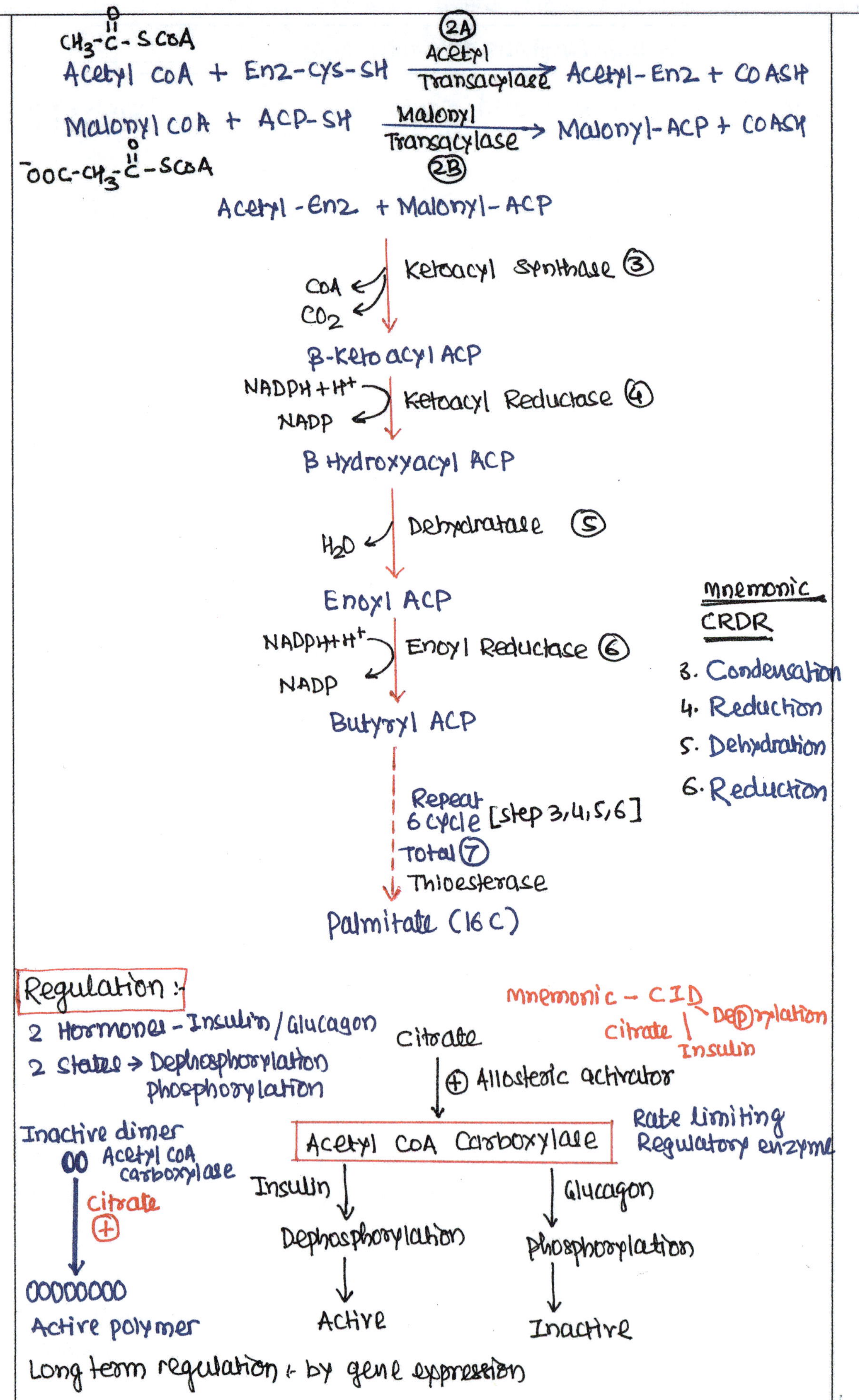

$CH_3-C(=O)-SCoA$
Acetyl CoA + Enz-CYS-SH → (2A) Acetyl Transacylase → Acetyl-Enz + COASH
Malonyl CoA + ACP-SH → Malonyl Transacylase (2B) → Malonyl-ACP + COASH
$^-OOC-CH_3-C(=O)-SCoA$
Acetyl-Enz + Malonyl-ACP
Ketoacyl Synthase (3)
CoA
CO_2
β-Keto acyl ACP
$NADPH+H^+$
NADP
Ketoacyl Reductase (4)
B Hydroxyacyl ACP
Dehydratase (5)
H_2O
Enoyl ACP
$NADPH+H^+$
NADP
Enoyl Reductase (6)
Butyryl ACP
Repeat 6 cycle [step 3,4,5,6]
Total (7)
Thioesterase
Palmitate (16 C)
Mnemonic
CRDR
3. Condensation
4. Reduction
5. Dehydration
6. Reduction
Regulation :-
2 Hormones - Insulin / Glucagon
2 States → Dephosphorylation
Phosphorylation
Mnemonic - C I D
Citrate
Insulin
Dephosphorylation
Citrate
⊕ Allosteric activator
Acetyl CoA Carboxylase
Rate limiting
Regulatory enzyme
Inactive dimer
00 Acetyl CoA carboxylase
Citrate ⊕
00000000
Active polymer
Insulin
Dephosphorylation
Active
Glucagon
Phosphorylation
Inactive
Long term regulation :- by gene expression

CHOLESTEROL

- Derived lipid [$C_{27}H_{46}O$]
- 27-carbon compound, OH at 3rd position, '=' betn 5-6 carbon
- Animal sterol
- Precursor for bile acids, vit D, steroid hormone
- Contain cyclopentanoperhydrophenanthrene nucleus (CPPP)
- Cholesterol is a component of plasma membrane and required for transport across the cell membrane.

LAQ:- Biosynthesis of Cholesterol c̄ its regulation.

Site: Liver, Intestine, Adrenal cortex, All nucleated cells
Cytosol + Endoplasmic reticulum

5 steps

① formation of HMG CoA
② formation of Mevalonate
③ conversion of Mevalonate to IPP
④ formation of squalene (30C)
⑤ conversion of squalene to cholesterol

Regulation :- ① Covalent Modification

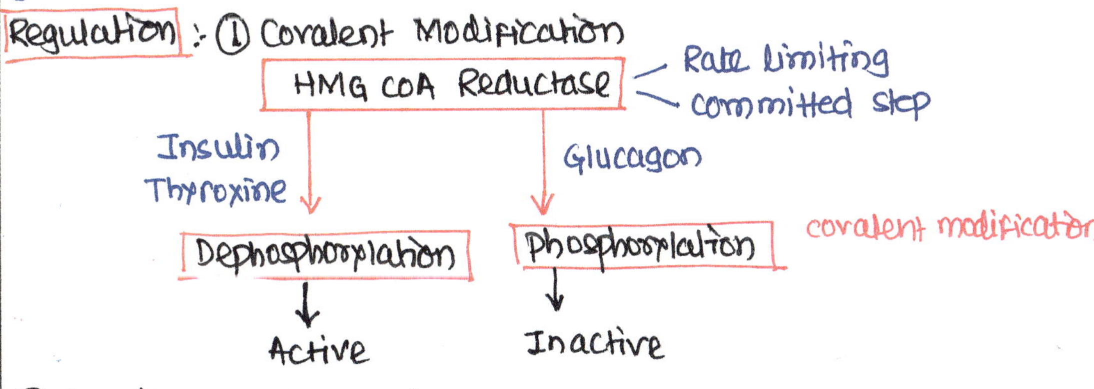

② feedback Regulation

↓ Genetic level

↓↓ Transcription of gene for synthesis of HMG CoA Reductase

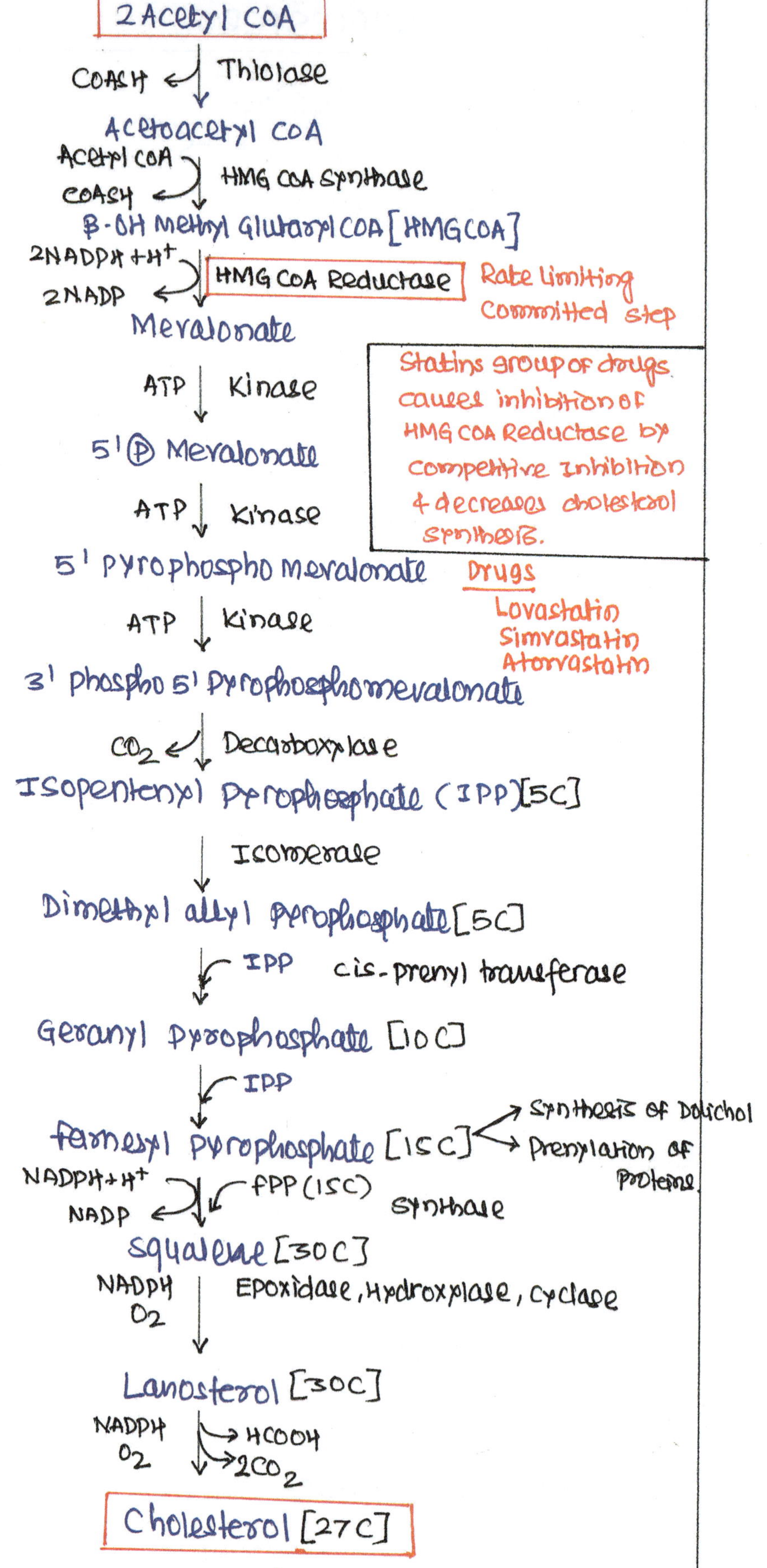
2 Acetyl CoA
CoASH
Thiolase
Acetoacetyl CoA
Acetyl CoA
CoASH
HMG CoA synthase
β-OH Methyl Glutaryl CoA [HMG CoA]
2NADPH + H+
2NADP
HMG CoA Reductase
Rate limiting
Committed step
Mevalonate
Statins group of drugs causes inhibition of HMG CoA Reductase by competitive inhibition & decreases cholesterol synthesis.
ATP
Kinase
5' (P) Mevalonate
ATP
Kinase
5' pyrophospho mevalonate
Drugs
Lovastatin
Simvastatin
Atorvastatin
ATP
Kinase
3' phospho 5' Pyrophosphomevalonate
CO2
Decarboxylase
Isopentenyl Pyrophosphate (IPP) [5C]
Isomerase
Dimethyl allyl pyrophosphate [5C]
IPP
cis-prenyl transferase
Geranyl Pyrophosphate [10C]
IPP
Farnesyl Pyrophosphate [15C]
Synthesis of Dolichol
Prenylation of Proteins
NADPH+H+
NADP
FPP (15C)
Synthase
Squalene [30C]
NADPH
O2
Epoxidase, Hydroxylase, Cyclase
Lanosterol [30C]
NADPH
O2
HCOOH
2CO2
Cholesterol [27C]

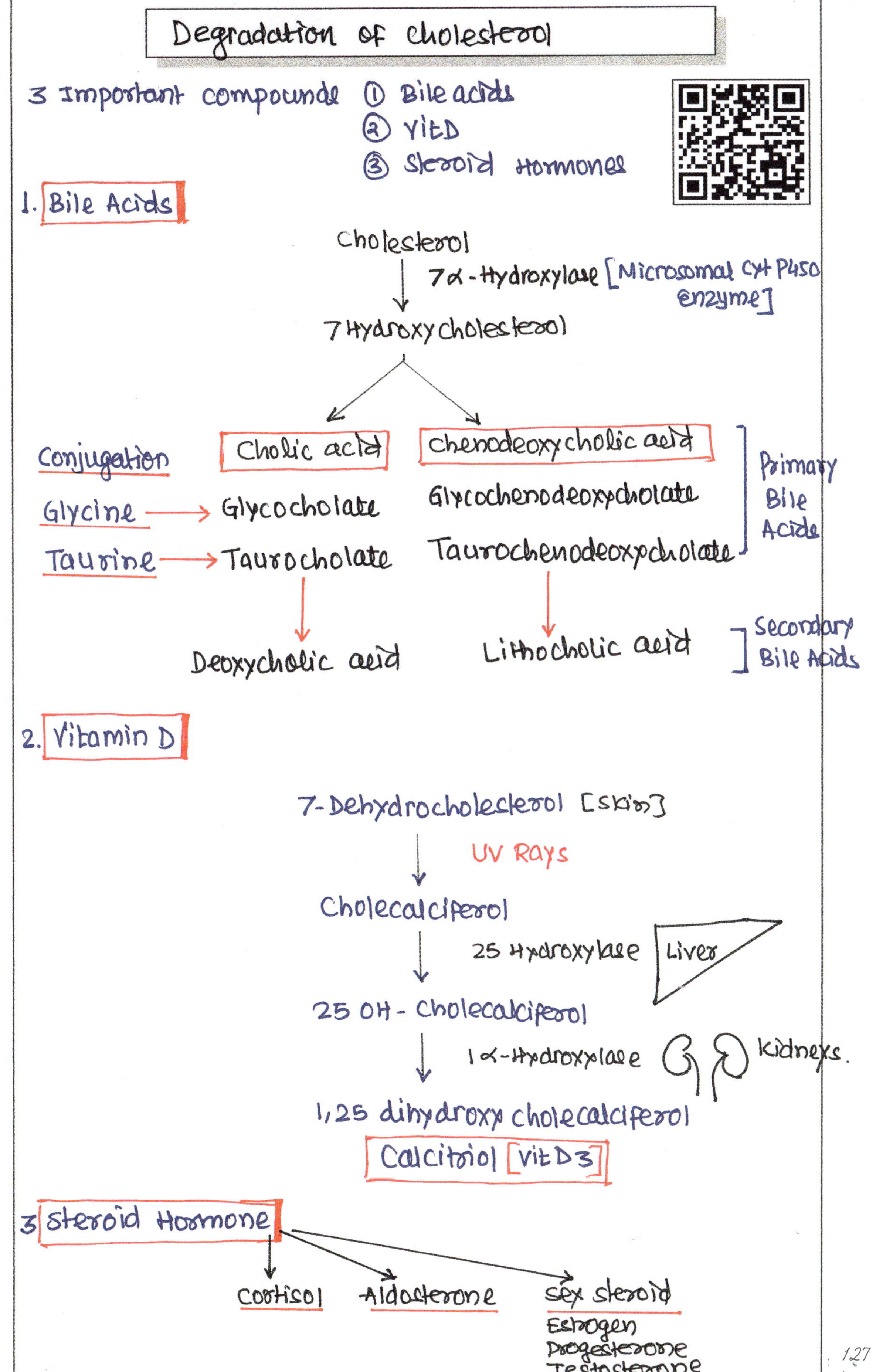

Degradation of cholesterol
3 Important compounds
① Bile acids
② Vit D
③ Steroid Hormones
1. Bile Acids
Cholesterol
7α-Hydroxylase [Microsomal Cyt P450 enzyme]
7 Hydroxy cholesterol
Cholic acid
Chenodeoxycholic acid
Conjugation
Glycine
Taurine
Glycocholate
Glycochenodeoxycholate
Taurocholate
Taurochenodeoxycholate
Primary Bile Acids
Deoxycholic acid
Lithocholic acid
Secondary Bile Acids
2. Vitamin D
7-Dehydrocholesterol [Skin]
UV Rays
Cholecalciferol
25 Hydroxylase
Liver
25 OH - Cholecalciferol
1α-Hydroxylase
Kidneys.
1,25 dihydroxy cholecalciferol
Calcitriol [Vit D3]
3 Steroid Hormone
Cortisol
Aldosterone
Sex steroid
Estrogen
Progesterone
Testosterone

Ketone bodies Synthesis and Utilization

Names ① Acetone ② Acetoacetate] True ketone bodies

③ β-OH butyrate

$CH_3-\overset{O}{\overset{\|\|}{C}}-CH_3$	$CH_3-\overset{O}{\overset{\|\|}{C}}-CH_2-COO^-$	$CH_3-\overset{OH}{\overset{\|}{CH}}-CH_2-COO$
Acetone (III)	Acetoacetate (I)	B-OH-Butyrate (II)

Clinical significance of ketone bodies

1. **Uncontrolled DM**

Insulin deficiency ⟶ ↓ uptake of glucose by cells
↓
starved for glucose
↓
lipolysis
↓
fatty acid ↑↑
↓ degradation
Acetyl CoA
↓
Ketone bodies synthesis.

2. **Starvation**

↓ Glucose
↓
↑ fatty acid consumption
↓
Acetyl CoA [Acetyl CoA Not handled by the TCA Cycle due to deficiency of Oxaloacetate]
↓
Ketone body synthesis.

Ketone body synthesis

Mitochondria
Liver

Acetyl COA + Acetyl COA

↓ Thiolase

Acetoacetyl COA

Acetyl COA ↓ HMG COA synthase [Regulatory]

OH Methyl Glutaryl COA
[HMG COA]

↓ HMG COA lyase

Acetoacetate [Primary ketone body]

CO_2 spontaneous → Acetone

Exhaled via lung

$NADH+H^+$ → NAD^+, β-OH Butyrate DH → β-OH Butyrate

Utilization of ketone bodies

By extrahepatic tissues

β-OH-Butyrate

NAD^+ → $NADH+H^+$ ↓ β-OH Butyrate DH

Acetoacetate

Succinyl COA → Succinate ↓ Thiophorase [Absent in liver]

Acetoacetyl COA

COASH

Acetyl COA   Acetyl COA

Energetics :- 2 Acetyl COA 2×10 = 20
NADH 2.5 = 2.5
−1 = −1
21.5 ATP by β-OH Butyrate

(19) ATP by Acetoacetate

No GTP production when succinyl COA ⟶ succinate in TCA cycle

Ketonemia + Ketonuria = Ketosis

[↑ K.B. in blood] [↑ K.B. in urine] C/s DKA Starvation

Rothera's test

Procedure	Observation	Principle
Saturate 3ml urine with Ammonium sulfate, Add 2 drops of Na-nitroprusside and 3ml Liquor ammonia slowly by the side of test tube.	purple ring at the Junction	Acetone & Acetoacetate react with Na-nitroprusside and Ammonia to give purple ring at the junction.

C/S:- Rothera's test is positive in patient c̄ DKA and starvation.

LAQ:- Describe the steps in ketone body metabolism. Add a note on significance of ketone body production.

Clinical case

51-year-old man was brought to the hospital in confused and semicomatose state. Several days before he was complaining of undue thirst and also used to get up several times during the night for Urination. His breath had fruity odour. Following is a data ~~for~~ of his Laboratory investigation.

Plasma glucose Random - 580 mg/dl
Rothera's test – purple ring
Benedict's test – Brick Red PPt

What is your probable diagnosis?

Ans → Diabetic Ketoacidosis

↓ Insulin OR Insulin Resistance → ↑ Gluconeogenesis, ↑ Glycogenolysis → Hyperglycemia

↓ Insulin OR Insulin Resistance → ↑ Lipolysis → FA + Glycerol

↑ K.B. Synthesis
↑ TAG Synthesis

CHYLOMICRONS

Density: < 0.95

Protein 1%, Lipid 99% [TAG, PL, Cholesterol]

Apoprotein:- ApoB48, ApoA, ApoC, ApoE

Site of synthesis - Intestine

Function:- Transport of exogenous [dietary] TAG from GIT to muscle and adipose tissue

Metabolism of CM

1. Synthesis of nascent CM occurs in intestine, from dietary TAG, cholesterol and phospholipid and apoprotein B-48. Then through lymphatics enters into circulation.

2. It matures by taking apo CII & apo E from HDL

3. ApoC II activates LPL (lipoprotein lipase) in Skeletal muscle, Adipose tissue & Heart

TAG —LPL→ Free FA + Glycerol
Free FA ↓ Peripheral tissues
Glycerol ↓ Liver

4. CM remnant is taken up by the liver and acted by Hepatic lipase.

Hepatic lipase <
- Facilitates uptake
- Hydrolyses remnant TAG & PL

VLDL, IDL and LDL

VLDL- Very low density lipoprotein [0.95–1.006 g/L]

Apoprotein :- Apo B100, Apo C I, II, III

Site of synthesis : Liver

function :- Transport of endogenous TAG from Liver to peripheral tissues.

LDL :- Low density lipoprotein [1.019 – 1.063 g/L]

Apoprotein :- Apo B100

function : Transport of Cholesterol from Liver to peripheral tissues.

Metabolism

1. Synthesis of nascent VLDL in the liver

2. formation of mature VLDL by acquiring apo CII & E from HDL

3. The loss of TAG by action of lipoprotein lipase and return of Apo CII to HDL causes formation of VLDL remnant or IDL

4. formation of LDL contain cholesterol and Apo B100

5. uptake of LDL by the liver & peripheral tissues

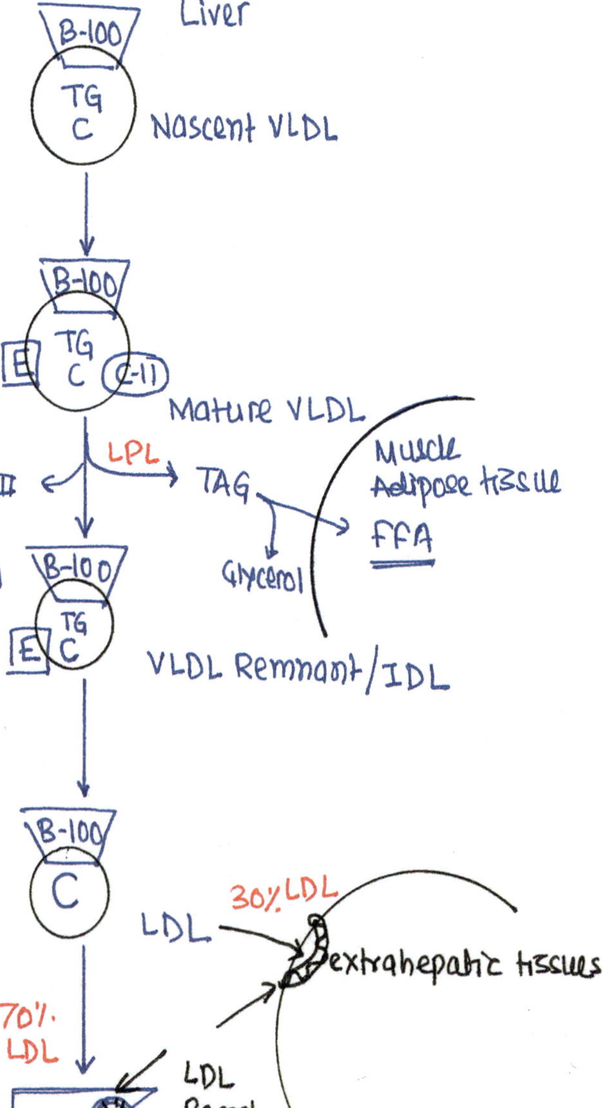

	VLDL	IDL	LDL
Protein	7-10%	11%	21%
TAG	60%	30%	10%
Cholesterol	20%	30%	50%
Phospholipid	15%	20%	20%
Apoprotein	B-100 APO C APO E	B-100 APO E	B-100

Uptake of LDL by Receptor mediated endocytosis

30% extrahepatic tissues

LDL

LDL Receptors

LDL-LDL Receptor complex formation

clathrin coated pits

Endocytosis of LDL-LDL Receptor

Endosomes

Shedding of clathrin & reutilization of receptors.

fusion c̄ lysosome to form endolysosomes [degradation]

Cholesterol

- Stored in the tissues
- Plasma membrane
- Inhibit cholesterol synthesis

HDL [High density Lipoprotein]

Also known as Good cholesterol

HDL- Density 1.063 - 1.125 g/L

Apoproteins:- Apo A I, II, IV, Apo C I, II, III, Apo D, Apo E

protein: 30-60%, lipid: 40-70%. [Phospholipid, cholesterol]

Function:- Reverse transport of cholesterol, from peripheral tissues to liver

Metabolism of HDL

liver → PL, C, A-1, LCAT ← Small intestine

Discoidal HDL

Apo A1 Activates LCAT - Lecithin cholesterol acyl transferase

Transfer of cholesterol from peripheral tissues to HDL

LCAT causes esterification of cholesterol to nonpolar cholesteryl esters

this nonpolar cholesteryl ester moves to the interior of HDL

Thus, nonpolar core is generated forming Spherical HDL

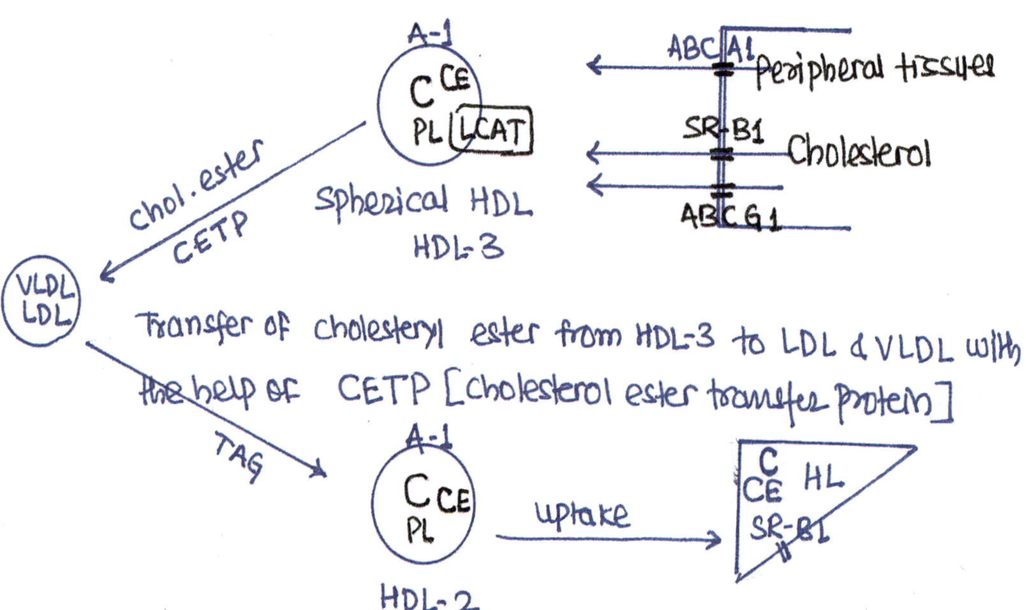

Transfer of cholesteryl ester from HDL-3 to LDL & VLDL with the help of CETP [Cholesterol ester transfer protein]

Cholesterol rich HDL is taken up by the liver via SR-B1 receptor mediated endocytosis.

Hyperlipoproteinemia

Hyperlipidemia
Fredrickson Classification

Type	Disease	Defect	Features
Type I	familial chylomicronemia	LPL deficiency OR Apo C II defect	↑ CM, ↑ VLDL ↑ TAG Milky plasma (↑ CM) Eruptive Xanthomas Pancreatitis No Risk of CAD
Type IIa	familial hypercholesterolemia	LDL Receptor defect OR Apo B100 defect	↑ LDL ↑ Cholesterol Tendon Xanthomas Atherosclerosis Risk of CAD
Type IIb	familial combined dyslipidemia		↑ LDL ↑ VLDL
Type III	familial dysbetalipoproteinemia [Broad beta disease]	Apo E defect	↑ Remnant VLDL Remnant CM ↑ cholesterol
Type IV	familial hypertriglyceridemia	overproduction of VLDL often associated with Glucose intolerance and increased insulin	↑ VLDL, ↑ Chol Associated c̄ CAD, Alcoholism, Obesity, DM
Type V	familial hypertriglyceridemia	Apo A-V defect	↑ VLDL, ↑ CM ↑ TAG Eruptive Xanthomas Pancreatitis.

Hypolipoproteinemia

NO - β-lipoprotein

Abetalipoproteinemia

Due to mutation in the gene encoding microsomal triglyceride transfer protein.

CM, VLDL and LDL are Not produced.

α-lipoprotein deficiency

1. **Tangiers disease**

   Mutation in the gene ABCA1
   - transport protein for transport of cholesterol from cell to HDL
   - Low levels of HDL and APOA1

   C/F:- enlarged yellow-orange tonsils, hepatosplenomegaly & peripheral neuropathy

2. **Fish eye disease** → partial LCAT deficiency
3. **Norum's de** ↳ complete LCAT defect

↑ Cholesterol	Tendon Xanthoma	Type IIa
↑ Triglyceride	Eruptive Xanthoma	Type I IV V
↑ Chylomicrons	Milky plasma	Type I
↓ HDL	Yellow-orange tonsils	Tangiers disease

↑ Triglyceride → Risk of Pancreatitis - Type I, IV, V

↑ Cholesterol → Risk of Coronary Artery de - Type IIa

Clinical cases

A patient has multiple tendon xanthomas. Serum cholesterol was 396 mg/dl and LDL was 232 mg/dl. Statins were given to the patient. He also told H/o recent heart attack.

What is your probable diagnosis?

→ Type IIa - familial hypercholesterolemia

Defect - LDL Receptor defect

Lab Δ :- ↑ Cholesterol
↑ LDL

T/t :- Statins [Lovastatin] → competitive inhibitor of enzyme HMG CoA Reductase
↓ cholesterol synthesis.

Clinical case-2

A child was brought to the hospital with pain in the abdomen. Ultrasonography revealed episode of acute pancreatitis. When phlebotomy was done, his serum showed milky appearance. Serum triacylglycerol levels were very high (1200 mg/dl).

What is your probable diagnosis?

→ Type I - familial chylomicronemia

Defect :- Lipoprotein lipase deficiency OR APO CII defect

Lab Δ :- ↑ TAG
↑ CM → gives milky plasma

Clinical case-3

A child was brought to the hospital with yellow-orange tonsils, hepatosplenomegaly and peripheral neuropathy. On investigation - low levels of HDL found

What is your probable diagnosis?

Lab Δ → Absent OR ↓ HDL

- familial Alpha lipoprotein deficiency - Tangier's disease

Defect :- Mutation in the transporter ABCA1 [Transport cholesterol from cells to HDL]

FATTY LIVER

It is a reversible condition in which large vacuoles of triacylglycerol accumulates in liver cells.

Alcoholic fatty liver

- due to heavy alcohol consumption
- In the process of breaking alcohol liver generates harmful substances, that damage liver cells, promote inflammation, ↓ immune mechanism.

Nonalcoholic fatty liver

- Not related c̄ Alcohol consumption.

Risk factors

Type 2 DM
Obesity
High blood pressure
Steroid therapy
Metabolic syndrome
Hepatic infection
Toxins

C/F :- discomfort in upper rt. side of abdomen
Enlarged liver
Nausea, loss of appetite or weight loss
Jaundice, weakness

Diagnosis :-

γ-Glutamyl transferase - specific for Alc. liver ds.
USG liver OR CT scan
Liver biopsy
Serum TG ↑↑, LDL cholesterol ↑↑, HDL ↓↓

Stages of liver damage

Fatty liver → Fibrosis → Cirrhosis

- Fatty liver: Accumulation of TGs
- Fibrosis: Scar tissue begins to replace healthy tissue
- Cirrhosis: Loss of liver cells and irreversible scarring

Causes

1. **Alcoholism** :- Most common cause of fatty liver in India. Alcohol metabolism increases NADH, which converts oxaloacetate to malate and decreases oxaloacetate level. Acetyl CoA is not metabolised by the TCA cycle due to ↓ oxaloacetate. Acetyl CoA is used for fatty acid synthesis and TG synthesis.

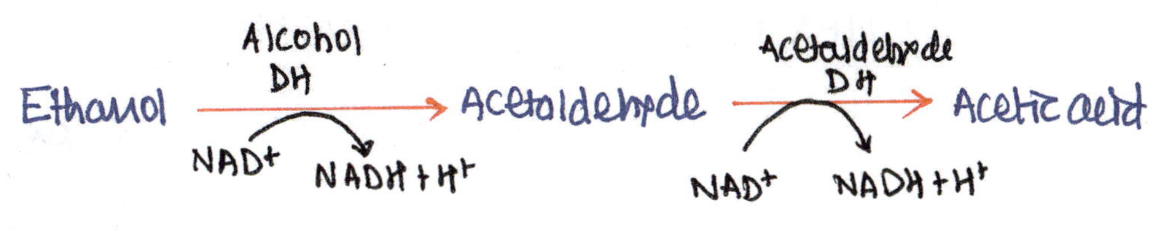

Oxaloacetate $\xrightarrow{NADH+H^+}$ Malate

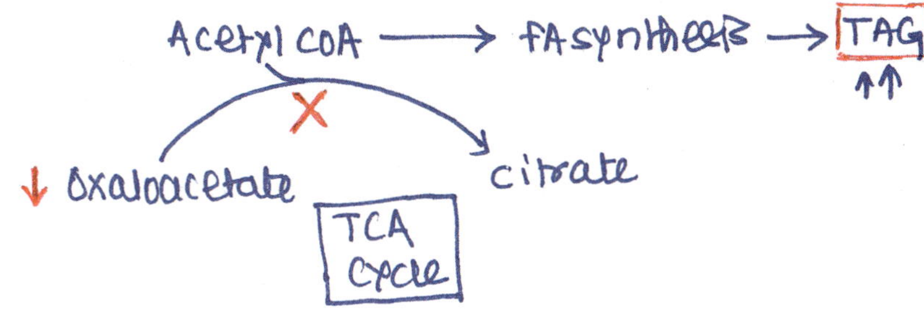

② **Increased TG synthesis**

DM, Alcoholic, high fat diet and starvation is associated c̄ increased fat mobilization from adipose tissue to liver.

③ **Impaired lipoprotein synthesis**

Decreased phospholipid synthesis – deficiency of lipotropic factors.
Defect in formation and secretion of lipoproteins.

④ **Toxic injury to liver** – poisoning by CCl_4, Arsenic, Lead.

Treatment of FATTY liver

① limit or Avoid alcohol
② Take a step to lose weight
③ Make changes in the diet – No excess calories, low saturated and trans fat in diet.
↑ consumption of PUFA, dietary fibres.
④ Moderate exercise
⑤ Medications for the disease

Lipotropic factors (Important BAQ) → Choline, Betaine, Methionine, Inositol, B-12, Folic acid, vit E, Selenium, Glycine, Serine

- Required for mobilization of fat from liver
- Deficiency of these factors causes deposition of Triacylglycerol (Fat) in liver.

CHEMISTRY OF AA & PROTEINS

MCI COMPETENCY BI 5.1

SAQ

1. Classify amino acids with suitable examples.
2. Functional classification of proteins
3. Secondary structure of proteins.
4. Plasma proteins.
5. Biologically important peptides c̄ their functions.
6. Immunoglobulins: structure and functions

LAQ:- Describe the structural organization of proteins. Give example of each structure.

Clinical conditions related c̄ abnormal electrophoresis.

- Multiple Myeloma
- Cirrhosis of liver
- Nephrotic syndrome

Multiple Myeloma :- MCQ + viva

- Plasma cell tumor
- Multiple punched out lesion in the bones
- Monoclonal gammopathy
- characteristic M-band on electrophoresis
- Excretion of light chain proteins (Bence Jones proteins) in urine
- Detected by classical Heat coagulation test.

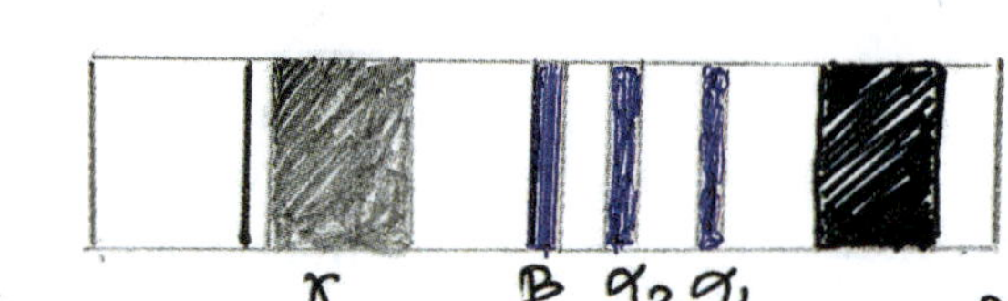

Electrophoresis pattern

Aminoacid classification

Classify aminoacids with suitable examples.

Total 20 Aminoacids incorporated in proteins

21st Aminoacid → Selenocysteine

22nd Aminoacid → Pyrrolysine

They are classified on the basis of 4 properties.

Aminoacids

① Based on structure

$NH_2 - \overset{H}{\overset{|}{C}} - COOH$, with (R) as the side chain

Structural classification

- Aliphatic side chain
  1. Simple
  2. Branched chain
  3. OH-group containing
  4. Sulfur containing
  5. Acidic AAs
  6. Amide group
  7. Basic AAs
- Aromatic AAs
- Imino acid

② Solubility

Polarity

- Polar
- Nonpolar

Nonpolar

- Aromatic-phe, Tyr, Try
- Branched- VIL
- S-containing-Met
- Simple-Alanine

③ Nutritional Requirement

- Essential
  1. Methionine
  2. Arginine
  3. Tryptophan
  4. Threonine
  5. valine
  6. Isoleucine
  7. Leucine
  8. Phenylalanine
  9. Histidine
  10. Lysine
- Nonessential

④ Metabolic fate

Glucose or FAT

- Glucogenic
- Ketogenic: Leucine, Lysine
- Both: Phe, Tyr, Try, Isoleucine

Simple AA	Branched chain	Amide group
Glycine Alanine	valine Isoleucine Leucine	Aspargine Glutamine

OH-group containing	Sulfur containing	Acidic AA	Basic AA
Tyrosine Serine Threonine	Methionine cysteine cystine	Aspartic acid Glutamic acid	Histidine Arginine Lysine

Aromatic :- Phenylalanine, Tyrosine, Tryptophan

Iminoacid: Proline

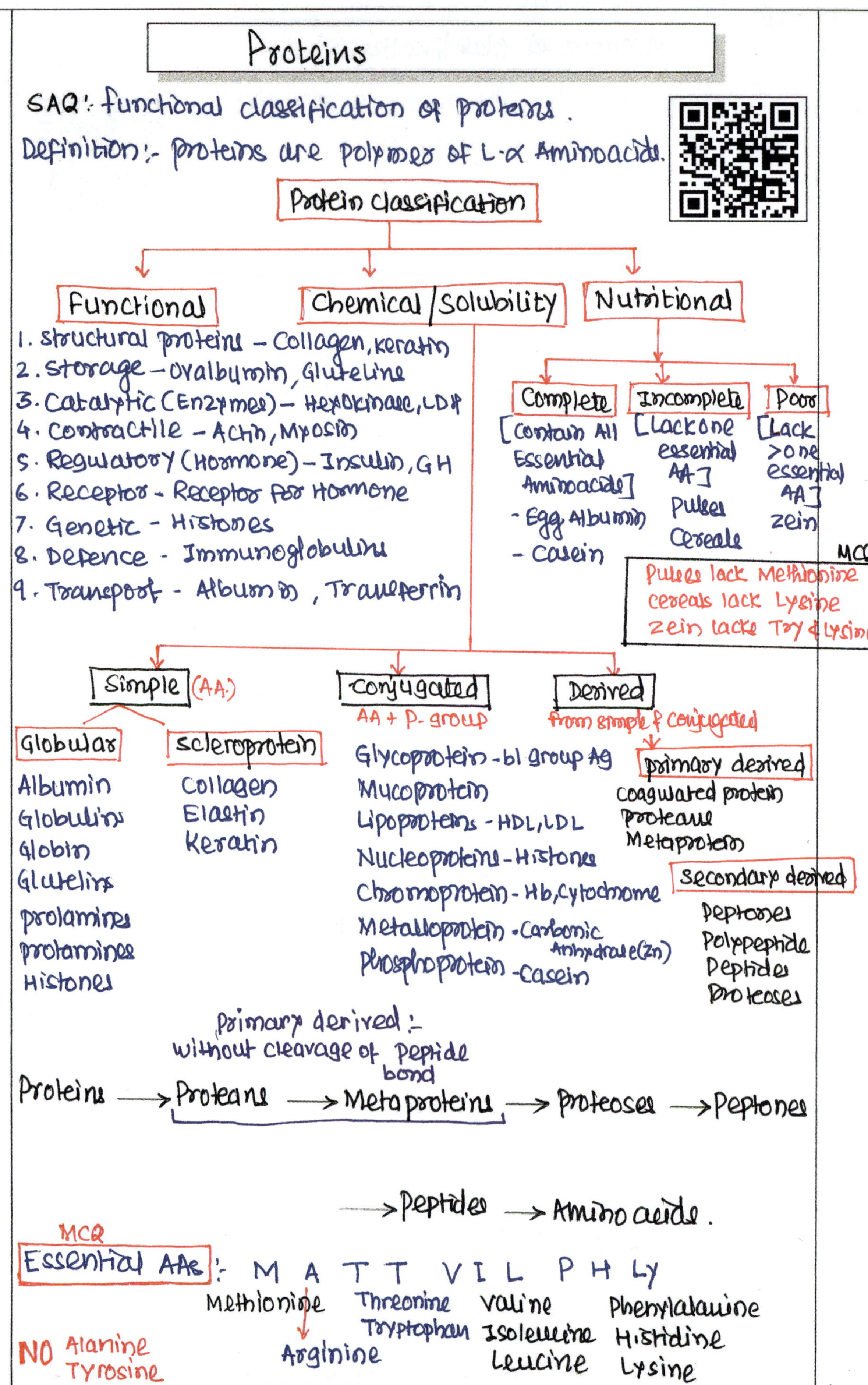
Proteins
SAQ:- functional classification of proteins.
Definition:- proteins are polymers of L-α Aminoacids.
Protein classification
Functional
Chemical / Solubility
Nutritional
1. structural proteins – Collagen, keratin
2. storage – ovalbumin, Gluteline
3. Catalytic (Enzymes) – Hexokinase, LDH
4. Contractile – Actin, Myosin
5. Regulatory (Hormone) – Insulin, GH
6. Receptor – Receptor for Hormone
7. Genetic – Histones
8. Defence – Immunoglobulins
9. Transport – Albumin, Transferrin
Complete
[Contain All Essential Aminoacids]
– Egg Albumin
– Casein
Incomplete
[Lack one essential AA]
Pulses
Cereals
Poor
[Lack >one essential AA]
zein
MCQs
Pulses lack Methionine
cereals lack Lysine
zein lacks Try & Lysine
Simple (AA)
Conjugated
AA + P. group
Derived
from simple & conjugated
Globular
Albumin
Globulins
Globin
Glutelins
prolamines
protamines
Histones
scleroprotein
Collagen
Elastin
Keratin
Glycoprotein – bl group Ag
Mucoprotein
Lipoproteins – HDL, LDL
Nucleoproteins – Histones
Chromoprotein – Hb, Cytochrome
Metalloprotein – Carbonic Anhydrase (Zn)
Phosphoprotein – Casein
primary derived
coagulated protein
Proteans
Metaprotein
Secondary derived
Peptones
Polypeptide
Peptides
Proteoses
Primary derived:- without cleavage of peptide bond
Proteins → Proteans → Metaproteins → Proteoses → Peptones
→ Peptides → Amino acids.
MCQ
Essential AAs :- M A T T V I L P H Ly
Methionine
Arginine
Threonine
Tryptophan
Valine
Isoleucine
Leucine
Phenylalanine
Histidine
Lysine
NO Alanine
Tyrosine

Structural Organization of Proteins

LAQ:- Describe the structural organization of Proteins give example of each structure.

④ levels of Organization

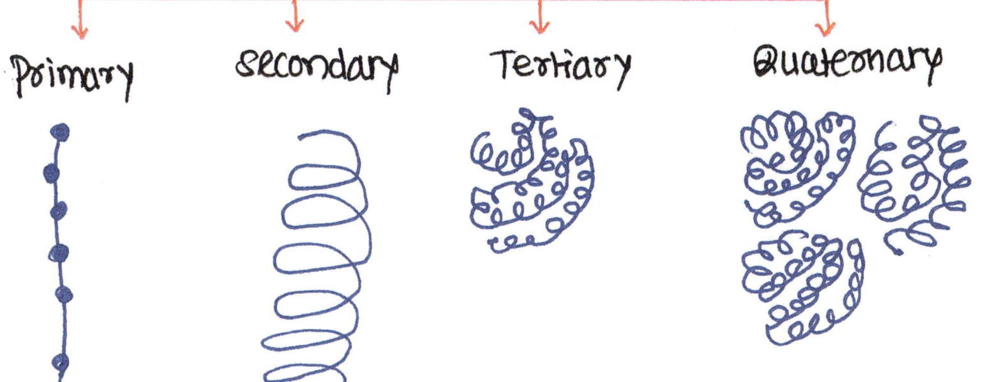

Primary: AAs are aranged in a linear sequence. AAs connected by peptide bond to form Polypeptide

Secondary: Folding of Polypeptide chain upon itself

Tertiary: Folding of secondary structure to form 3D domain

Quaternary: two or more than two polypeptide chain kept together

Single peptide → I, II, III

Two peptide IV

Primary structure

depends on Number of Aminoacids
Sequence of Aminoacids

AAs are connected by peptide bond and occasionally by disulfide bond.

Peptide bond: partial double bond in character betn carboxyl group of 1 AA and Amino group of 2 AA

N-Terminal (Amino) $\overset{+}{N}H_3$ – CH(R1) – [CO–NH] – CH(R2) – [CO–NH] – CH(R3) – [CO–NH] – CH(R4) – COO^- (Carboxy C-Term...

Example :- Insulin

sequenced by fredrick sanger [By Sanger Method]

first isolated in pure form by Banting and Best

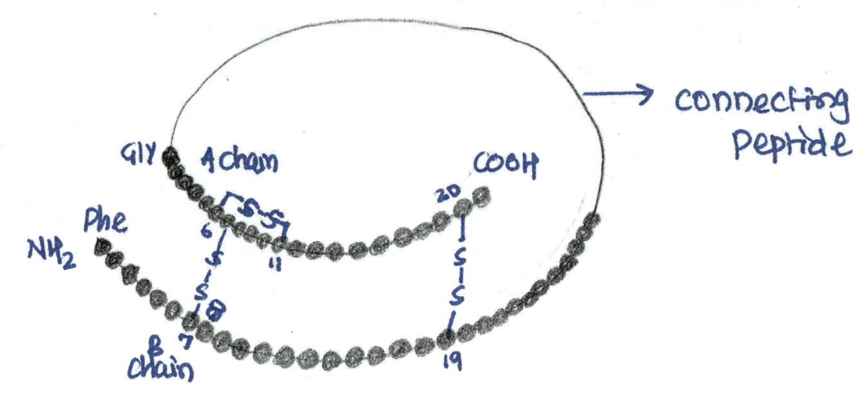

structure of Inculin

Secondary structure → α Helix
β Pleated sheets
Turns
Bends
Loops

α - Helix

- Most common
- stable conformation
- spiral
- Right handed
- Most common AA in α Helix is Methionine

Bonds :- H-Bonds - Intrachain
Hydrophobic interaction
Electrostatic forces
Vanderwaal forces.

Stabilized mainly by intrachain H Bond **H Bond** [betn 1st & 4th AA] [Carbonyl oxygen & Amide nitrogen]

N
H
O
C

Each Turn
3.6 AAs
pitch of turn 0.54 nm
0.15 nm distance betn two AA

destabilized by

Proline
Glycine
series of Acidic (-vely) charged AA [Glutamic & Aspartic Acid]
series of Basic (+vely) charged AA [Lysine, Arginine, Histidine]

Example :- α Helix is found in structure of Myoglobin & Hemoglobin.

β-Pleated sheets Known as β, because it is 2nd in discovery by Pauling and Corey

- These are extended in zigzag manner
 Two types - c̄ reference to N & C terminal

① Parallel ② Antiparallel

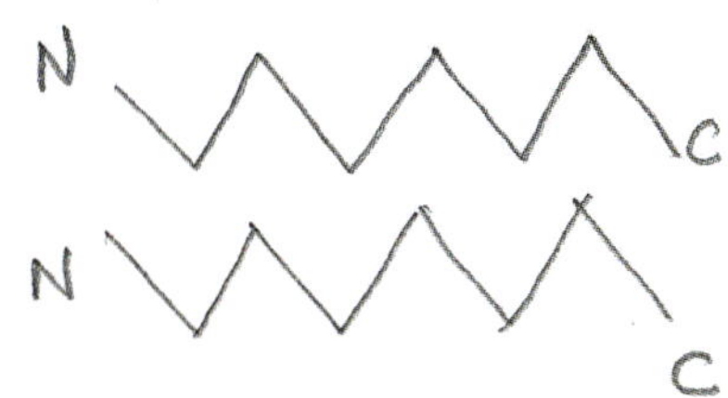

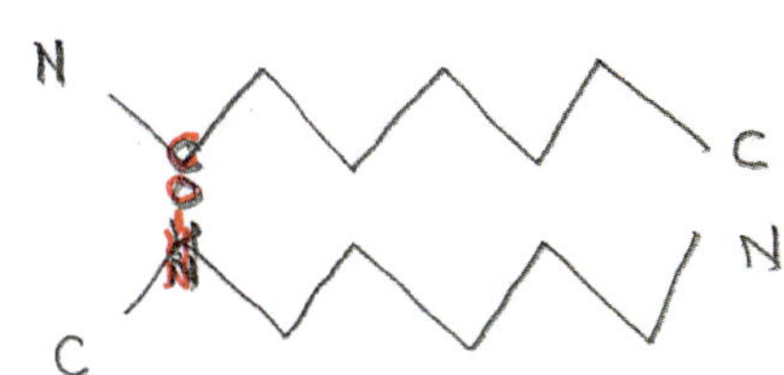

Stabilized by H-Bond betn carbonyl oxygen and amide nitrogen. (Interchain H-Bonds)

Example: ① Carbonic anhydrase - contain both parallel and antiparallel β-sheets.

② Silk fibroin

Turns } Involved in a joining of two secondary structures
Bends } short chain peptides
Loops } Long chain

Super secondary structures - Zinc finger motif
Leucin zipper motif
Helix turn helix motif

Tertiary structure

Folding of secondary structure to form 3 dimensional structure

It is functional unit

Example : **Myoglobin**

Myoglobin: Made up of single polypeptide chain, found in higher concentration in muscle. It is made up of 8 α Helices c̄ Iron in the centre.
It stores oxygen in the muscle, which can be utilized as energy.
It is 3 dimensional and functional protein.

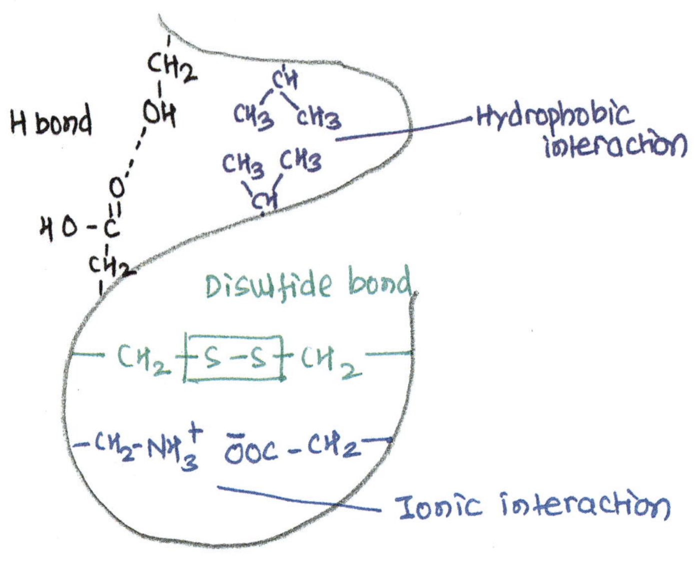

Tertiary structure

stabilized by - Hydrophobic interactions
H-Bonds
Ionic interaction
vander waal forces

	Tertiary structure	Quaternary structure
	Myoglobin	Hemoglobin
Primary structure	153 AAs	141 AAs - α chain 146 AAs - β chain
secondary structure	8 α Helices	8 α Helices for each α chain & β chain
Tertiary structure	folding of α Helices c̄ heme Iron in centre	folding of α Helices c̄ heme
Quaternary structure	—	Tetramer - 2α, 2β units

Quaternary structure

Formed by two or more polypeptide chains.
It is a association of several polypeptide chains or subunits into closely packed arrangement.
Each subunit has its own primary, secondary & tertiary level of organization.

Example :- Hemoglobin < 2α chains, 2β chains

Made up of 4 polypeptide subunits, two α and two β each subunit contains Heme (contains Iron). Each Iron atom binds to one oxygen molecule. Thus one Hb binds with 8 atoms of oxygen.

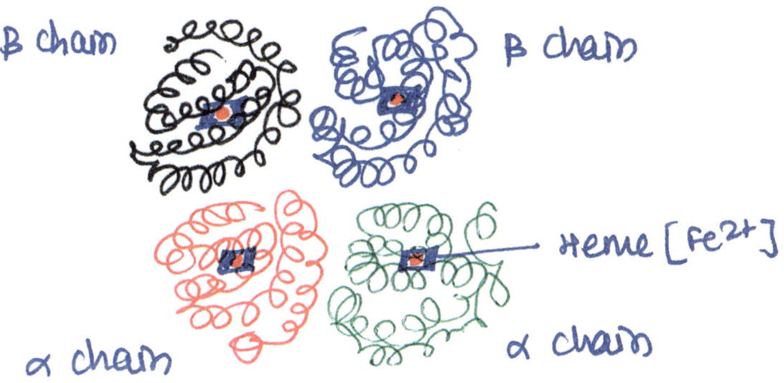

3° and 4° structures stabilized by – Hydrophobic interactions, H-Bonds, Electrostatic forces, Van der waal forces

Plasma Proteins

Classified on the basis of electrophoresis
- Albumin $3.5-5.0$ g/dl
- Globulins $2.5-3.5$ g/dl
- Fibrinogen 200-300 mg/dl

Albumin

(+) (−) Serum Protein Electrophoresis

α_1 α_2 β γ Globulins

α_1 [0.3-0.5 g/dl]	α_2 0.4-0.8 g/dl	β 0.6-1.1 g/dl	γ 0.8-1.8 mg/dl
α_1-antitrypsin	α_2-Macroglobulin	β-lipoprotein (LDL)	IgG
α_1-lipoprotein (HDL)	Haptoglobin	Transferrin	IgA
Orosomucoid	Prothrombin	Hemopexin	IgM
Thyroid binding globulins	Ceruloplasmin	Plasminogen	IgD
Cortisol binding globulin			IgE

Functions of Albumin

1. Maintenance of colloidal osmotic pressure
2. Nutrition
3. Buffering action
4. Transport of free fatty acids, steroids, Hormones, ions, bilirubin etc.

α_1
- α_1-antitrypsin :- Inhibition of trypsin
- HDL :- Reverse transport of cholesterol [from tissues to liver]
- Retinol binding protein :- transport of Retinol
- Thyroid binding globulin :- transport of T3 & T4
- Cortisol binding globulin :- transport of cortisol

α_2
- α_2 Macroglobulin :- It has antitrypsin and antiplasminogen activity
- Haptoglobin :- bind with heme and prevent excretion of Heme.
- Prothrombin :- Clotting factor
- Ceruloplasmin :- transport of copper (Cu^{2+})
  Conversion of $Fe^{2+} \rightarrow Fe^{3+}$

β **β-lipoprotein [LDL]** :- Transport of cholesterol from liver to peripheral tissues

Transferrin :- transport of Iron

Hemopexin :- transport of Heme

plasminogen :- fibrinolytic

γ **IgG** :- Responsible for secondary immune response compliment fixation and opsonization

IgA :- found in secretions [responsible for mucosal immunity]

IgM :- Primary immune response compliment fixation

IgD :- Acts as a B-cell receptor

IgE :- Responsible for allergic response.

Abnormal electrophoretic Pattern

1. Multiple Myeloma - Prominent M-Band
2. Cirrhosis of liver - decreased Albumin, increased γ-globulins β-γ-bridging.
3. Nephrotic syndrome. All proteins lost in urine except α_2 prominent α_2 band, due to retained α_2 macroglobulin, because of larger size.

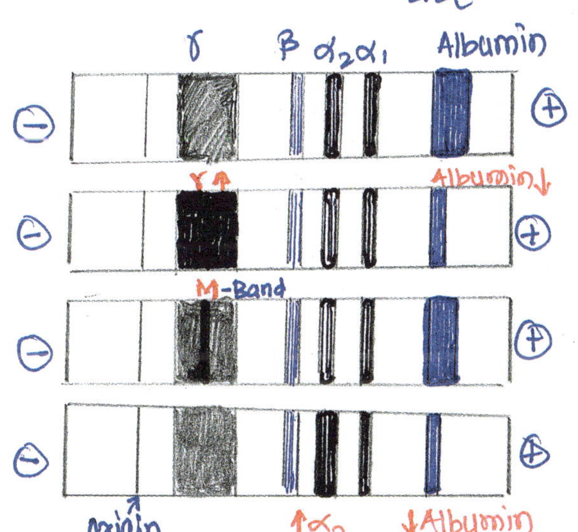

Normal serum electrophoresis

Cirrhosis of Liver
↓ Albumin, ↑ γ-Globulin

Multiple Myeloma
prominent M-band in γ region

Nephrotic syndrome
↓ Albumin, ↑ α_2 globulin

IMMUNOGLOBULINS

SAQ:- Write a note on structure and functions of Immunoglobulins.

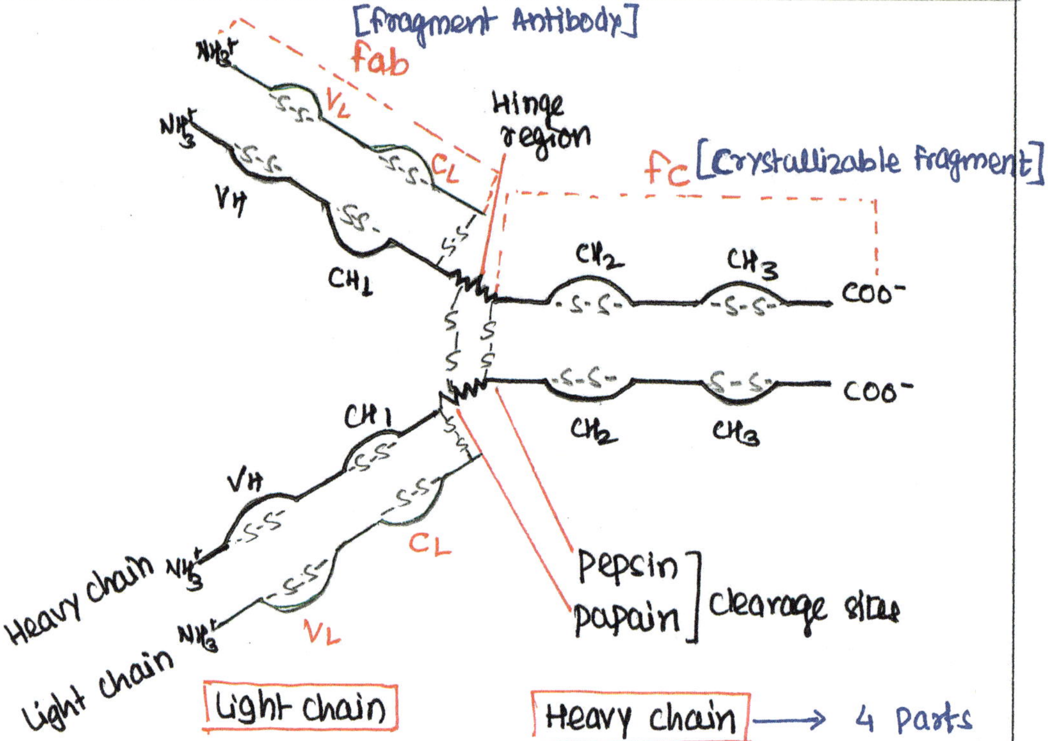

[Fragment Antibody]
Fab

Hinge region

Fc [Crystallizable fragment]

NH_3^+ — V_L — C_L

NH_3^+ — V_H — CH_1

CH_2 — CH_3 — COO^-

CH_2 — CH_3 — COO^-

CH_1 — V_H — NH_3^+ Heavy chain

C_L — V_L — NH_3^+ Light chain

Pepsin / papain] cleavage sites

Light chain

K (60%) OR

λ (40%)

Heavy chain → 4 Parts

γ → IgG

α → IgA

μ → IgM

δ → IgD

ε → IgE

Constant region ③

- CH_1
- CH_2
- CH_3

Variable region ①

- V_H

Fc: crystallizable fragment - Role in complement fixation

Fab: fragment antibody - Antibody binding

MCQ papain:- cleave beyond the disulfide linkage [Fab + Fab + Fc]

pepsin:- cleave before the disulfide linkage [Joint Fab + Fc]

On the basis of Heavy chains - 5 Types

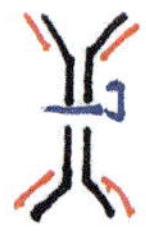

IgG	IgA	IgM	IgD	IgE
Monomer	Monomer/ Dimer	Pentamer	Monomer	Monomer

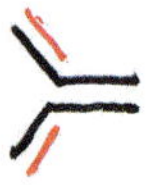

	IgG	IgA	IgM	IgD	IgE
serum concentration	75% Max	15%	9%	0.2%	0.004%
In mg/dl	1000	200	120	3	0.05
Sedimentation coefficient	7S	7S-11S	19S Max	7S	8S
Mol. weight	1,50,000	1,70,000 — 400000	900000 Max	1,80000	1,90,000
MCQ Placental transfer	⊕ Except G2	No	No	No	No
MCQ Compliment fixation	⊕	By Alternate Pathway	⊕	⊖	⊖
MCQ Opsonisation	⊕				
Special features	secondary immune response Types IgG1, G2, G3, G4	Present in secretions. J-chain - Mucosal immunity - Tears, Saliva colostrum Milk	- J-chain primary Immune Response - B-cell receptor	B-cell receptor	Immediate hypersensitivity react. MCQ Allergies Worm infection.

Biologically Important Peptides

SAQ:- Enumerate different biologically important peptides. Write their functions in our body.

Protein :- Polymer of Lα-amino acid connected by peptide linkage

Polypeptide:- Constituent AAs are less in number as compared to proteins and molecular weight is < 1000

Peptide :- Polymer of amino acid (<10 AAs) connected by peptide linkage (-CO-NH-)

$$H_2N-\overset{R_1}{\underset{H}{C}}-\boxed{CO-NH}-\overset{R_2}{\underset{H}{C}}-COOH$$

Amino terminal — Carboxyl terminal

Dipeptide [2 AAs c̄ one peptide linkage]

Tripeptide :- 3 AAs c̄ two peptide linkages.
Tetrapeptide, Pentapeptide, Hexapeptide, Heptapeptide
Octapeptide, Nonapeptide, Decapeptide.

Important Peptides

Dipeptides :- Carnosine, Anserine, Aspartame (Asp + Phe) [Artificial sweetner]

Tripeptide :- ① Glutathione [γ-Glutamyl cysteinyl glycine]

Functions:- Redox buffer
- Important in maintenance of –SH group, so imp. for activation of enzymes.
- Prevent RBC hemolysis
  - Antioxidant
  - Keep Iron Fe^{2+} state
- Detoxification of H_2O_2
- Transport of AA (Meister cycle) in kidneys and intestine

2 GSH
Reduced Glutathione
↓
G-S-S-G
Oxidized Glutathione

② TRH [Thyrotropin releasing Hormone]

Imp. in release of thyroid hormones.

Pentapeptide :- Encephalins - Natural analgesic
acts like opiates
Relieves pain

Octapeptide :- Angiotensin II — ↑ Blood pressure
Stimulate release of Aldosterone

Angiotensinogen —Renin→ Angiotensin I —ACE→ Angiotensin II

⊖ Captopril, Enalapril — ACE Inhibitors used in t/t of Hypertension

Nonapeptide (9 AAs) :- Vasopressin [ADH]
Secreted by posterior pituitary gland
Anti diuretic Hormone - Retention of water by the kidneys
Increases blood pressure [vasoconstriction in blood vessels]

Oxytocin
produced in hypothalamus and secreted by post. pituitary gland.
Responsible for contraction of uterus during child birth. Also used as a drug during labour.
During the breastfeeding it promotes the movement of milk through the ducts, and help the baby to get the milk easily.

Decapeptide :- Kallidin is converted into Bradykinin (Nonapeptide) and causes vasodilation.
Angiotensin I - precursor for Angiotensin II

Other peptides :- Gastrin, Secretin } Gastrointestinal Hormones.
Antibiotics :- Bacitracin
Actinomycin
Gramicidin

METABOLISM OF AA & PROTEINS

MCI COMPETENCY BI 5.3- 5.5

LAQs

1. Describe formation, transport, storage, functions and toxicity of Ammonia.
2. Urea cycle - steps, energetics, regulation & disorders.
3. Write about phenylalanine and tyrosine metabolism with inborn errors.
4. Describe the specialized products given by Phe & Tyr.
5. Tryptophan metabolism with specialized product.
6. Glycine metabolism with specialized product.
7. Metabolism of sulfur containing Amino acids

SAQs

① Transamination

② Transmethylation

③ Digestion & Absorption of proteins

④ Specialized product → Tyrosine, Tryptophan, Glycine, Methionine

⑤ Biological Amines

Clinical cases

① Phenylketonuria → Mousy odour

② Alkaptonuria → Urine turns black on standing

③ Albinism → Hypopigmentation, Photophobia

④ Homocystinuria → subluxation of lens [Ectopia lentis]

⑤ Urea cycle disorders → ↑ NH_3

⑥ Maple syrup urine disease → smell of burnt sugar to urine

⑦ Primary hyperoxaluria → ↑ urinary oxalate, oxalate stones

⑧ Pheochromocytoma → Hypertension, Headache, ↑ VMA in Urine

⑨ Parkinson's disease → Mask face, rigidity, involuntary movements

⑩ Carcinoid syndrome → flushing, diarrhea

⑪ Hartnup's disease → skin rash (pellagra), Ataxia

Digestion and Absorption of Proteins

Hydrolases/Peptidases

- **Endopeptidases [Proteases]**
  - Attack on internal peptide bond
  - Pepsin, Trypsin
- **Exopeptidases**
  - **Carboxypeptidases**
    - Act on C-terminal AA
    - Carboxypeptidase A & B
  - **Aminopeptidases**
    - N-terminal AA.
    - Aminopeptidase

In stomach: Gastric juice contain HCl - Denaturation of peptide
Pepsinogen
Rennin in infant

Pepsinogen —(HCl / Autocatalysis (Pepsin))→ **Pepsin**
cleaves peptide bond of Aromatic & Acidic AA

Pancreatic enzymes

Intestine: Enteropeptidase → Pancreas: Trypsinogen → Trypsin

① Attack protein escape from pepsin
② Act on carbonyl group of Lysine, Arginine

Chymotrypsinogen → Chymotrypsin

Proelastase → Elastase

Procarboxypeptidase → Carboxypeptidase A & B

Chymotrypsin → Act on carboxyl group of Aromatic AA, Leucine, Methionine Asp, Histidine

Elastase → Act on Glycine, Alanine, Serine

Absorption:- ① Na^+ dependent active transport - symport
② Na^+ independent system
③ γ-glutamyl cycle OR Meister cycle [Intestine, Brain & kidney]
involves Glutathione, 3ATP used for single AA
For transport of Cysteine & Glutamine

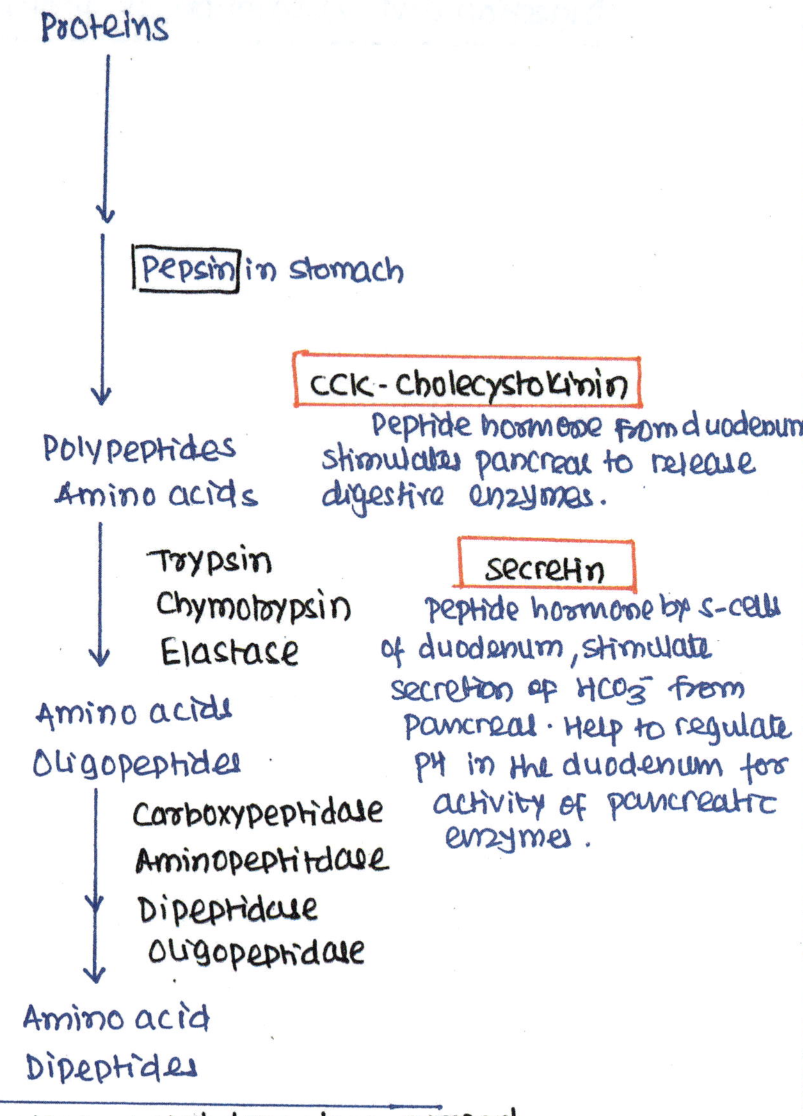

Absorption – Na+ dependent symport
Na+ independent
γ-Glutamyl Cycle

Disorders related ē digestion & absorption

1. Hartnups disease - Tryptophan absorption defect
2. Celiac disease - Lack of N-terminal Glu. splitting enzyme
   Gluten sensitive enteropathy
   Intolerance to wheat, Barley, Oat
3. Cystinuria :- defect in transporter - cysteine, ornithine, Lysine, Arginine
4. Pancreatic ds - Pancreatitis, cystic fibrosis
5. Iminoglycinuria - Inability to absorb imino acid & glycine
   defect in transporter

AMMONIA METABOLISM

LAQ :- Describe formation, transport, storage, function and toxicity of Ammonia

Sources :- ① Transamination & deamination
② Purines & pyrimidines
③ Biogenic amines
④ By the action of intestinal bacteria
Urea ⟶ NH_3

Formation :-

Transamination :- It is a process of transfer of amino group ($-NH_2$) from an aminoacid to a ketoacid to generate a New Ketoacid and amino acid

Glutamate is the central acceptor of NH_3

Aspartate + αKG —SGOT OR AST / PLP⟶ **Glutamate** + oxaloacetate

Alanine + αKG —SGPT OR ALT / PLP⟶ **Glutamate** + Pyruvate

Oxidative deamination :- oxidation coupled with liberation of ammonia.

occurs in mitochondria

Glutamate —mt Glutamate Dehydrogenase (NAD^+ → $NADH+H^+$)⟶ α-Ketoglutarate + **NH_3**

Storage and transport

Glutamate + **NH_3** —Glutamine synthase⟶ **Glutamine**

Major transport & storage form of ammonia from **Brain** & **Intestine**

MCQ

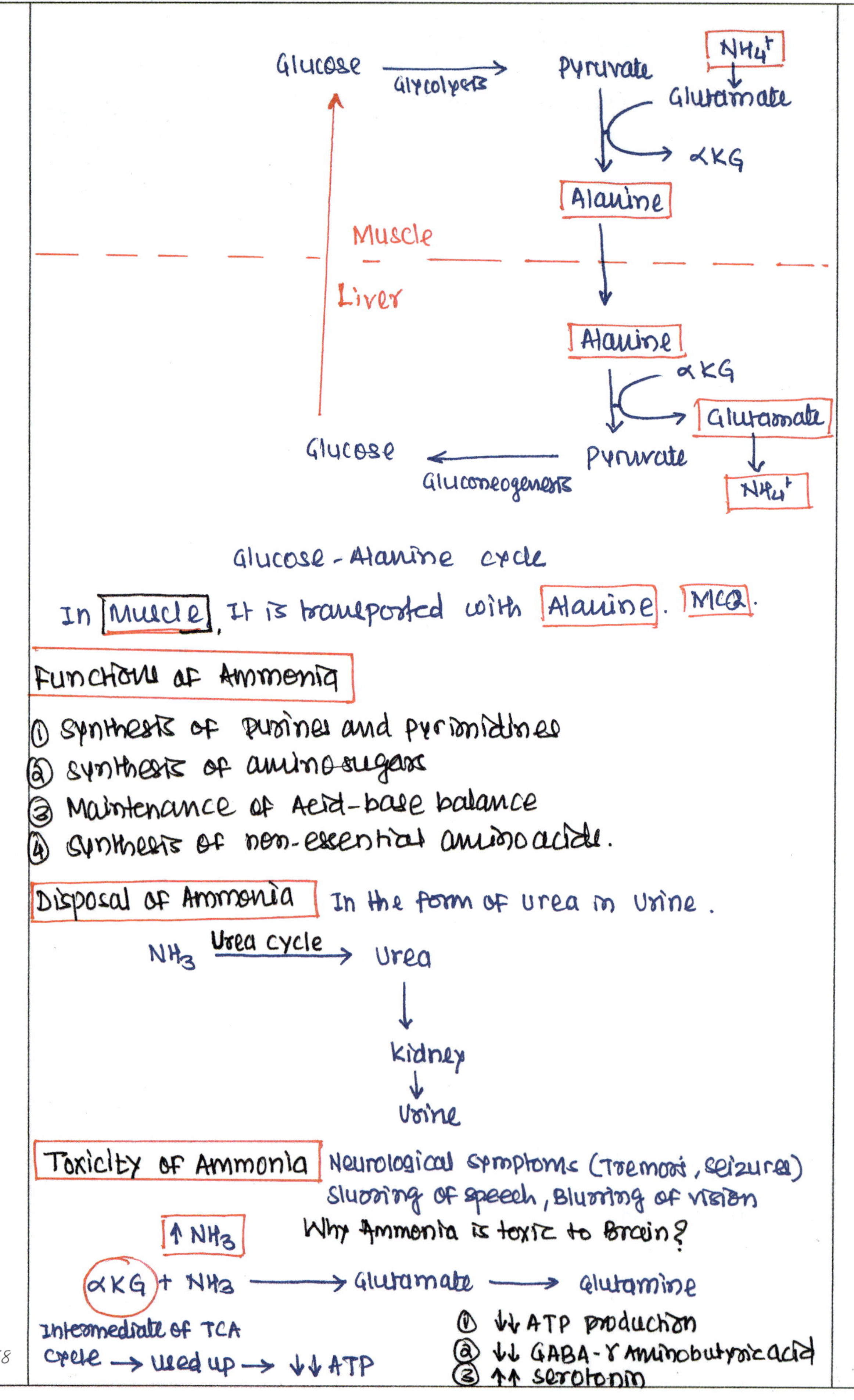

Glucose - Alanine cycle

In Muscle, It is transported with Alanine. MCQ.

Functions of Ammonia

1. Synthesis of purines and pyrimidines
2. Synthesis of amino sugars
3. Maintenance of Acid-base balance
4. Synthesis of non-essential amino acids.

Disposal of Ammonia

In the form of urea in Urine.

NH_3 —Urea cycle→ Urea → Kidney → Urine

Toxicity of Ammonia

Neurological symptoms (Tremors, seizures) Slurring of speech, Blurring of vision

↑ NH_3

Why Ammonia is toxic to Brain?

αKG + NH_3 → Glutamate → Glutamine

αKG: Intermediate of TCA cycle → used up → ↓↓ ATP

1. ↓↓ ATP production
2. ↓↓ GABA-γ Aminobutyric acid
3. ↑↑ serotonin

UREA CYCLE

Ornithine cycle / Krebs Henseleit Cycle

Site : Liver [Arginase present only in liver]

1st two Reactions → Mitochondria

Next three steps → cytosol

5 Important enzymes

One Activator → NAG [Allosteric activator of CPS I]

5 Disorders + HHH Syndrome

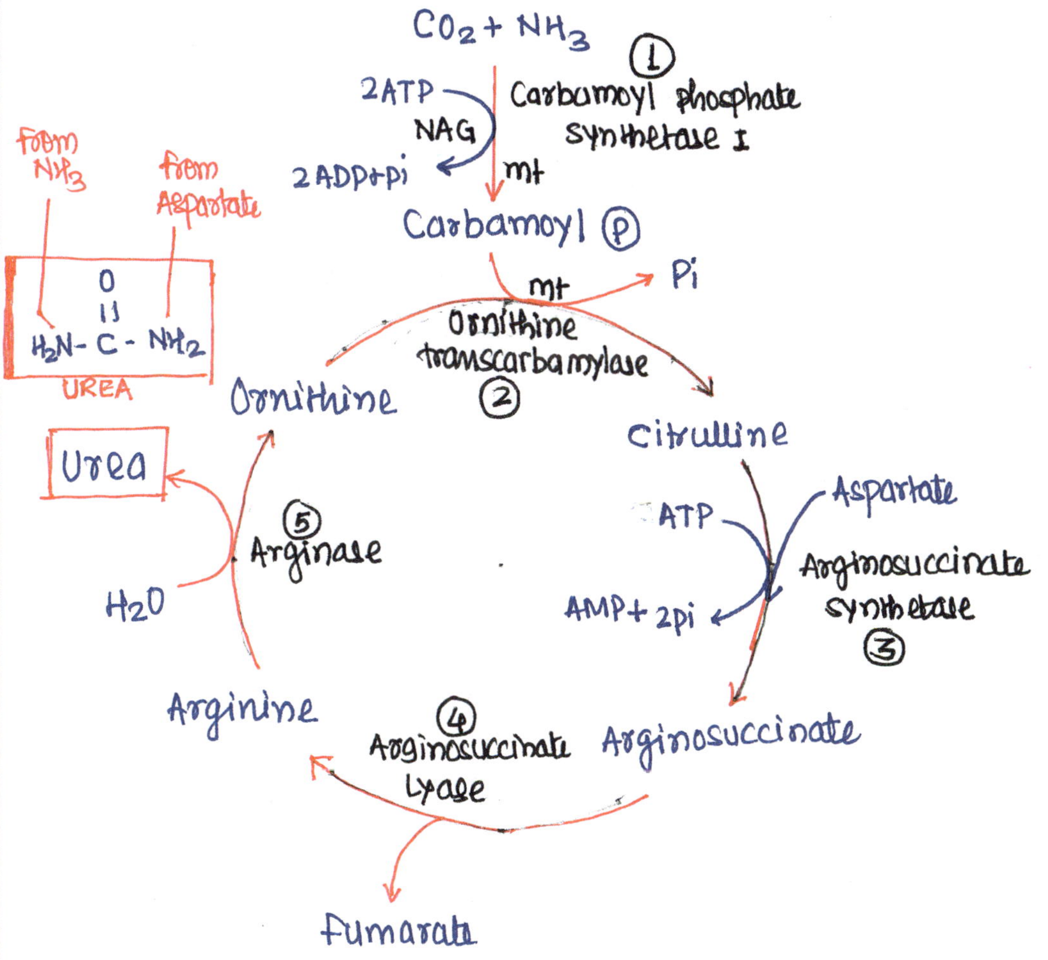

Regulation: NAG – N-Acetyl Glutamate is allosteric activator of carbamoyl Ⓟ synthetase I

1st two steps occurs in the mitochondria and Next three in the cytosol [compartmentalization]

Acetyl CoA + Glutamate $\xrightarrow[\oplus \text{ Arginine}]{\text{NAG synthase}}$ N-Acetyl glutamate

Energetics of urea cycle

1st step (CPS I) → 2ATP

3rd step (Argininosuccinate synthetase) → 2 ATP
Utilization of 2 high energy ⓟ bond

Total = 4 ATP
− 2.5 ATP fumarate → Malate
↓
NADH → 2.5 ATP

Net consumption = 1.5 ATP

Urea cycle disorders

↑ NH_3 → Brain toxicity
Respiratory Alkalosis → ↓ GABA
Loss of inhibition on Resp. centres

These are genetic diseases caused due to deficiency of enzymes of urea cycle.

These enzymes are responsible for removing ammonia from the blood. Deficiency of any enzyme causes elevation of **Ammonia** level and its toxic effects.

Disorders	Defective enzyme	Product Accumulated
Hyperammonemia type I	CPS-I	Ammonia
Hyperammonemia type 2	Ornithine transcarbamylase	Ammonia, orotic acid
Citrullinemia	Arginosuccinate synthetase	Citrulline
Arginosuccinic aciduria	Arginosuccinate lyase	Arginosuccinate
Argininemia	Arginase	Arginine
HHH syndrome [blocked transport of ornithine from cytosol to mt]	Due to defect in ORNT1 gene [ornithine translocase]	Ammonia Ornithine Citrulline

Clinical case

1. A neonate developed hyperammonemia on the 6th day of life. Also had poor feeding, Lethargy, hypothermia, seizures, apnoea & coma. He died due to sepsis.
   Diagnosis :- Urea cycle disorder [Hyperammonemia type I or II]

2. A Male infant with H/o Lethargy, poor feeding and seizures brought to the Hospital. Blood showed increased levels of citrulline and ammonia.
   Diagnosis :- Citrullinemia

3. A 3 month old infant developed poor feeding, hepatosplenomegaly, convulsion, respiratory failure. Blood showed raised NH_3 & Arginosuccinic acid. Diagnosis- Arginosuccinic aciduria.

Phenylalanine and Tyrosine

Phenylalanine

- Essential AA
- Nonpolar
- Both G & K
- Aromatic AA

Tyrosine

- Aromatic AA
- phenolic OH group
- Nonessential
- Relatively polar among Nonpolar
- Both G & K

Phenylalanine
→ (O_2 → H_2O; Tetrahydrobiopterin BH_4 → Dihydrobiopterin) **Phe Hydroxylase** — PKU
Tyrosine
→ (αKG → Glutamate) **Transaminase** — Tyrosinemia II
P-OH-Phenylpyruvate
→ (Ascorbate → Dehydro Ascorbate + H_2O + CO_2) **Hydroxylase** — Neonatal tyrosinemia
Homogentisate
→ (O_2) **Homogentisate Oxidase** — Alkaptonuria
4 Maleyl acetoacetate
→ **Maleyl acetoacetate Isomerase** — Tyrosinemia I
4 fumaryl acetoacetate
→ **Hydrolase** — Tyrosinemia I
→ Fumarate → Glucose
→ Acetoacetate → Fat

Phenylalanine → (Phenylalanine Hydroxylase) → Tyrosine

- Phenylalanine → Protein
- Tyrosine → Protein
- Tyrosine → Glucose
- Tyrosine → Fat
- Tyrosine → Melanin
- Tyrosine → Dopamine
- Tyrosine → Norepinephrine, Epinephrine
- Tyrosine → T3 & T4

Specialized Products of PHE & TYR

① Melanin [Skin, Hair, Eyes]
② Dopamine [CNS]
③ Norepinephrine & Epinephrine [Adrenal medulla]
④ Thyroid Hormones [T3 & T4] [Thyroid gland]

Melanin

Tyrosine
↓ Tyrosinase ✗ Albinism
DOPA
↓ Tyrosinase
Dopaquinone → (Cysteine) → Red Melanin
↓
Leucodopachrome
↓
5,6 Dihydroxyindole
↓
Indole 5,6-quinone
↓
Melanochrome
↓
Black Melanin

Leucoderma/Vitiligo

- Localised Hypopigmentation due to defect in tyrosinase
- White patches.

Albinism:- Inborn error of Phe and Tyr Metabolism
Lack of synthesis of pigment Melanin
Autosomal recessive
Incidence 1 in 20,000
Most common cause is defective **Tyrosinase**

C/F: Sensitive to sunlight, photophobia
susceptible to skin cancer, Generalised Hypopigmentation

Synthesis of Dopamine, Norepinephrine, Epinephrine

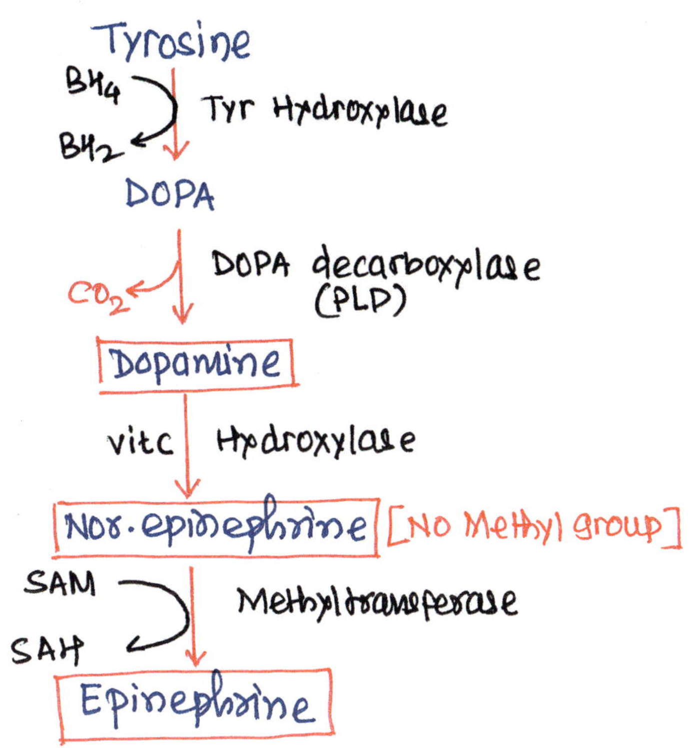

Synthesis of Thyroid Hormones [T3 & T4]

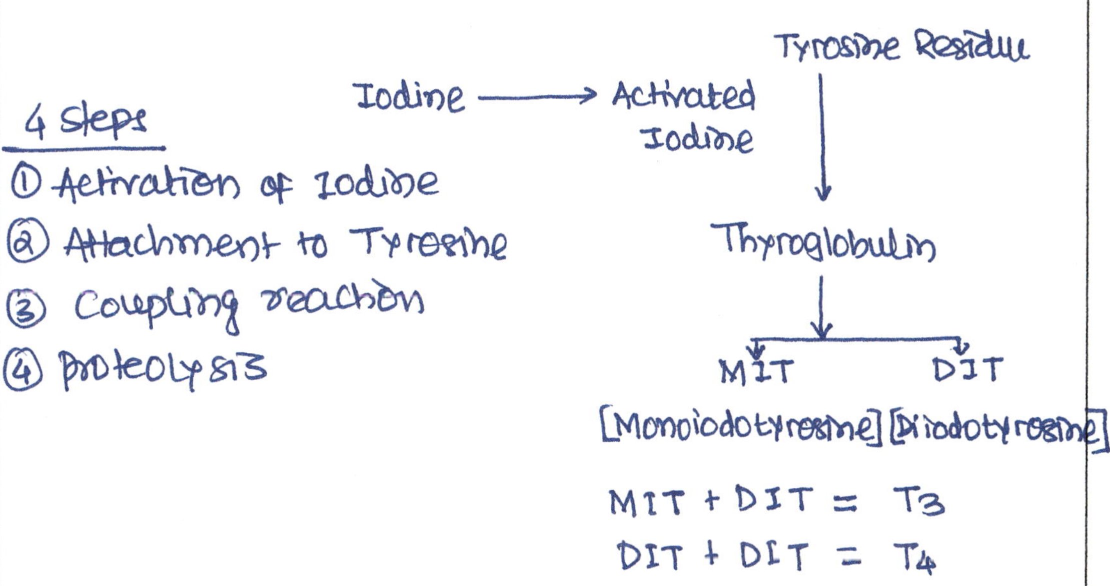

Phenylketonuria [PKU]

It is an inborn error of amino acid (Phen & Tyr) metabolism.

Autosomal recessive disorder

Deficient enzyme :- Phenylalanine hydroxylase

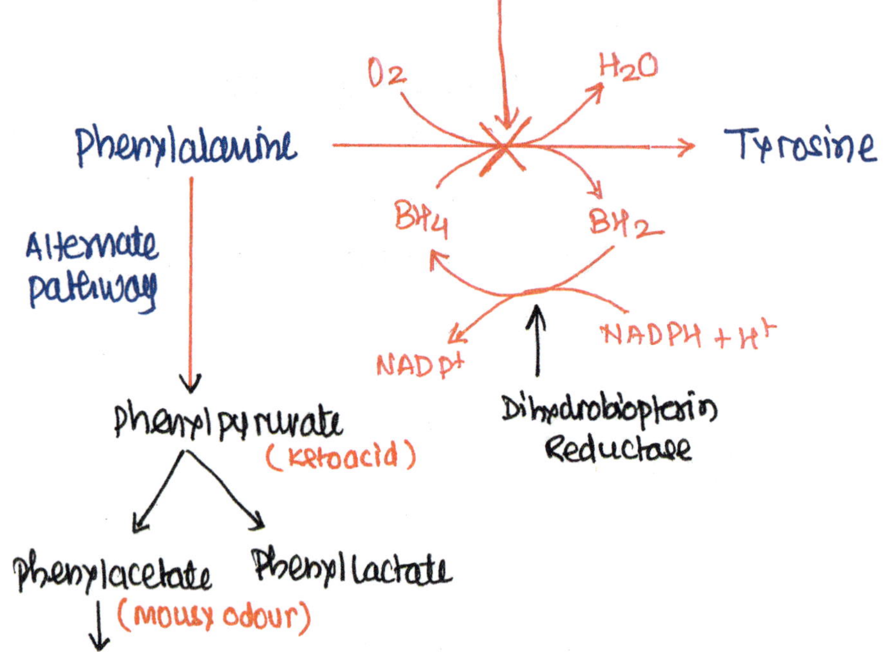

Due to deficiency of phenylalanine hydroxylase, phenylalanine is diverted to alternate pathway and gives phenylpyruvate, phenylacetate, phenyllactate and phenyl acetylglutamine.

Phenylpyruvate is a ketoacid hence the name PKU.

Mousy odour is due to phenylacetate and other byproducts.

Types

Type 1 classical phenylketonuria :- Phenylalanine hydroxylase

Type 2 & 3 :- defect in dihydrobiopterin reductase

Type 4 & 5 :- defect in dihydrobiopterin biosynthesis

Clinical manifestations.

Neurological manifestations :- due to decreased synthesis of Dopamine, Norepinephrine, epinephrine which are neurotrans-mitters in CNS.

Also due to decreased serotonin production and defective myelin formation: (↑ Phe, interferes with Tryptophan metabolism hence, ↓ serotonin)

Hypopigmentation :- due to inhibition of tyrosinase involved in melanin formation. [light skin colour, fair hair, Blue eyes]

C/F :- Mental retardation
- Failure to walk or talk
- Growth failure
- Seizures and tremors
- Generalised hypopigmentation / Localised hypopigmentation
- Fair hair, Blue eyes.
- Mousy odour of urine

Lab diagnosis

screening test :- Guthrie test MCQ
Bacterial inhibition assay - bacillus subtilis overcome the growth by using Phenylalanine on agar plate.

confirmatory test :- Tandem mass spectrometry MCQ
Chromatography

prenatal diagnosis :- DNA probes

Other test :- ferric chloride test

Treatment

Avoid Phenylalanine in diet (Avoid meat, Poultry products)
Tyrosine becomes essential aminoacid, has to be supplied in the diet.
Phenylalanine free commercial preparations
supportive t/t :- behavioural therapy.

ALKAPTONURIA
SAQ Clinical case

Autosomal recessive
1 in 25000
Defect in **Homogentisate Oxidase**

Benzoquinone Acetate ← (Polyphenol Oxidase) — Homogentisate

Benzoquinone Acetate ↓ Polymerisation → Alkapton ↓ deposit in cartilage, Bone, tendons, joint → Ochronosis

Homogentisate ↓ ✗ Homogentisate oxidase → 4 Maleyl acetoacetate

- Homogentisate accumulates in tissues and blood and excreted in urine.
- Homogentisate on standing, gets oxidized to corresponding quinones which polymerises to give black or brown color.
- Change in colour of urine on standing to brown or dark.
- Asymptomatic in child hood
- Tendency towards Arthritis in adulthood - Ochronosis.

C/F :- Urine turns dark or brown on exposure to air, but often goes unnoticed because this change takes time.
Dark staining of diper in infancy.
Skin, cartilage, sclera discolored - blue, grey or black
Accumulation of Alkapton in tendons, bones cause tendonitis & bone pain.
Arthritis of large joints and intervertebral disc.

Diagnosis :- urine gives positive test c̄ $FeCl_3$ & Silver Nitrate due to strong reducing activity of Homogentisate.
Confirmatory test :- Tandem Mass Spectrometry

T/t :- No specific t/t, Low phe in diet, symptomatic t/t for Arthritis.

Tyrosinemia type 2

Also known as Richner Hanhart Syndrome

Defective enzyme :- Tyrosine transaminase

Accumulation and excretion of Tyrosine and metabolites.

C/F :- Dermatitis
Eye lesions
Rarely Mental retardation

Tyrosine
αKG → Glutamate (Tyr transaminase ✗)
↓
P-OH phenylpyruvate
↓ ✗
Homogentisate
↓ oxidase
4 Maleylacetoacetate
↓ ✗
4 Fumaryl acetoacetate
↓ ✗
Fumarate + Aceto acetate

Neonatal tyrosinemia

Defective enzyme :- P-OH phenylpyruvate hydroxylase

Tyrosinemia type 1

Also known as Tyrosinosis

Defect in enzyme - Maleylacetoacetate isomerase
Fumarylacetoacetate hydrolase

C/F :- diarrhea
vomiting
Cabbage like odour from body secretions

PARKINSON'S DISEASE

Associated with ↓↓ dopamine in brain, due to loss of neurons in basal ganglia

Affect near about 1% population above 60 yrs of age.

C/F :- Muscle rigidity
Mask like face
Tremors
Lethargy
Involuntary movements

T/t :- L-DOPA and Carbidopa is given
In brain L-DOPA is converted into Dopamine
carbidopa prevents peripheral conversion of Dopa into dopamine and prevents side effects.

TRYPTOPHAN METABOLISM

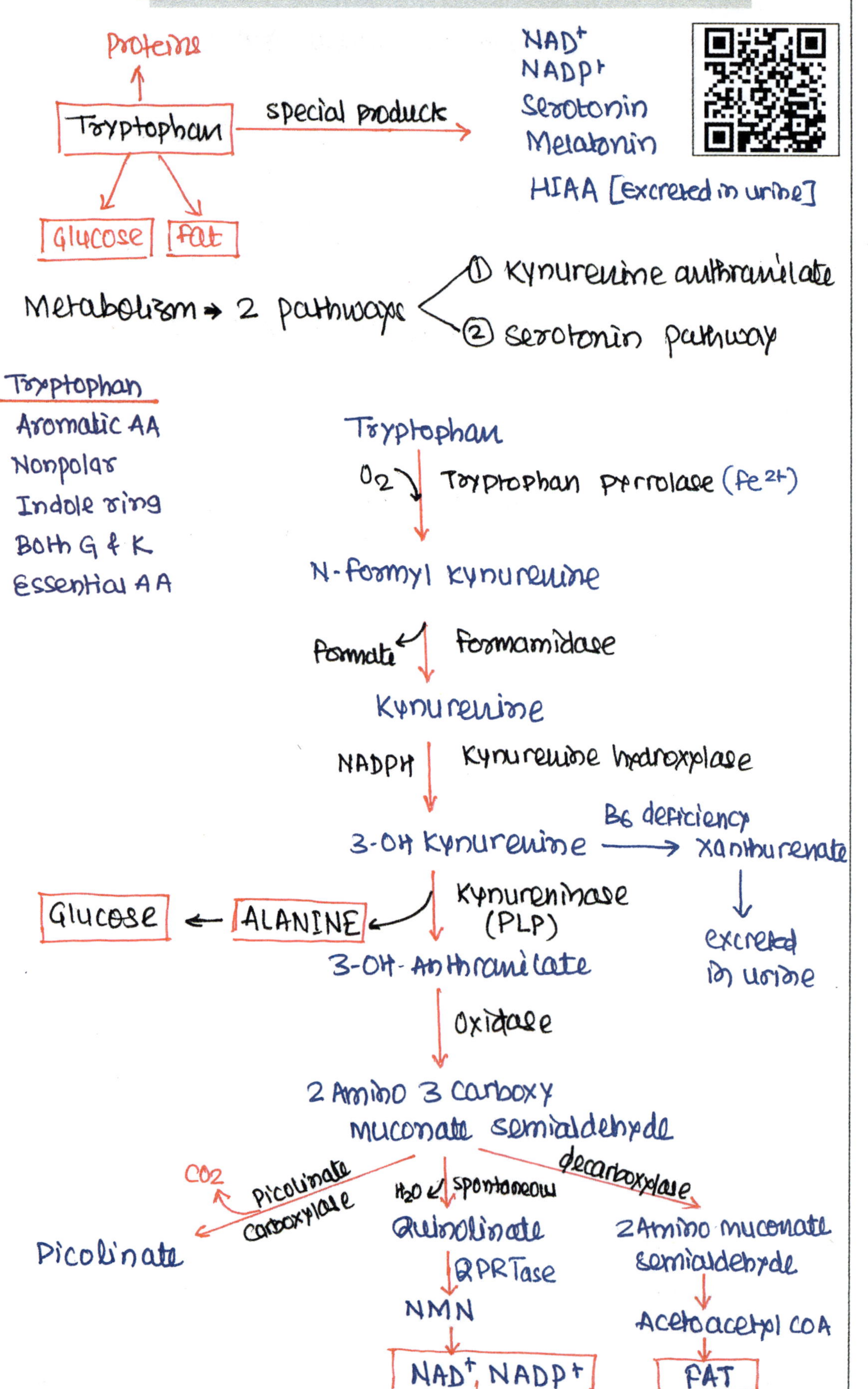

Serotonin pathway

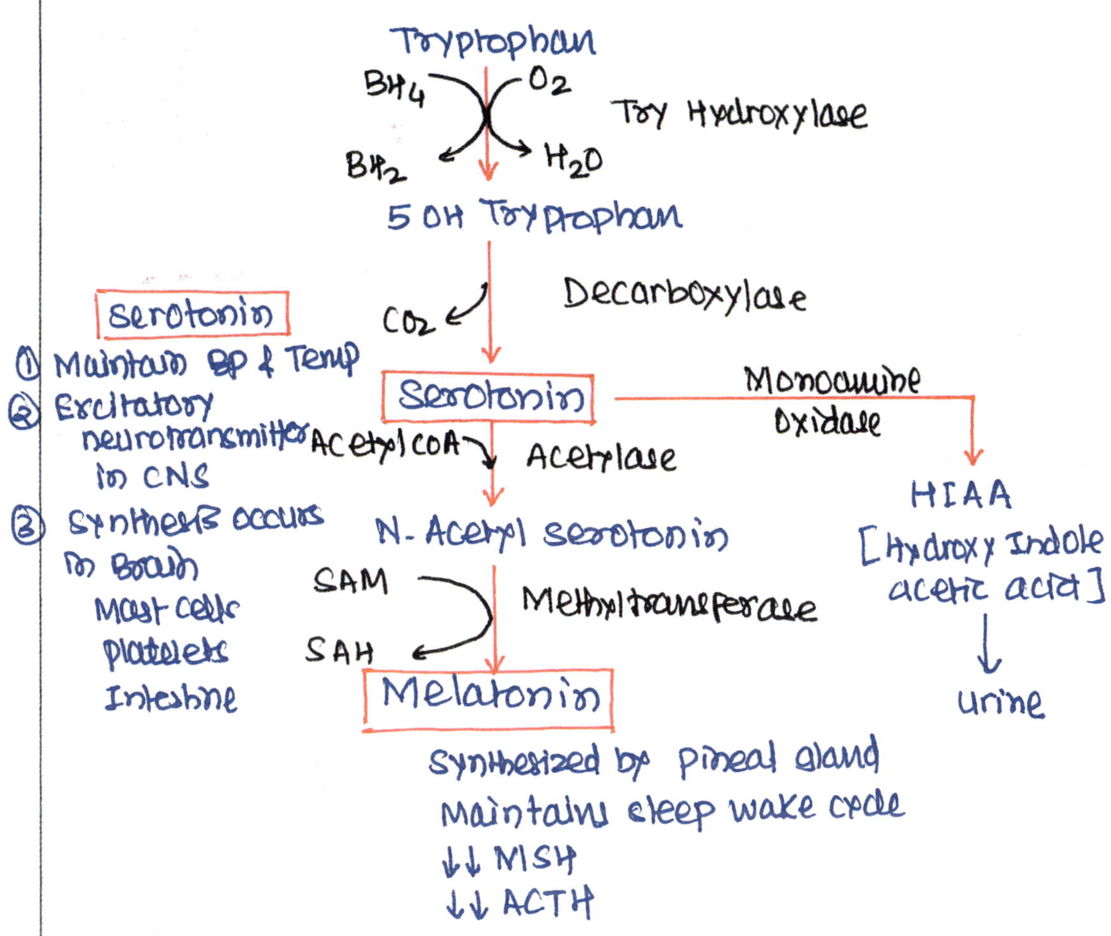

Synthesized by pineal gland
Maintains sleep wake cycle
↓↓ MSH
↓↓ ACTH

Disorders

① **Hartnups disease**

defect in absorption of Tryptophan from intestine and renal tubules. C/F : skin Rash, Ataxia, CNS symptoms
pellagra like symptoms → diarrhea, dementia, dermatitis
60 mg Tryptophan ⟶ 1 mg Niacin

② **Carcinoid syndrome / Argentaffinoma**

Arises from argentaffin cells of intestine (tumor)
60% Tryptophan ⟶ serotonin [↑ serotonin] → skin flushing
facial skin lesions
Diarrhea
Rapid breathing and Heart rate

③ **Xanthurenate in Urine**

In B6 deficiency (PLP)
defective kynureninase

GLYCINE METABOLISM

Nonessential Amino acid

Optically inactive

Gives Glucose hence glucogenic AA

LAQ: Glycine metabolism with specialized products given by glycine.

Glycine → Proteins (Collagen) (30% glycine)

Glycine → Glucose

Glycine → Heme

Glycine → Creatine

Glycine → Glutathione

Glycine → Conjugation reactions

Glycine → C_4, C_5, N_7 of purine ring

Synthesis

from ① Threonine

② Serine

③ Glycine synthase

④ Glyoxylate

① Threonine —Aldolase→ Glycine

② Glyoxylate ←←← Serine → Pyruvate → Glucose

Serine ⇄ Glycine (Serine hydroxymethyl transferase)

③ $NH_3 + CO_2 + N^5N^{10}$ Methylene THF ⇄ Glycine (Glycine synthase)

Glycine —Oxidase→ Glyoxylate

④ Glyoxylate —Transaminase→ Glycine

✗ Primary Hyperoxaluria [Protein targetting defect]

Formation of oxalate stones

Glyoxylate → Oxalate → Urine

Glyoxylate → Formate [One C-unit]

Degradation

① $NH_3 + CO_2 + THF$

② Glyoxylate → Oxalate

③ Serine → Glucose

Specialized products of Glycine

① Glutathione [Glutamyl-cysteinyl-glycine] tripeptide → Maintains RBC integrity

② Glycine + Succinyl CoA

↓ ALA synthase

δ ALA

↓

↓

↓

Heme

1st step in the Heme synthesis

③ Glycine + Arginine —Liver→ Guanidoacetate

↓ SAM, Muscle

Creatine

↓

Creatine Ⓟ

↓ (H_2O)

Creatinine (In urine)

④ Benzoyl CoA + Glycine ⟶ Hippuric acid ⟶ Urine

[Conjugation Reaction]

Glycine + Cholic acid ⟶ Glycocholate

Glycine + Chenodeoxycholic acid ⟶ Glycochenodeoxycholate

Disorders

① Glycinuria :- It is due to defective renal reabsorption of glycine

② Primary hyperoxaluria :- due to defective protein targetting of Glycine transaminase [found in mitochondria instead of peroxisomes]

Methionine, Cysteine, Cystine

LAQ: Metabolism of sulfur containing aminoacids with inborn errors.

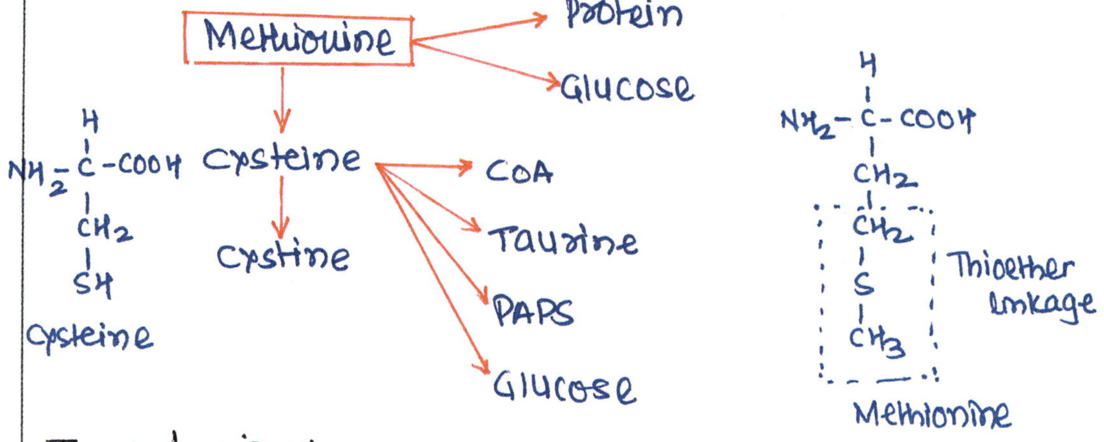

Three steps in the Metabolism of Methionine

① Transmethylation reaction: Methionine ⟶ Homocysteine

② Cysteine synthesis

③ Degradation of Cysteine

① Methionine → Homocysteine

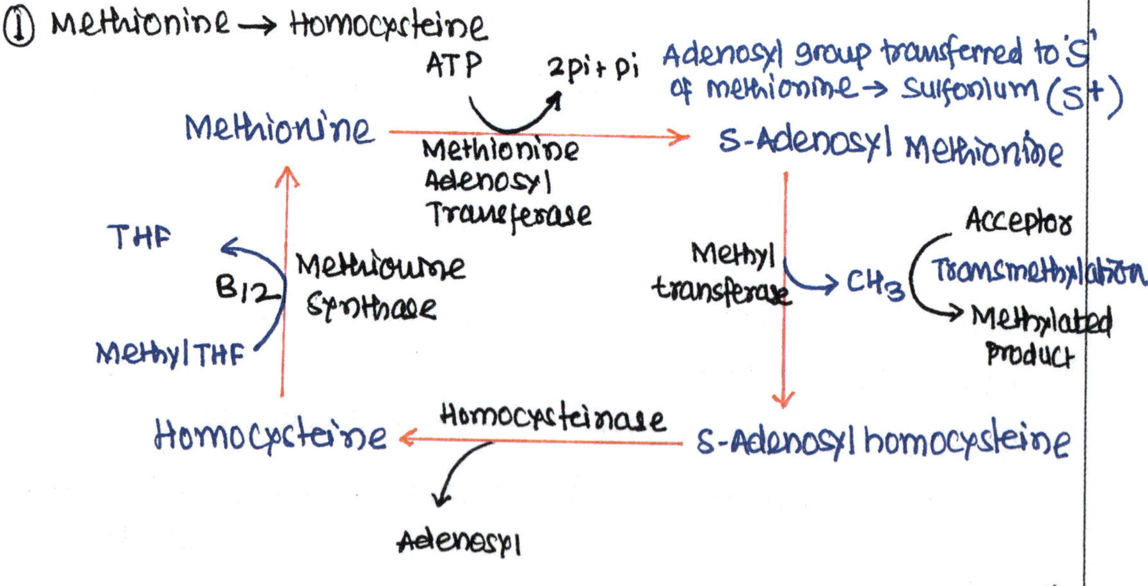

Transmethylation :- Methyl group is transferred from one compound to other.

Guanidoacetate $\xrightarrow{CH_3}$ Creatine

Norepinephrine $\xrightarrow{CH_3}$ Epinephrine

Ethanolamine $\xrightarrow{CH_3}$ Choline

N-Acetyl serotonin $\xrightarrow{CH_3}$ Melatonin

Homocysteine $\xrightarrow{CH_3}$ Methionine

Epinephrine $\xrightarrow{CH_3}$ Metanephrine

Carnosine $\xrightarrow{CH_3}$ Anserine

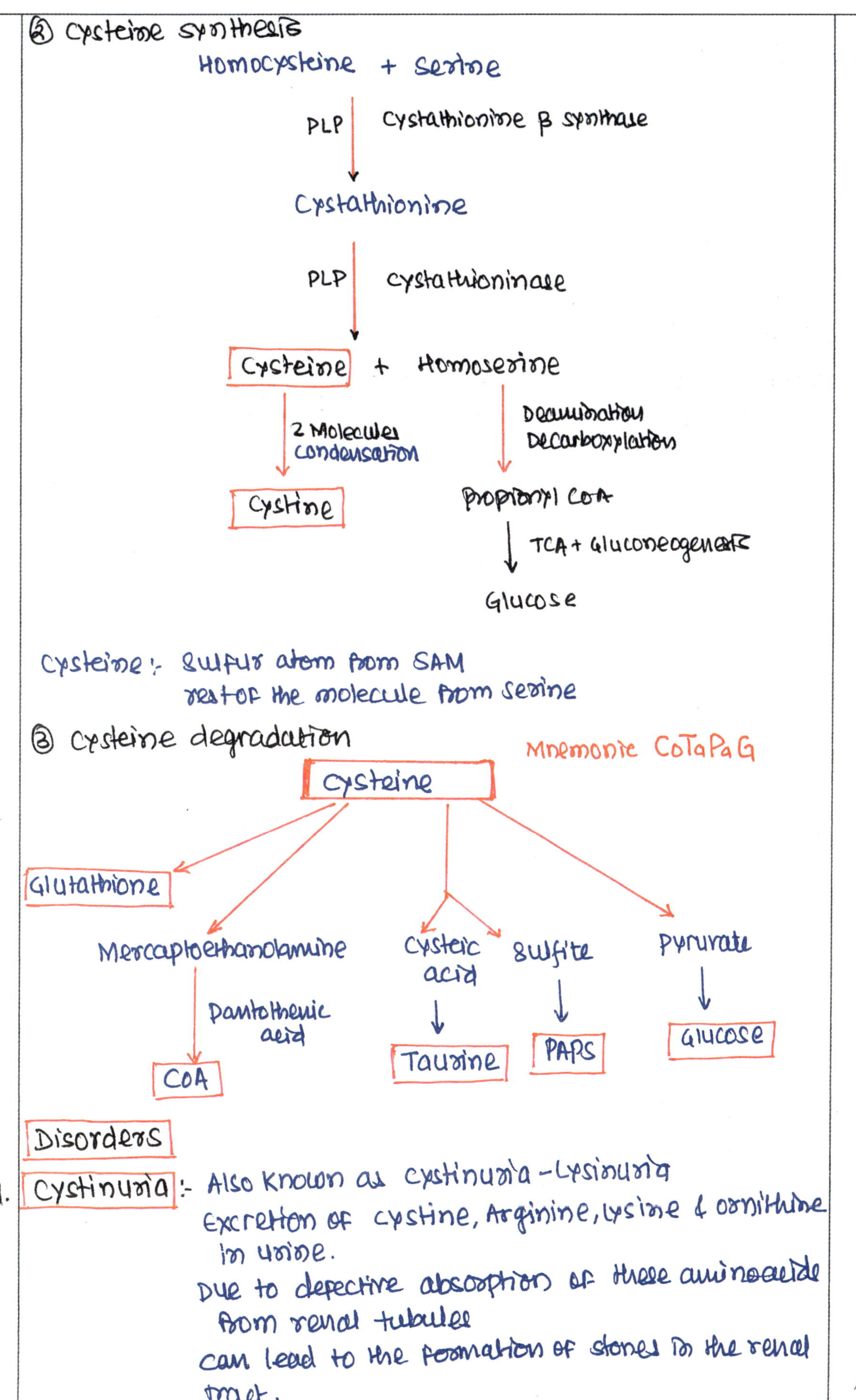
② Cysteine synthesis
Homocysteine + Serine
PLP
Cystathionine β synthase
Cystathionine
PLP
Cystathioninase
Cysteine
+ Homoserine
2 Molecules condensation
Deamination Decarboxylation
Cystine
Propionyl CoA
TCA + Gluconeogenesis
Glucose
Cysteine :- Sulfur atom from SAM
rest of the molecule from serine
③ Cysteine degradation
Mnemonic CoTaPaG
Cysteine
Glutathione
Mercaptoethanolamine
Pantothenic acid
CoA
Cysteic acid
Taurine
Sulfite
PAPS
Pyruvate
Glucose
Disorders
1. Cystinuria :- Also known as cystinuria - Lysinuria
Excretion of cystine, Arginine, lysine & ornithine in urine.
Due to defective absorption of these aminoacids from renal tubules
can lead to the formation of stones in the renal tract.

2. **Cystinosis**

Due to accumulation of cystine in the tissues & organs of the body.

Accumulation of cystine in Lysosomes → Blood
→ Reticuloendothelial system
spleen liver Bone marrow

3. **HOMOCYSTINURIA**

due to deficiency of Cystathionine β-synthase
Accumulation of Homocysteine in blood and excretion in Urine.

Homocysteine + Serine
↓ PLP ✕ cystathionine β synthase
cystathionine

Clinical features

- Myopia (Nearsightedness)
- Dislocation of lens
- Abnormal blood clotting
- brittle bones that are prone to fracture
- developmental delay
- Learning problem
- failure to thrive
- Megaloblastic anemia

↑ Homocysteine
↓
Interferes c̄ collagen formation
Responsible for aggregation of LDL
↓
Coronary Artery diseases.

Lab diagnosis

① Homocystein level in Blood & urine
② Enzyme assay
③ Cyanide-Nitroprusside test - rapid, simple, qualitative test for determination of cystine concentration :- Red colored complex

Treatment :- Vit B6 supplementation, cysteine is given
Dietary restriction of methionine
Vit B12 & folic acid

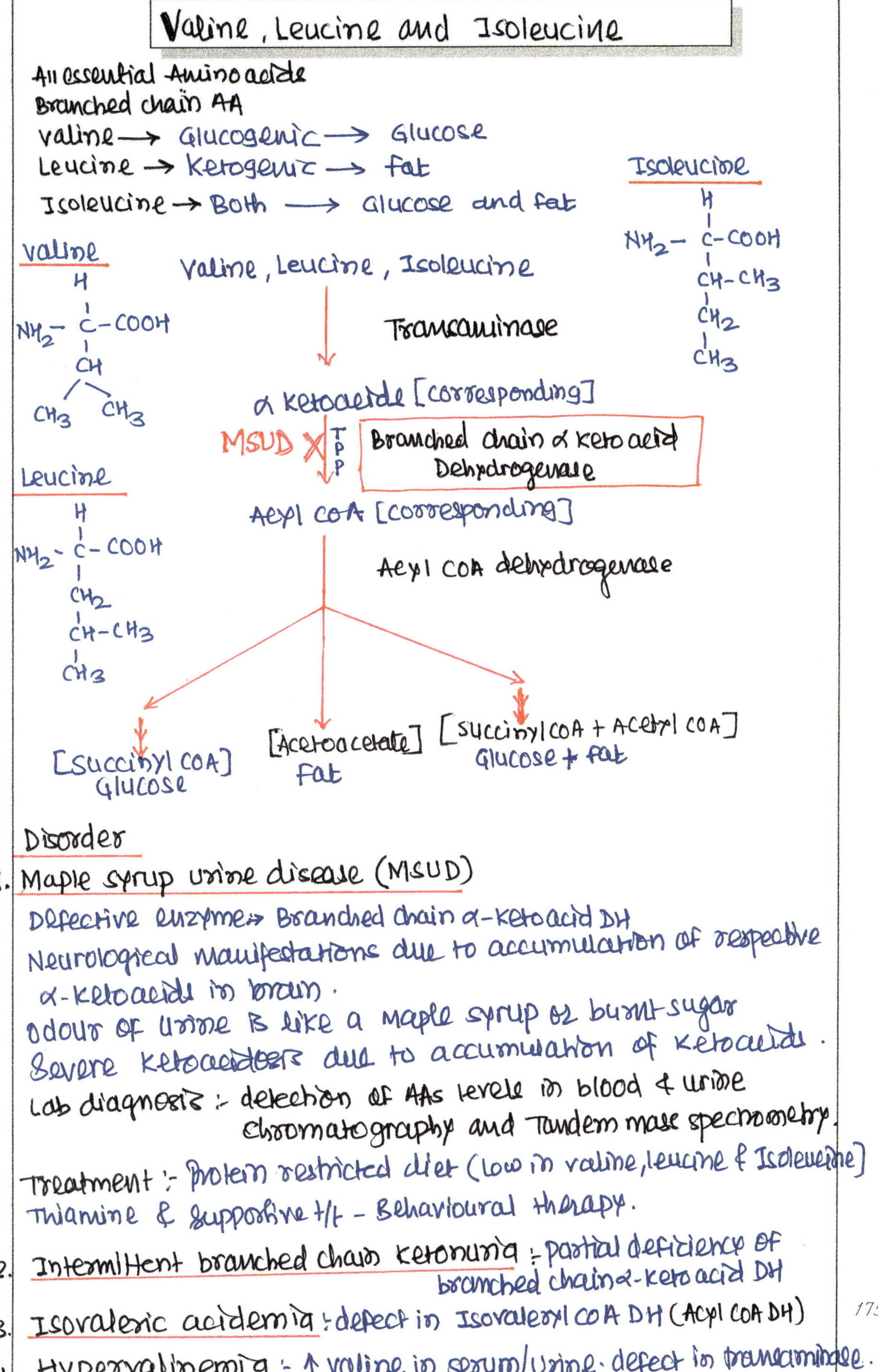

Valine, Leucine and Isoleucine

All essential Amino acids
Branched chain AA

valine → Glucogenic → Glucose
Leucine → Ketogenic → fat
Isoleucine → Both → Glucose and fat

Valine

```
        H
        |
NH2 - C - COOH
        |
       CH
      /  \
   CH3    CH3
```

Leucine

```
        H
        |
NH2 - C - COOH
        |
       CH2
        |
       CH - CH3
        |
       CH3
```

Isoleucine

```
        H
        |
NH2 - C - COOH
        |
       CH - CH3
        |
       CH2
        |
       CH3
```

Valine, Leucine, Isoleucine
↓ Transaminase
α ketoacids [corresponding]
↓ TPP — Branched chain α keto acid Dehydrogenase (MSUD ✗)
Acyl CoA [corresponding]
↓ Acyl CoA dehydrogenase

- [Succinyl CoA] Glucose
- [Acetoacetate] Fat
- [Succinyl CoA + Acetyl CoA] Glucose + fat

Disorder

1. **Maple syrup urine disease (MSUD)**
   Defective enzyme → Branched chain α-ketoacid DH
   Neurological manifestations due to accumulation of respective α-ketoacids in brain.
   Odour of urine is like a maple syrup or burnt sugar
   Severe ketoacidosis due to accumulation of ketoacids.
   Lab diagnosis :- detection of AAs levels in blood & urine
   chromatography and Tandem mass spectrometry.
   Treatment :- Protein restricted diet (low in valine, leucine & Isoleucine)
   Thiamine & supportive t/t – Behavioural therapy.
2. **Intermittent branched chain ketonuria** :- partial deficiency of branched chain α-keto acid DH
3. **Isovaleric acidemia** :- defect in Isovaleryl CoA DH (Acyl CoA DH)
4. **Hypervalinemia** :- ↑ valine in serum/urine. defect in transaminase.

Biogenic Amines and Polyamines

Nitrogenous compounds formed mainly by decarboxylation of amino acids or by amination or by transamination of aldehyde or ketones.

Glutamate $\xrightarrow{PLP}$ γ-aminobutyric acid [GABA] ($\downarrow CO_2$)

Tryptophan ⟶ 5HT $\xrightarrow{PLP}$ Serotonin ⟶ Melatonin ($\downarrow CO_2$)

Histidine $\xrightarrow{PLP}$ Histamine ($\downarrow CO_2$)

Tyrosine ⟶ DOPA $\xrightarrow{PLP}$ Dopamine ⟶ NE ⟶ E ($\downarrow CO_2$)

Amines :- GABA → Inhibitory Neurotransmitter

Serotonin → Excitatory Neurotransmitter

Melatonin → Sleep and wake process

Histamine → vasodilator, promotes gastric HCl

Dopamine → Increases blood pressure

Norepinephrine → Neurotransmitter, ↑ BP

Epinephrine → Neurotransmitter, ↑ BP

Taurine → constituent of bile acid

Polyamines :- Biological amines with multiple amino groups

Examples – Putrescine, Spermidine, Spermine

Arginine

↓

Ornithine

↓ Ornithine decarboxylase (CO_2)

Putrescine

↓ Spermidine synthase (decarboxylated SAM)

Spermidine

↓ Spermine synthase (decarboxylated SAM)

Spermine

Functions :- They are involved in regulation of transcription and translation – Growth and differentiation

- Imp. for diagnosis and treatment of certain cancers.

Clinical cases

A 12-year-old child was brought to the pediatric OPD with chief complaints of joint pain. Routine investigations were unremarkable. Urine examination showed - black color on standing. What is your probable diagnosis?

- Alkaptonuria

Defective enzyme - Homogentisate oxidase

Lab Δ :- Benedict's test - positive

$FeCl_3$ test - positive

Homogentisate —(Polyphenol oxidase)→ Benzoquinone acetate —(Polymerisation)→ Alkapton deposit in cartilage, bone, tendons, joints ⇓ Ochronosis

Homogentisate ↓ ✗ Homogentisate oxidase → 4 Maleylacetoacetate

Confirmatory test :- Tandem mass spectrometry

HPLC

T/t :- Low protein diet [Restrict Phe & Tyrosine]

Analgesic for joint pain

Clinical case 2

A 55-year-old male patient came with chief complaints of severe headache and palpitation. On examination BP was very high and heavy sweating. Urine examination showed increased excretion of VMA [vanilylmandelic acid]. Physician asked for CT-Abdomen for suspected adrenal gland tumor.

What is your probable diagnosis?

Ans :- Pheochromocytoma - Benign tumor of adrenal gland

↑ Epinephrine, Norepinephrine causes symptoms.

Lab Δ :- 24 Hour urinary VMA & catecholamines → Increased.

Symptoms :- Headache, Sweating, ↑ Heart rate, ↑ BP

Clinical Case - 3

A 4-year-old child was brought to the clinic with chief complaints of failure to thrive and developmental delay. The child was tall, thin with elongated limbs and digits. Eye examination revealed ectopia lentis (subluxation of ocular lens). Urine showed positive cyanide-nitroprusside test, and also elevated levels of methionine & homocysteine.

Homocysteine + serine
↓ Cystathionine β synthase ✕
Cystathionine

What is your probable diagnosis?

Ans:- Homocystinuria

Defective enzyme:- Cystathionine β-synthase

Why do increased homocysteine levels causes atherosclerosis & coronary artery diseases?

↑ Homocysteine

↓

Interfere c̄ lysyl residue of collagen cross links

↓

forms homocysteine thiolactone (free radical)

↓

OX-LDL formation

↓

Aggregation of LDL and taken up by Macrophages, foam cells

↓

Atherosclerosis & CAD

Lab Δ:- Cyanide-Nitroprusside test for cysteine-SH group - Red color
Confirmation by - Tandem mass spectrometry OR HPLC

T/t:- Restrict Methionine and give Cysteine
Vit B6, B12, folic acid

Clinical case-4

A neonate was brought to the hospital with poor feeding, convulsions and vomiting. The urine smells of burnt sugar. Biochemical analysis suggestive of increased branched chain aminoacids and their ketoacids in urine. ABG analysis shows decrease HCO_3^- levels.

What is your probable diagnosis?

- Maple syrup urine disease

Defective enzyme :- Branched chain α-Ketoacid dehydrogenase

Clinical case-5

A 6-year-old boy was brought to the hospital with chief complaints of mental retardation, failure to thrive, hypopigmented patches all over body and mousy odour of urine. He had episode of seizures in the past.

What is your probable diagnosis?

- Phenylketonuria

Enzyme deficient :- Phenylalanine hydroxylase

Mousy odour :- Phenylacetate & other byproducts.

T/t :- Restrict Phe and give Tyrosine

Clinical case-6

A 30-year-old man came to hospital with complaints of Photophobia and nystagmus. He has hypopigmented skin, hairs and eyes.

Diagnosis - Albinism

Deficient enzyme :- Tyrosinase

Tyrosine —(Tyrosinase)✗→ DOPA —✗→ Dopaquinone ———→ Melanin

Integration of Metabolism & Starvation

MCI BI 6.1

Starve Feed cycle – divided into 5 stages

1. Well fed state (up to 4 Hrs after food)
2. Early fasting (4 to 16 Hrs)
3. fasting (16 to 48 Hrs after food)
4. Starvation (2-5 days without food)
5. prolonged starvation (>5 days without food)

1. Well fed state (up to 4 Hrs after food)

↑ Glycogenesis INSULIN ↑
↑ Lipogenesis
↑ Protein synthesis

Glucose is the major fuel

Insulin promotes utilization of glucose and storage into glycogen

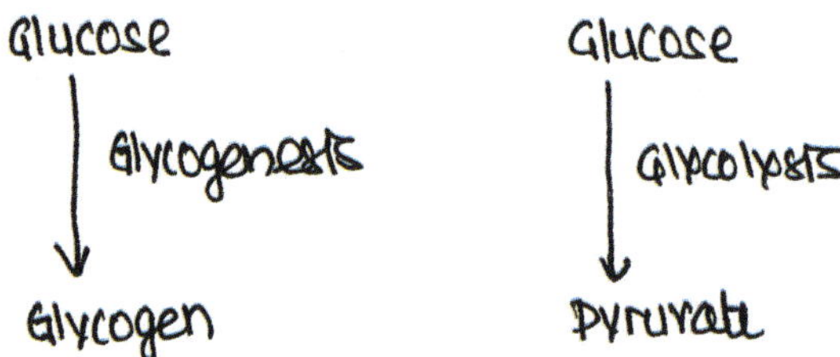

↑ uptake of glucose into the cell

Fat metabolism → ↑ FA synthesis , ↑ TG synthesis

Adipose tissue → ↑ uptake via insulin mediated GLUT-4
Inhibit Lypolysis

Muscle → ↑ uptake of glucose via GLUT4
Glucose ⟶ Glycogen (stored in muscle)

2. Early fasting (4 to 16 Hrs)

– Glucose provided by breakdown of liver glycogen.

Glycogen
↓ Glycogenolysis [In liver]
[Glucose level in between meal is maintained by Glycogenolysis]
Glucose

Decreased glycogenesis and lipogenesis

3. **Fasting [16-48 Hrs after Food]**

- Hepatic Gluconeogenesis is stimulated
- By 12-16 Hrs - Liver glycogen level decreases
- Muscle proteins are degraded - specially alanine reaches the liver for gluconeogenesis

Glucogenic amino acid, Glycerol

↓ Gluconeogenesis [Synthesis of Glu from Noncarbohydrate precursor]

Glucose.

4. **Starvation** (2-5 days without food)

- ↑ Activity of Hormone sensitive lipase
- ↑ lipolysis, free FA levels increases
- ↑ Ketone bodies synthesis

Liver :- Utilizes Amino acids and free FA

Heart :- Ketone bodies (Acetoacetate, β-OH-Butyrate)

Brain :- 80% Glucose, 20% Ketone bodies

Muscle :- Ketone bodies

Hormones :- Glucagon, GH, Epinephrine, ACTH, Cortisol

Triacyl glycerol → (Hormone sensitive lipase (HSL)) DAG + FA

DAG → (HSL) MAG + FA

MAG → (Lipase) Glycerol + FA

FA → Acetyl CoA → Ketone body synthesis by liver

Ketogenesis Liver

2 Acetyl CoA

↓

Acetoacetyl CoA

↓

HMG CoA

↓

Acetoacetate

↓

β-OH-Butyrate

Ketolysis Extrahepatic tissues.

β-OH-Butyrate (KB)

↓ NAD → NADH + H⁺

Acetoacetate (KB)

↓

Acetoacetyl CoA

↓

2 Acetyl CoA

Stage 5. Prolonged starvation (> 5 days without food)

Increased muscle protein breakdown

Brain is completely shifted to ketone bodies for energy production

Liver utilizes free fatty acids & amino acids

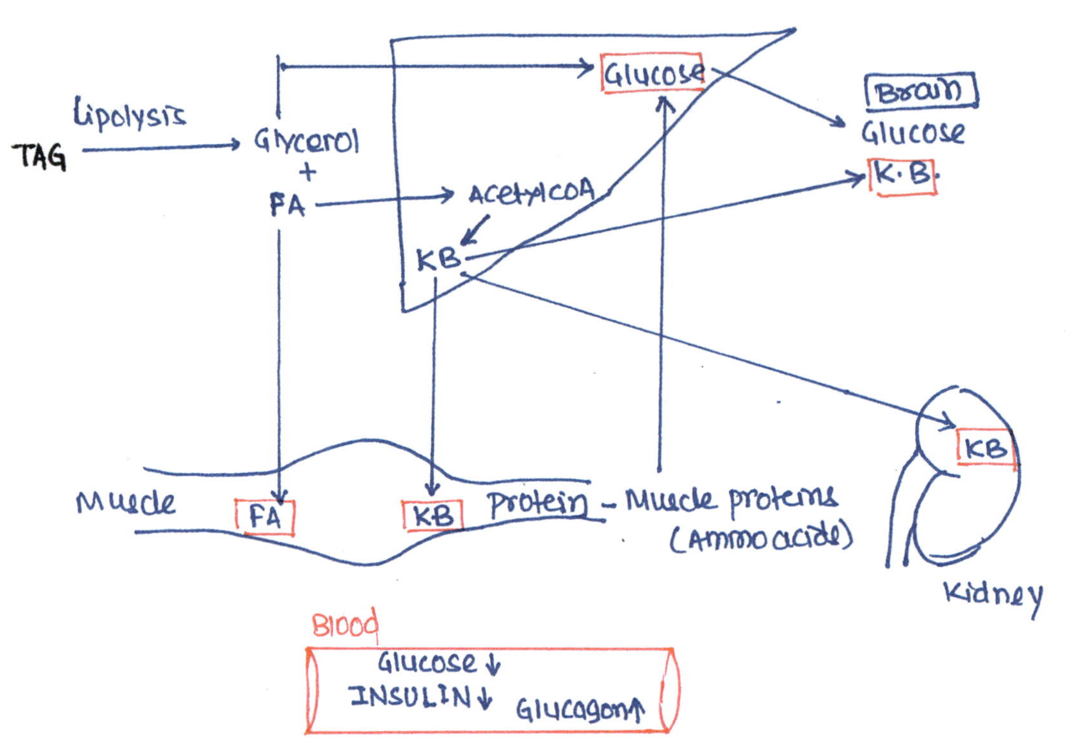

Major fuel for organs

Liver → Glucose in fed state, free FA, Amino acids No KB - Absent Thiophorase

Brain → Glucose, Ketone bodies in prolonged starvation

Heart → Ketone bodies, free FA

Kidney → free FA, Glucose

RBC's → Glucose

Muscle → Glucose, Glycogen, Free FA, Ketone bodies

Major metabolic pathways

Well fed state → Synthesis of Glycogen, lipid & protein

In betn Meal → Glycogenolysis up to 12-16 Hrs

Fasting → Gluconeogenesis from glycerol & Alanine

Starvation → Fatty Acid oxidation, KB synthesis

Prolonged starvation → Muscle protein breakdown

BIOLOGICAL OXIDATION

MCI COMPETENCY BI 6.6

SAQ

Electron transport chain

Chemiosmotic hypothesis

Oxidative phosphorylation

Inhibitors of ETC

Uncouplers

Redox potential

High energy compounds

Substrate level phosphorylation

Shuttle pathway for NADH

Diagram :- Electron transport chain

Chemiosmotic hypothesis

ATP synthase

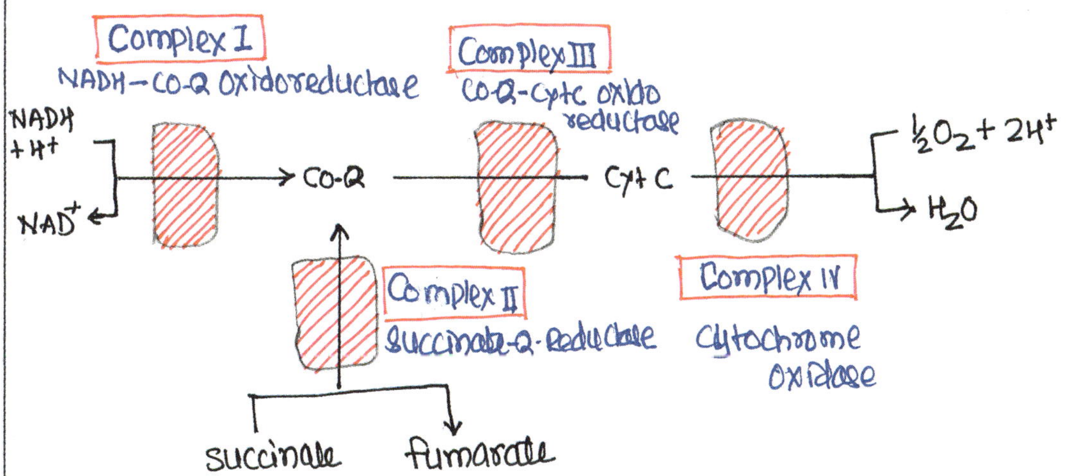

Diagram - Electron Transport Chain

LAQ :- Write in detail about ETC and inhibitors of ETC

SAQ :- Enumerate uncouplers and their role.

SAQ :- Enumerate Inhibitors of ETC & oxidative phosphorylation.

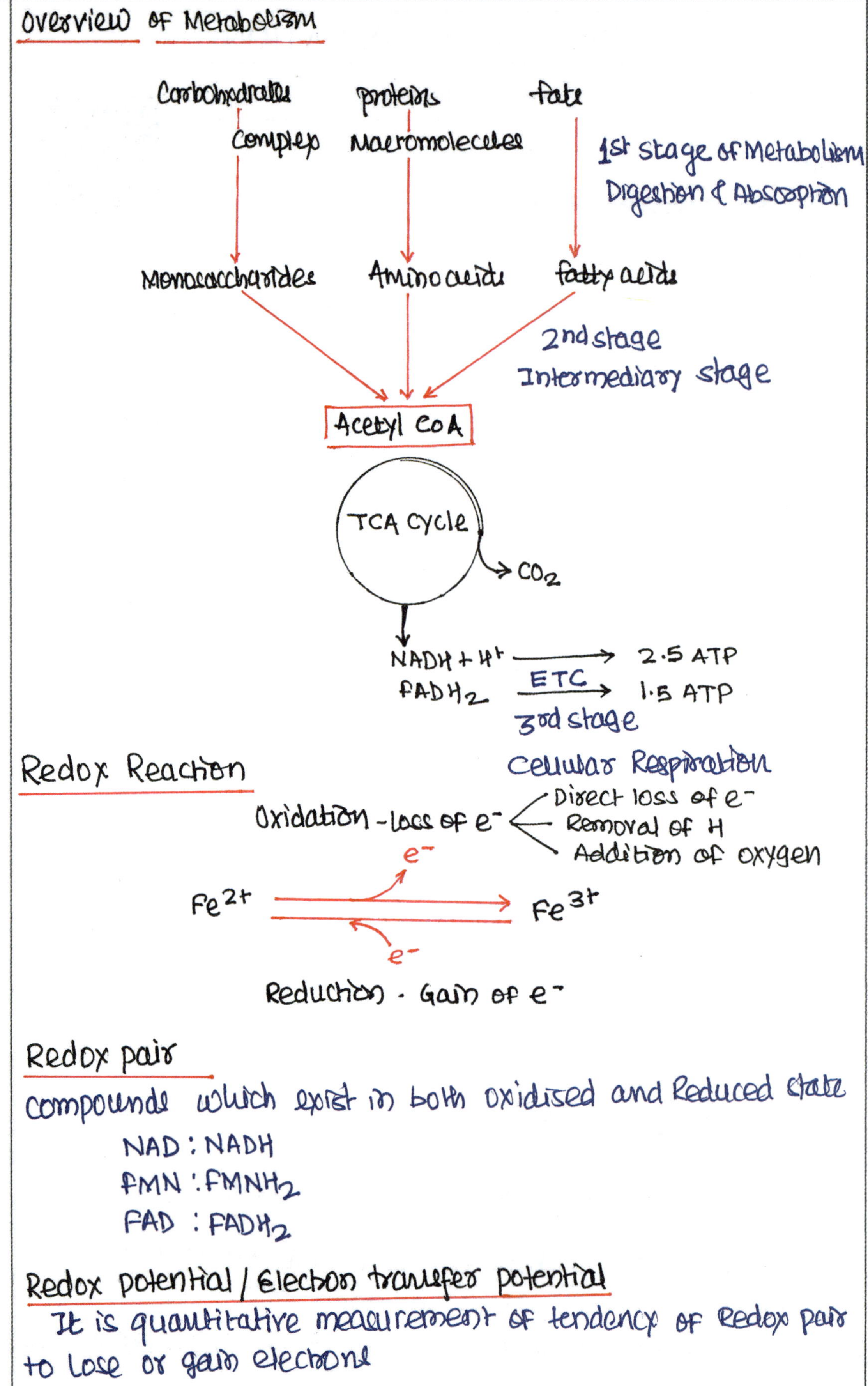

Overview of Metabolism

Redox Reaction

Oxidation - loss of e^-
- Direct loss of e^-
- Removal of H
- Addition of oxygen

Reduction - Gain of e^-

Redox pair

compounds which exist in both oxidised and Reduced state

NAD : NADH

FMN : $FMNH_2$

FAD : $FADH_2$

Redox potential / Electron transfer potential

It is quantitative measurement of tendency of Redox pair to lose or gain electrons

NAD : NADH → −0.32
FMN : $FMNH_2$ → −0.30
FAD : $FADH_2$ → −0.22

Negative Redox potential means tendency to lose electrons

Ubiquinone : Ubiquinol +0.045
Cytochrome b → +0.077
Cytochrome C_1 → +0.22
Cytochrome c → +0.25
Cytochrome a → +0.29
Cytochrome a_3 → +0.35
$\frac{1}{2}O_2 : H_2O$ → +0.82

Positive Redox Potential means tendency to gain electrons

In ETC, they are arranged in ascending order of Redox potential

Electrons transport from Negative redox potential to positive redox potential.

Electron Transport Chain

Site :- Inner mitochondrial membrane

Sequence :-

$NAD^+ \rightarrow CoQ \rightarrow cyt\,b \rightarrow cyt\,c_1 \rightarrow cyt\,c \rightarrow a.a_3 \rightarrow H_2O$

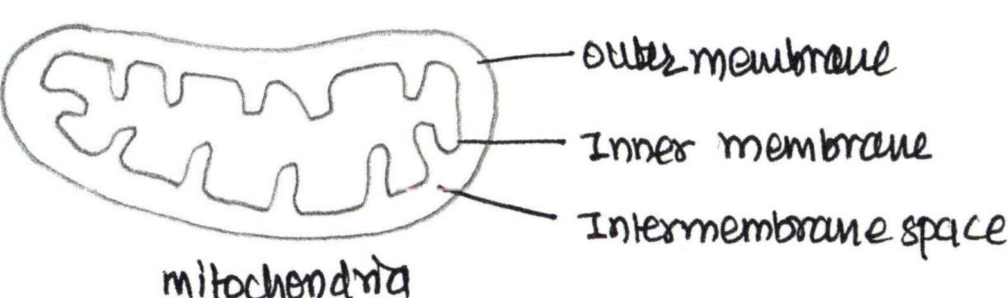

Complexes of ETC

① Complex I : NADH - CoQ - Oxidoreductase

FMN is prosthetic group, Also contain FeS complex

2. Complex II : Succinate DH complex OR Succinate-CoQ Reductase

Succinate ⟶ FAD^+ ⟶ FeS ⟶ CoQ
↓
QH_2

3. Complex III : CO-Q-Cyt C oxidoreductase

Contain cytb, cytc (Fe^{2+}/Fe^{3+})

Rieske Iron sulfur complex

CoQ ⟶ Cytb ⟶ FeS ⟶ $CytC_1$ ⟶ CytC

4. Complex IV : Cytochrome c oxidase

[Cu^{2+}/Cu^{+}]

[Fe^{2+}/Fe^{3+}]

Cyt C ⟶ Cyt $a.a_3$ ⟶ $\frac{1}{2}O_2$ ⟶ H_2O

Mobile electron carrier — ① Coenzyme Q
Ubiquinone [lipid with isoprenoid unit]

② Cytochrome C

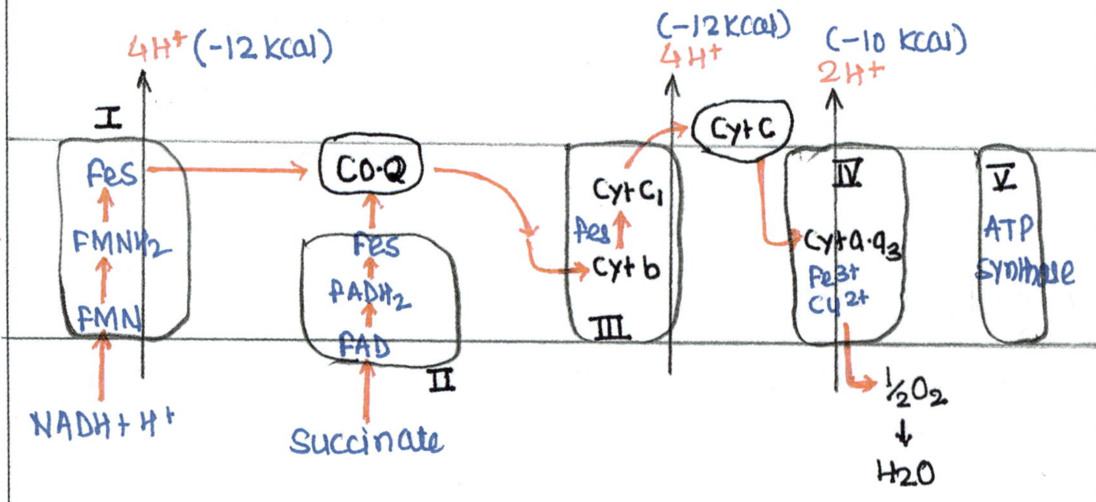

Substrate for NADH – Glyceraldehyde 3 Ⓟ
Pyruvate
Isocitrate
α Ketoglutarate
Malate
β-OH-Acyl CoA

Substrate for $FADH_2$ – Succinate
Fatty acyl CoA
Glycerol 3 Ⓟ

Oxidative phosphorylation

When ATP synthesis is linked with ETC, it is known as oxidative phosphorylation

P:O Ratio :→ It denotes the number of inorganic phosphate molecules consumed for ATP production per molecule of O_2 consumed

In simple term, it denotes ATP production from

NADH ⟶ 2.5 ATP
$FADH_2$ ⟶ 1.5 ATP

ATP synthase

Fo and F1 subunit
Smallest motor
Proposed by Paul Boyer 1964
Responsible for ATP synthesis

Mechanism - chemiosmotic hypothesis
– development of proton gradient across the mt. membrane
Fo → Inhibition by Oligomycin
F1 → 1st factor discovered in mt.

Chemiosmotic Hypothesis

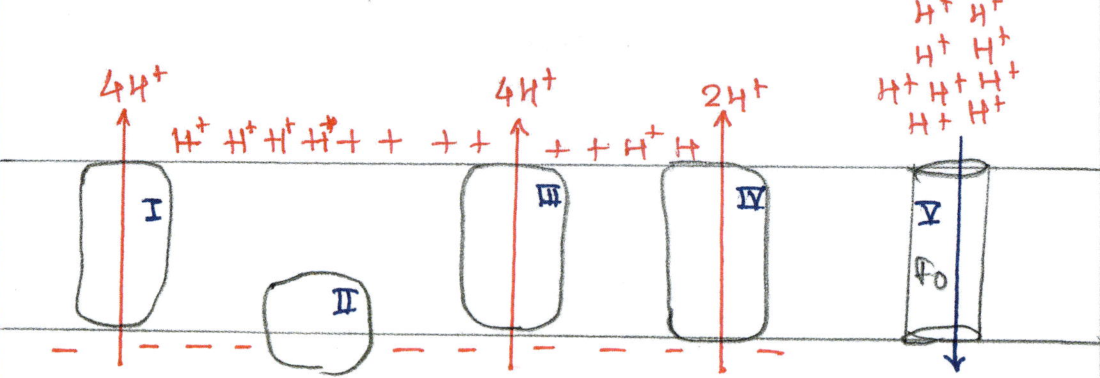

During the transfer of electron through the inner mt membrane complexes, protons are pumped out

4 protons from complex I

4 protons from complex III

2 protons from complex IV

transfer of protons causes generation of proton gradient across the membrane because it is impermeable to ions. The only route is through 'Fo'-subunit of ATP complex.

when protons moves through fo-subunit, it causes rotation of γ-handle of F_1 complex. Rotation of γ-handle causes conformational change in β-subunit known as binding change mechanism.

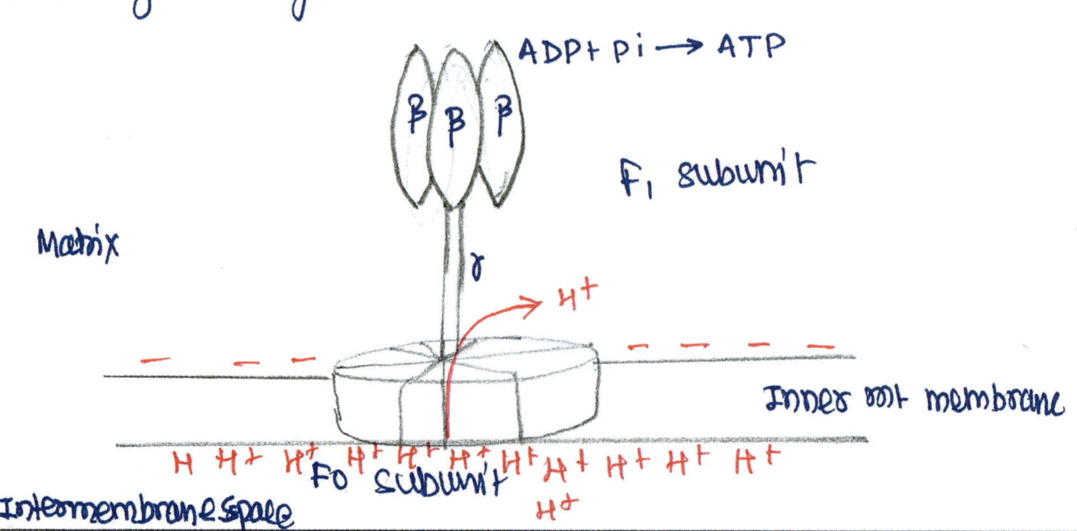

Conformation of B-subunit

Open – O — Release of ATP

Loose - L — Binding of ADP + Pi

Tight - T - Synthesis of ATP

Change in conformation of B- subunit causes ATP formation.

Inhibitors of ETC

Complex	Inhibitors
Complex I	Piercidine Guanithidine Rotenone (fish poison) Barbiturate
Complex II	Malonate Carboxin TTFA [Thenoyltrifluoro-acetone]
Complex III	Antimycin British Antilevisite Phenformin
Complex IV	Carbon monoxide Azide Cyanide Hydrogen sulfide

Inhibitors of Oxidative Phosphorylation

① Atractyloside → Inhibit translocation of ADP & Pi

② Oligomycin → Inhibit fo subunit of ATP synthase

③ Ionophores → Valinomycin, Gramicidin, Nigericin
K^+
disrupt the proton gradient across the membrane

UNCOUPLERS

Mechanism :- Increase permeability of inner mitochondrial membrane

↓

Delink oxidative phosphorylation from ETC

↓

Production of Heat
No ATP production

Two Types

Physiological

1. Thyroxine
2. Long chain FFA
3. Unconjugated bilirubin
4. Thermogenin [Brown adipose tissue] [Newborn & Hibernating animals]

Chemical

1. 2,4 Dinitrophenol (DNP)
2. 2,4 Dinitrocresol (DNC)
3. Pentachlorophenol
4. FCCP [trifluoromethoxy phenylhydrazone]
5. Aspirin (High doses)
6. calcium
7. Dicumarol

High energy compounds

compounds that have high negative free energy change means → -7.0 kcal/mol or more on hydrolysis

1. Pyrophosphates :- ATP, GTP, CTP, UDP-Glucose
2. Phosphoenol pyruvate :-
3. Acyl phosphates :- Carbamoyl phosphate, Bisphosphoglycerate
4. Acyl thioesters :- Acetyl COA, succinyl COA, SAM
5. Creatine phosphate

Substrate level phosphorylation

Synthesis of ATP without electron transport chain is known as substrate level phosphorylation.

It is due to hydrolysis of high energy compound

Examples

① 1, 3 bisphosphoglycerate

ADP → ATP (Phosphoglycerate Kinase)

3 phosphoglycerate

② Phosphoenol pyruvate

ADP → ATP (pyruvate Kinase)

pyruvate

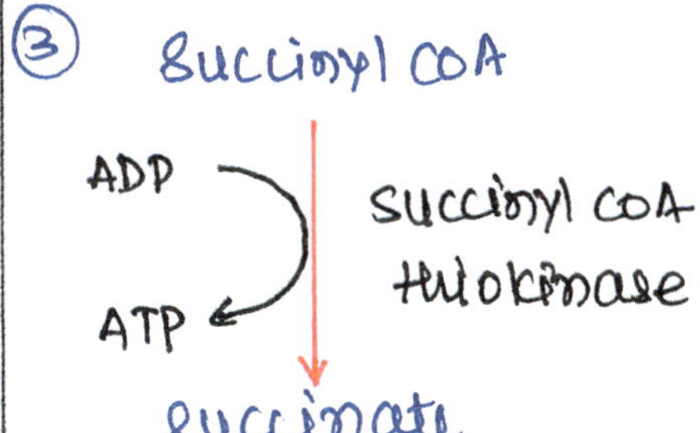

Shuttle pathway for transport of cytosolic NADH

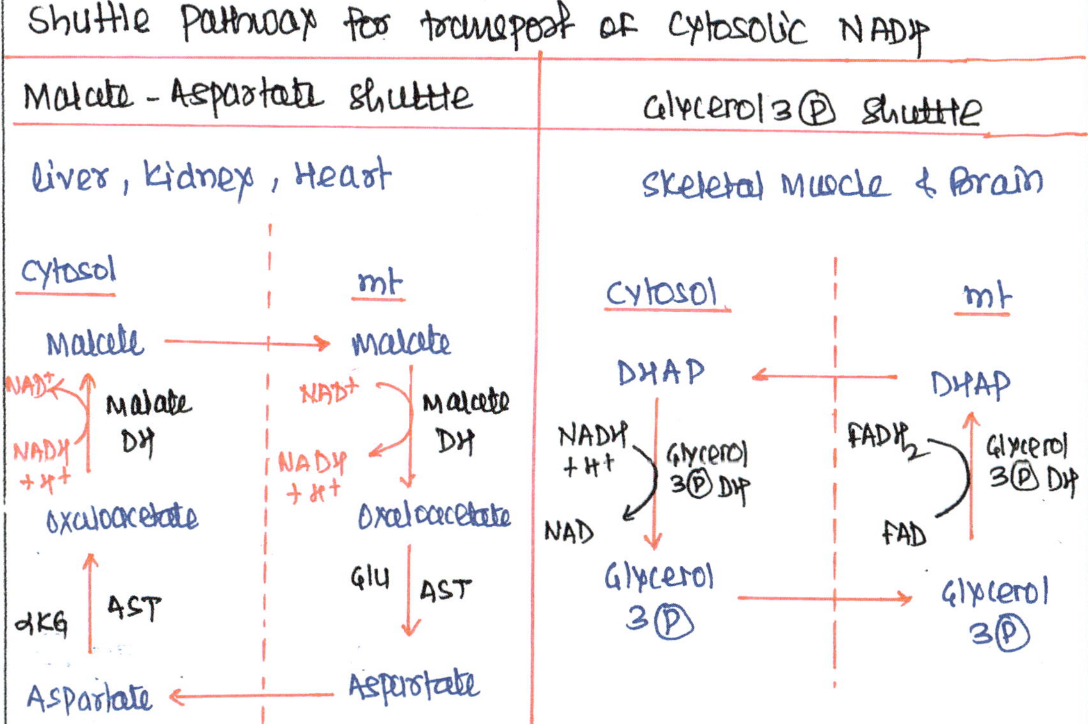

ENZYMES INVOLVED IN BIOLOGICAL OXIDATION

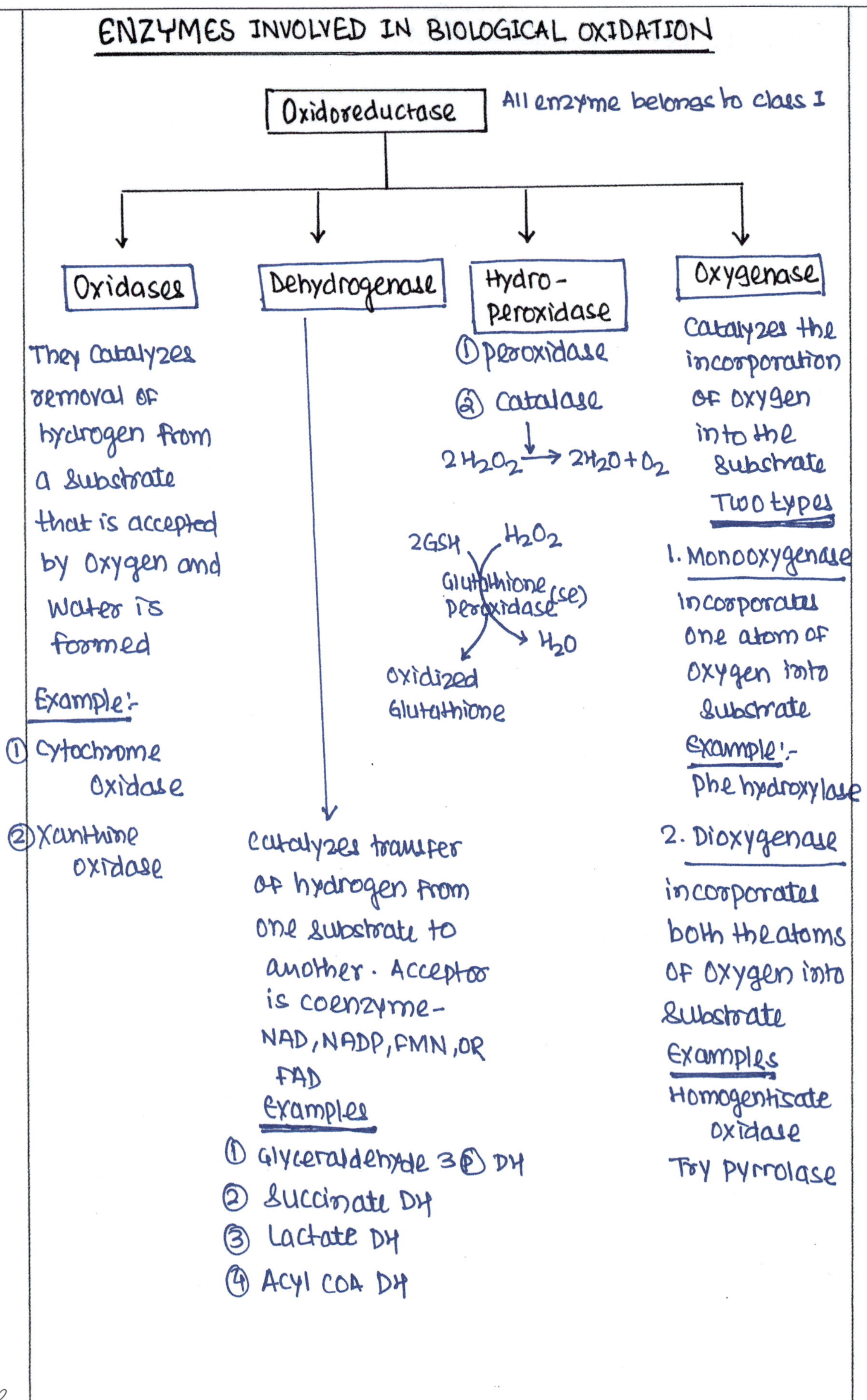

NUCLEOTIDE CHEMISTRY AND METBOLISM

MCI COMPETENCY BI 6.2-6.4

LAQ:- Denovo synthesis of purine nucleotide
Pyrimidine nucleotide synthesis & degradation

SAQ:-

Purine salvage pathway
Biologically important nucleotides
Synthetic nucleotide analogs
Causes of Primary & secondary Gout
purine NT degradation
Pyrimidine NT degradation

Clinical cases

1. Gout :- 1st Metatarsophalyngeal joint involved, ↑ uric acid level
Aspiration of crystals & microscopy

2. Lesch Nyhan syndrome – Self mutilating activity, MR
↑ uric acid level

3. Orotic aciduria :- Anemia, orotate crystal formation, Growth retardation

4. SCID - severe combined immunodeficiency
→ Both T & B cell affected, recurrent infections, ↓ uric acid, Adenosine deaminase deficiency

Clinical case

A 12-year-old boy was brought to the hospital with chief complaints of mental retardation, joint pain and irresistable urge to bite his finger and lips. Also having head banging activity. The serum uric acid was 11 mg/dl.

What is your probable diagnosis?

Ans :- Lesch Nyhan syndrome
Pathway – purine salvage pathway
Deficient enzyme :- HGPRTase

Biologically Important Nucleotides

SAQ:- Enumerate biologically important Nucleotide and write about their functions.

① Important in DNA and RNA synthesis

Nucleoside [Ribose + N-Base]

Nucleotide [Ribose + N-Base + Ⓟ]
OR Nucleoside Monophosphate

N-Base	Nucleoside	Nucleotide
Adenine	Adenosine	Adenylate (5'AMP)
Guanine	Guanosine	Guanylate (5'GMP)
Cytosine	Cytidine	Cytidylate (5'CMP)
Thymine	Thymidine	Thymidylate (5'TMP)
Uracil	Uridine	Uridylate (5'UMP)

② They act as a coenzymes - NAD^+, $NADP^+$, FAD, FMN, COA

③ Second messengers in hormone action - cAMP - Group II A
cGMP - Group II B

④ Metabolic intermediates - UDP-Glucose - Glycogen metabolism
CDP-Acylglycerol - Phospholipid metabolism.

⑤ Methyl group donor - S-Adenosyl methionine [SAM]

⑥ High energy compounds - ATP, GTP - 7 Cal/Mol

⑦ Important in the synthesis of Glycoproteins - UDP-Glucose
UDP-Galactose

⑧ Regulates Metabolic pathways - cAMP

NAMES

ATP GTP CTP UDP-Glucose PAPS : Active sulfate donor
ADP GDP CDP UDP-Glucuronic acid : conjugation of Bilirubin
AMP GMP CMP UDP-Galactose TMP Adenosyl B12
cAMP cGMP UTP TDP
UMP TTP

Synthetic Nucleotide Analogs

SAQ :- Write about synthetic nucleotide analogs & their role in cancer treatment.

- Nucleotide analogs are produced by altering heterocyclic ring OR deoxyribose
- They are Analogs to – Purines
  Pyrimidines
  Ribose
  Nucleosides
  Nucleotides
- They are useful in various disorders :- ① Cancer
  ② Gout
  ③ Neurological ds
  ④ AIDS

1. Anticancer agent

6 Mercaptopurine
5-Fluorouracil
6-Azaguanine
6-Azacytidine
6-Azauridine
Idouracil
Arabinosyladenine
Arabinosylcytidine

Mechanism

Similar in structure
↓
Incorporated in DNA
↓
Defective DNA formation
↓
Inhibit cell growth
Inhibit cell proliferation

2. Allopurinol – In t/t of Gout

Hypoxanthine
↓ xanthine oxidase ✗
Xanthine
↓ xanthine oxidase ✗
Uric acid

Allopurinol ↓ uric acid level by inhibiting xanthine oxidase enzyme by two mechanism ① competitive inhibition
② Suicide inhibition

Used in t/t of Gout, ↓ uric acid level, prevents urate crystal formation in joints

3. AIDS

Zidovudine (AZT) – Azidothymidine is a thymidine analog. Acts by inhibiting HIV reverse transcriptase. Decrease viral load in AIDS patient.

Dideoxyinosine

4. **Arabinosyladenine** :- Useful in neurological diseases and viral encephalitis.

5. **Azathioprine** :- Useful in graft rejection reactions

Purine Nucleotides

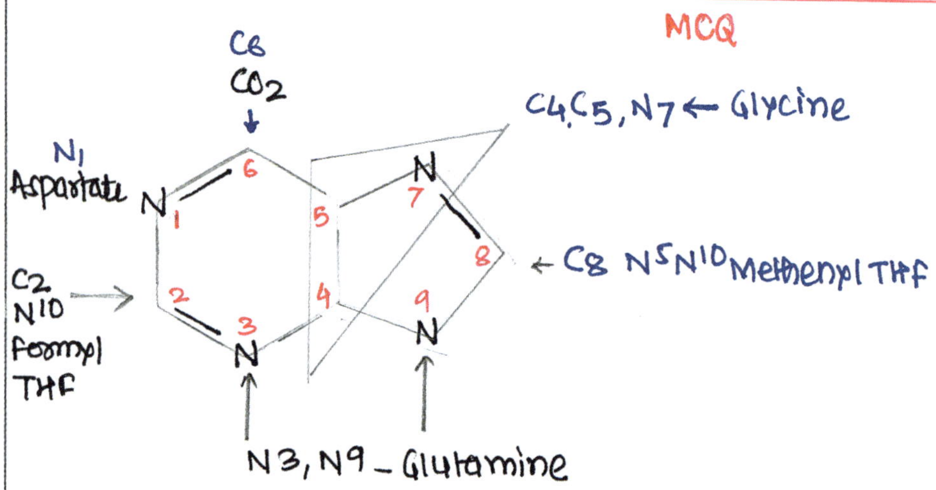

Source of C and N in the purine ring.

Mnemonic to remember denovo synthesis pathway.

Steps:-

ATP 1 — (2) Glutamine — ATP 3 Glycine

ATP 6 — ATP (5) — (4) THF

7 CO_2 — ATP 8 Asp — 9 × F — (10) — 11 × H_2O

1,3,5,6,8 — ATP synthetase

2,5 – Glutamine

3 – Glycine

4,10 – THF

7 – CO_2

8 – Aspartate

9,11 – Nothing added

Denovo Synthesis of Purine Nucleotides

built on preexisting Ribose 5 (P)

Ribose 5 (P)

↓ 1 ATP — PRPP synthetase ⊖ ADP, GDP

Phosphoribosyl pyrophosphate (PRPP)

↓ 2 Glutamine — Glutamyl amidotransferase ⊖ IMP, AMP, GMP

Phosphoribosylamine

↓ 3 Glycine, ATP — Synthetase

Glycinamide Ribosyl 5 (P)

↓ 4 N5, N10 methenyl THF — Transferase

Formyl glycinamide Ribosyl 5 (P)

↓ 5 Glutamine, ATP — Synthetase

Formylglycinamidine R5 (P)

↓ 6 ATP — Synthetase

Aminoimidazole R5 (P) [AIR]

↓ 7 CO_2 — Carboxylase

Aminoimidazole carboxylate R5 (P)

↓ 8 Aspartate, ATP — Synthetase

Aminoimidazole succinyl carboxamide R5 (P)

↓ 9 Lyase → Fumarate

Aminoimidazole carboxamide R5 (P)

↓ 10 Formyl THF — Transformylase

Formaminoimidazole carboxamide R5 (P)

↓ 11 Cyclase (Ring closure) → H_2O

IMP [Ionosine Monophosphate]

Three pathways for purine synthesis

1. De Novo Synthesis
2. purine salvage pathway
3. phosphorylation of purine nucleotides.

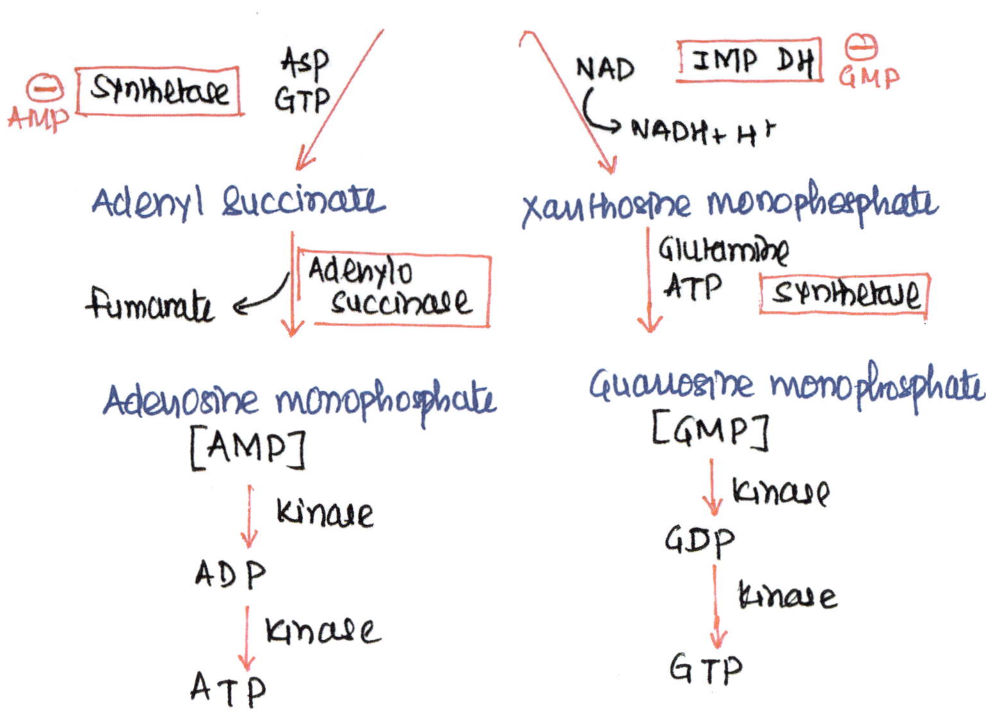

Ribonucleotide ADP, GDP —Ribonucleotide Reductase (Thioredoxin, NADPH)→ Deoxyribonucleotide dADP, dGDP

Regulation :- feedback Mechanism

PRPP synthetase :- Inhibited by ADP, GDP

Glutamyl amidotransferase :- Inhibited by IMP, AMP, GMP

Adenyl succinate synthetase :- Inhibited by AMP

IMP dehydrogenase :- Inhibited by GMP

Inhibitors of purine synthesis

- Sulfonamide
- Methotrexate
- 6 Mercaptopurine
- Azaserine
- Mycophenolic acid

Purine Salvage Pathway

SAQ: Write a note on purine salvage pathway and disorder related with it.

Adenine —(PRPP → PPi; Adenine Phosphoribosyl transferase)→ AMP

Hypoxanthine —(PRPP → PPi; HGPRTase: Hypoxanthine-Guanine Phosphoribosyl transferase)→ Ionosine Monophosphate [IMP]

Guanine —(PRPP → PPi; Hypoxanthine-guanine phosphoribosyl transferase, HGPRTase)→ GMP

Small part of nucleotides are synthesized from degraded DNA molecule is known as salvage pathway.

During degradation of cells free purines and pyrimidines are released. These purines are used in the synthesis of purine nucleotide in one step reaction.

Significance: Brain and RBC's are partly dependent on purine salvage pathway for synthesis of nucleotides.

Lesch Nyhan Syndrome

X-linked disorder – Affect only males

Defective enzyme – Hypoxanthine Guanine phosphoribosyl transferase [HGPRTase]

C/F: Aggressive behaviour
- Self mutilating activity [biting of Nails and lips]
- Mental retardation
- Hyperuricemia and Gout

Neurological symptoms: due to ↓ synthesis of purine nucleotides in brain

Hyperuricemia: ↓HGPRTase → ↑PRPP → Denovo synthesis ↑ → UA↑
Hypoxanthine and Guanine converted into Uric acid

Treatment: Allopurinol for Hyperuricemia, Behavioural therapy

Purine nucleotide degradation

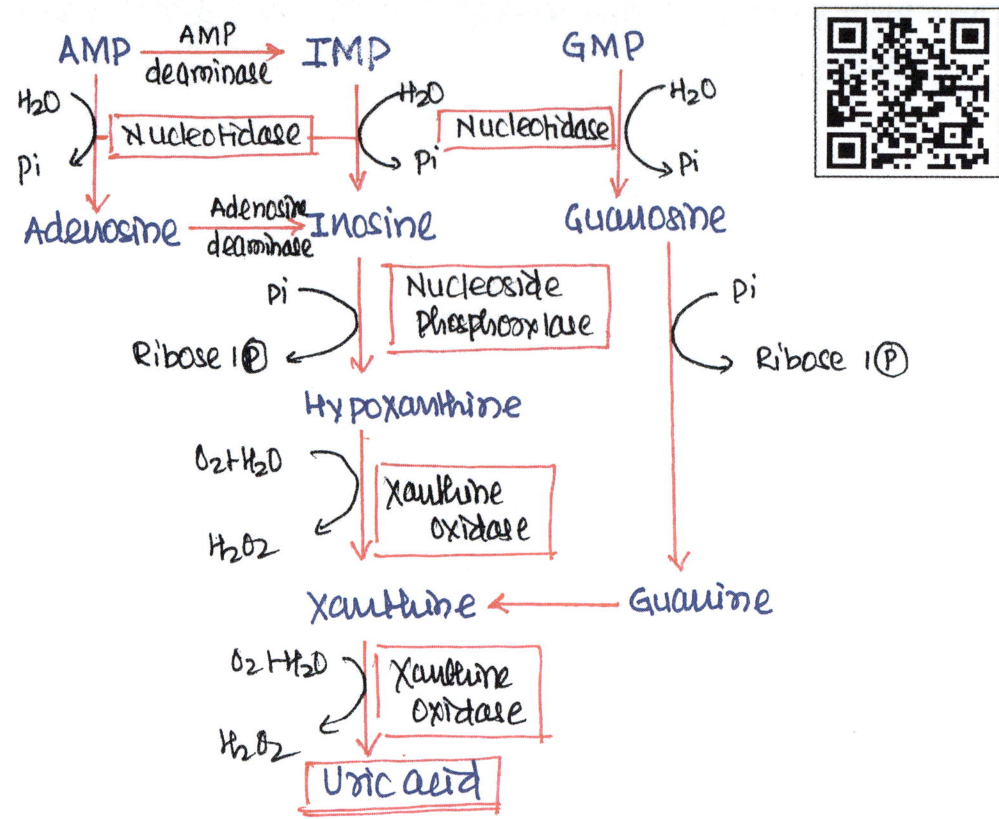

Normal values :- Males 3 – 7 mg/dl

females 2 – 5 mg/dl

Disorders of purine Nucleotide Metabolism

1. Gout :- ↑ uric acid
   Monosodium urate crystal deposition in joint
   MC Joint :- Metatarsophalangeal joint [Great toe]
   t/t :- Allopurinol.
2. Lesch-Nyhan syndrome – HGPRTase deficiency
   X-linked
   Neurological symptoms & ↑ UA
3. Severe combined Immunodeficiency disorder [SCID]
   due to deficiency of Adenosine deaminase
   affect B & T cell function, ↓ uric acid level
4. Purine nucleoside phosphorylase (PNP) deficiency
   ↓ Uric acid level, disorder of immune system
   Recurrent infections, neurological symptoms & autoimmune disorder.

GOUT

Disorder of purine nucleotide metabolism, due to excessive production of uric acid.

Most common site :- 1st Metatarsophalangeal joint
Great toe (90% cases)

Other sites - Knee joint, Ankle joint, Wrist joint etc.

Symptoms :- ① Swollen painful joints

② Excruciating pain - due to inflammation and arthritis

③ Podagra gout — Involvement of great toe

④ Renal stones

Normal serum uric acid in Males 3-7 mg/dl
Females 2-5 mg/dl

When Sr. uric acid > 7 mg/dl
↓
Uric acid precipitated

↓ Deposits in joint → ↓ Tophi → ↓ Inflammation, Intense pain

↘ Deposited in renal tubules → ↓ Renal stones

Causes of Gout

Primary - due to defect in Metabolic pathway

① ↑↑ Activity of PRPP synthetase
② ↑↑ Activity of Glutamyl amidotransferase
③ HGPRTase deficiency
④ Glucose 6 phosphatase deficiency
⑤ ↑↑ Activity of Glutathione reductase

Secondary - due to other diseases

① Leukemia
② Lymphoma
③ Chemotherapy
④ Psoriasis
⑤ Renal failure

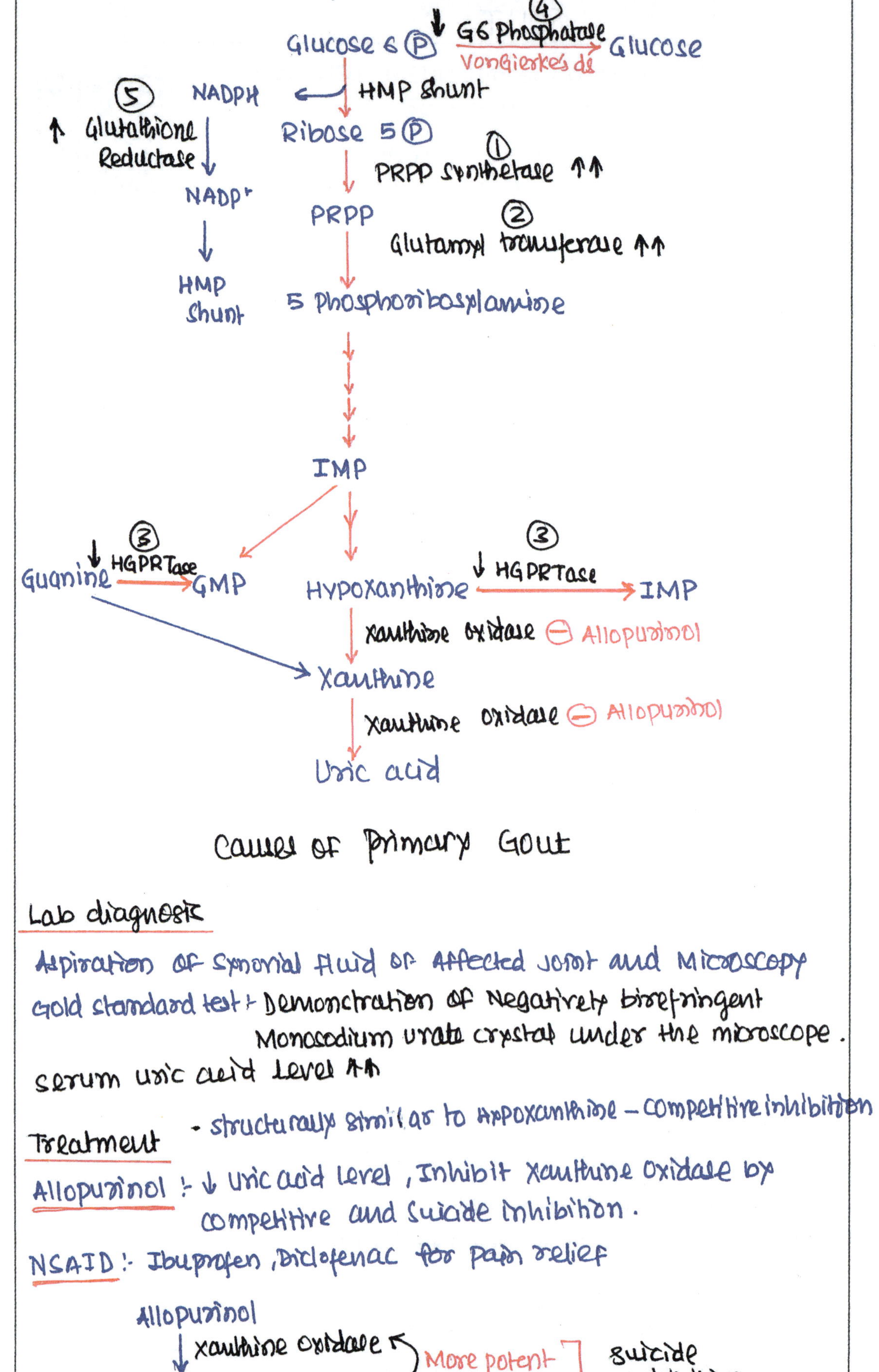
Glucose 6 (P)
↓ G6 Phosphatase
④
Vongierkes ds
Glucose
HMP shunt
NADPH
⑤
↑ Glutathione Reductase
NADP+
HMP shunt
Ribose 5 (P)
① PRPP synthetase ↑↑
PRPP
② Glutamyl transferase ↑↑
5 Phosphoribosylamine
IMP
③ ↓ HGPRTase
Guanine
GMP
Hypoxanthine
③ ↓ HGPRTase
IMP
Xanthine oxidase ⊖ Allopurinol
Xanthine
Xanthine oxidase ⊖ Allopurinol
Uric acid
Causes of Primary Gout
Lab diagnosis
Aspiration of synovial fluid of affected joint and Microscopy
Gold standard test :- Demonstration of Negatively birefringent Monosodium urate crystal under the microscope.
Serum uric acid level ↑↑
Treatment
- structurally similar to Hypoxanthine – competitive inhibition
Allopurinol :- ↓ uric acid level, Inhibit Xanthine Oxidase by competitive and suicide inhibition.
NSAID :- Ibuprofen, Diclofenac for pain relief
Allopurinol
xanthine oxidase
Alloxanthine
More potent Inhibitor
suicide inhibition

Why Alcohol precipitates attack of Gout?

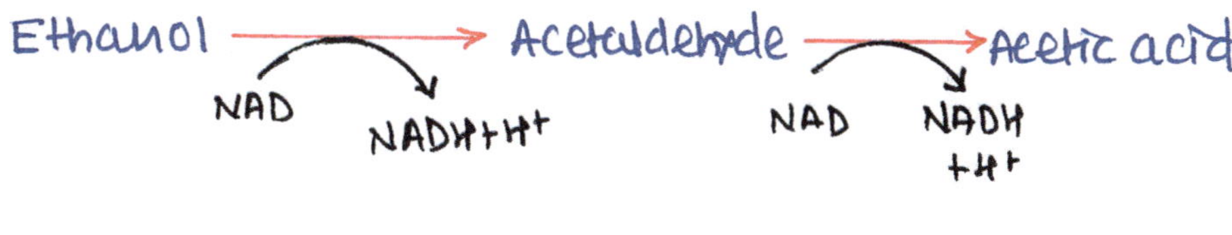

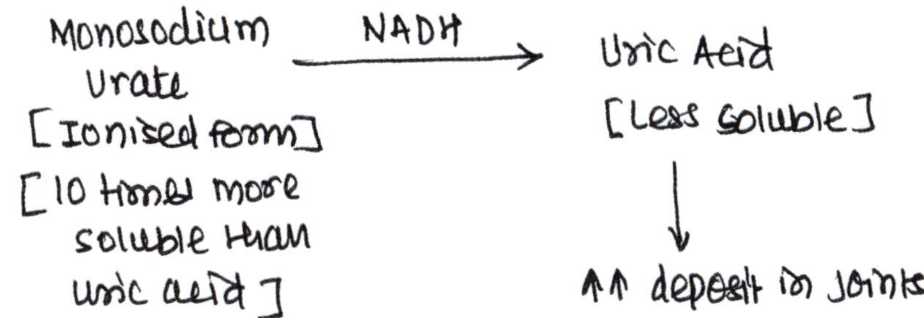

↓

ppt Gout

1. NADH favors Lactic acid formation. Lactic acid ↑ renal threshold for uric acid, hence retention of UA occurs.
2. NADH favours the conversion of monosodium urate ions into uric acid. Uric acid is less soluble and more precipitable when concentration is above 7 mg/dl. Hence alcohol ppt the attack of Gout.

Clinical Case

1. A 44 year old man, chronic alcoholic and smoker, came to Hospital with complaints of severe pain in great toe. On examination his 1st Metatarsophalangeal joint was inflammed. He has H/O similar episodes in the past.

   Diagnosis :- Gout

2. A 5 year old child was brought to the OPD with chief complaints of irresistable urge to bite his fingers and lips. Lab investigation showed ↑ uric acid level. What is your provisional diagnosis?

   Diagnosis :- Lesch Nyhan syndrome

3. A 42-year-old man, a chronic alcoholic, came to hospital c̄ complaint of severe pain in the great toe. The pain was severe and 1st MTP joint was inflamed. Serum uric acid was 11 mg/dl. Allopurinol was given in the treatment. What is your probable diagnosis?

Pyrimidine Nucleotide

Cytosine ⟶ CMP
Uracil ⟶ UMP
Thymine ⟶ TMP

Pyrimidine synthesis

LAQ :- Describe the steps in pyrimidine nucleotide synthesis along with regulation and disorders.

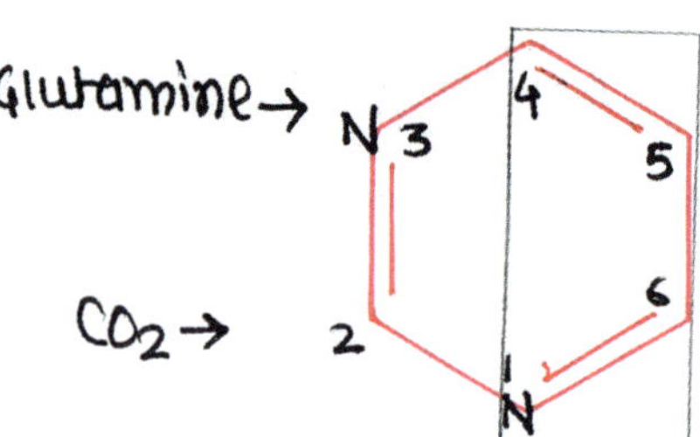

N1, C4, C5, C6 from Aspartate

Sources of carbon & Nitrogen in Pyrimidine ring

Enzymes involved in the Pyrimidine synthesis

- C Carbamoyl Ⓟ synthetase II
- A Aspartate transcarbamoylase
- D Dihydroorotase

} Multifunctional enzyme

- DH Dihydroorotate DH
- O Orotate Phosphoribosyl transferase
- O OMP decarboxylase

] Bifunctional enzyme

Disorder :- **OROTIC ACIDURIA** :- AR disease - defect in Pyr synthesis
Deficient. enzymes ① Orotate phosphoribosyl transferase
② OMP decarboxylase

C/F :- Excretion of orotate crystals in urine
Failure to thrive
Megaloblastic anemia
Immunodeficiency
Remission with oral Uridine

Orotic Acid
↓ ✗ OPRTase
OMP
↓ ✗ OMP decarboxylase
UMP

T/t :- Oral Uridine
Symptomatic T/t

Pyrimidine Nucleotide synthesis

CO_2 + Glutamine

↓ 2ATP — Carbomoyl Ⓟ synthetase II [CPS-II]

Carbamoyl Ⓟ

↓ Asp → Pi — Aspartate transcarbomylase [ATC]

Carbamoyl Aspartate

↓ H_2O — Dihydroorotase

Dihydro-orotate

↓ NAD^+ → $NADH + H^+$ — Dihydroorotate DH

orotate

↓ PRPP → PPi — OPRTase

Orotidine Monophosphate [OMP]

↓ CO_2 — OMP decarboxylase

Uridine Monophosphate [UMP]

↓ ATP — Kinase

UDP

- UDP → (ATP, Kinase) → UTP
  - UTP → (ATP, Glutamine; CTP Synthetase) → CTP
- UDP → (NADPH, Ribonucleotide Reductase) → dUDP
  - dUDP → (H_2O → Pi) → dUMP
  - dUMP → (N^5N^{10} Methylene THF; Thymidylate synthetase) → dTMP

Regulation:- CPS II Activated by PRPP & inhibited by feedback mechanism by UTP

ATC – Activated by ATP & inhibited by CTP

Both enzymes are Allosterically regulated.

CPS I	CPS II
Mitochondria	Cytosol
Urea cycle	Pyrimidine Synthesis
NAG is Activator	Activated by PRPP
NH_3 is a source of N	Glutamine is a source of N

Degrodation of Pyrimidine Nucleotides

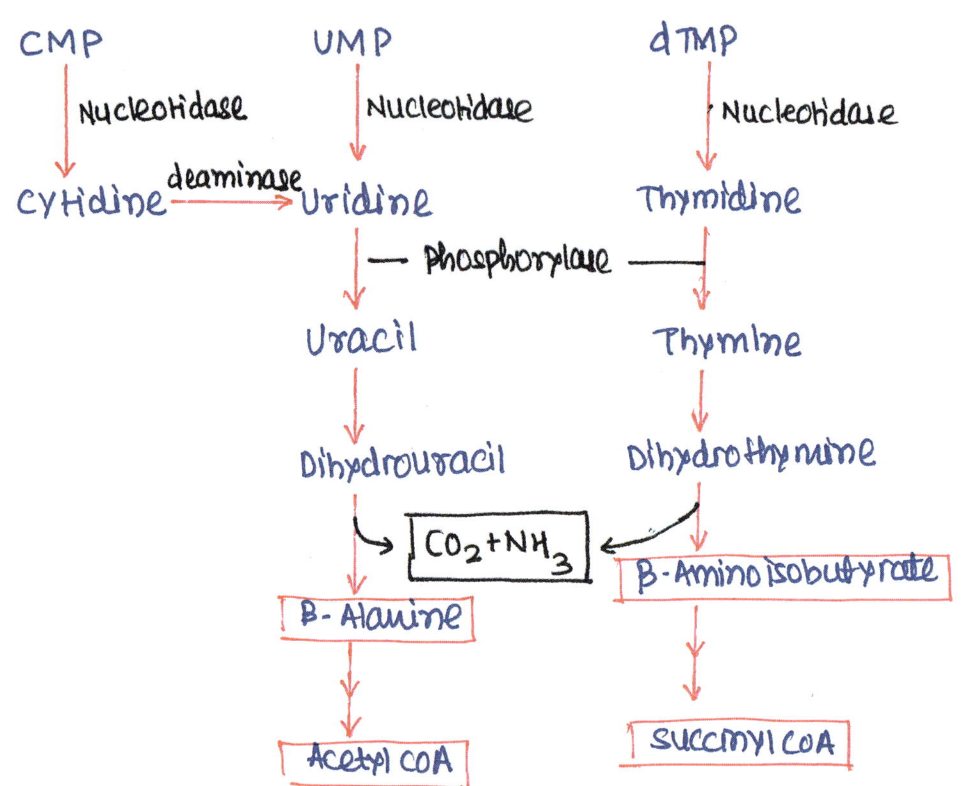

Why type-2 Hyperammonemia [Urea cycle ds] is associated with Orotic aciduria ?

Hyperammonemia-2 is an urea cycle ds due to deficiency of ornithine transcarbamoylase (OTC). In OTC deficiency accumulation of carbamoyl phosphate occurs, which is diverted for pyrimidine NT synthesis. Orotic acid is produced in pyrimidine NT synthesis.

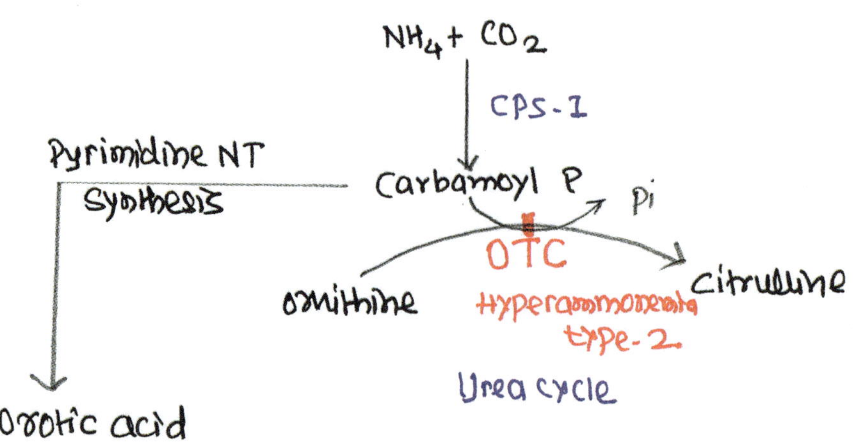

HEMOGLOBIN CHEMISTRY AND METBOLISM

MCI COMPETENCY BI 6.11-6.12

LAQ:- Heme synthesis pathway - steps, regulation, disorders
synthesis, uptake and conjugation of Bilirubin

SAQs

1. Hemoglobin derivatives - Normal, Abnormal
2. Sickle cell anemia
3. Alpha thalassemia
4. Beta thalassemia
5. porphyrias

Clinical cases

1. Acute intermittent porphyria - Abdominal pain, neurological symptoms
2. Porphyria cutanea tarda - Cutaneous manifestations
3. Congenital erythropoietic porphyria - Photosensitivity, skin lesions
4. Sickle cell anemia - Pain in extremities, HbS in electrophoresis
5. Thalassemia - Anemia, Jaundice, Hepatomegaly - β-thal-Major
6. Jaundic — Hemolytic → ↑ indirect bilirubin
   Hepatic → ↑ Both direct/Indirect, ↑↑↑ SGOT, SGPT
   Obstructive → ↑ direct bilirubin, ↑↑↑ ALP, GGT

7. Neonatal physiological Jaundice
   Yellowish discoloration of skin 2-3 days after birth
   due to inability of liver to conjugate bilirubin
   occurs in premature babies in which UDP-glucuronyl transferase
   activity is low due to immature hepatocytes
   ↑ in Indirect bilirubin
   T/t → Phototherapy (blue light 420-470 nm)
   Bilirubin absorb blue light and converted into Nontoxic
   Photoisomer - Lumirubin, it can be excreted from kidneys.

HEME SYNTHESIS

LAQ:- Write about steps, regulation and disorders related with heme synthesis pathway.

Site:- Mitochondria + cytosol

- Erythroid precursor cells (80%) in Bone marrow
- Liver

Steps ① Synthesis of Porphobilinogen

② PBG to Uroporphyrinogen (UPG)

③ UPG to Protoporphyrin

④ Incorporation of Iron in Protoporphyrin

Mnemonic:-

Mnemonic	Pathway	Disorder
Some + Great	Succinyl CoA + Glycine	
	↓ AS [ALA synthase]	① Sideroblastic Anemia
Doctors	δ ALA	
	↓ AD [ALA dehydratase]	② Plumboporphyria
Palpate	Porphobilinogen (PBG)	③ AIP Ac. Intermittent porphyria
	↓ U1 [UPG I synthase]	
Heart	Hydroxymethylbilane (HMB)	④ CEP congenital erythropoietic
	↓ U3 [UPG III cosynthase]	
Under	Uroporphyrinogen (UPG)	⑤ PCT Phorphyria cutanea Tarda
	↓ Ud [UPG decarboxylase]	
Cover	Coproporphyrinogen (CPG)	⑥ HC Hereditary coproporphyria
	↓ Co [CPG oxidase]	
Produces	Protoporphyrinogen (PPG)	⑦ VP Variegate Porphyria
	↓ Po [PPG oxidase]	
Pure	Protoporphyrin	
	↓ F [Ferrochelatase]	⑧ Protoporphyria
Heme	Heme	

HEME SYNTHESIS PATHWAY

Succinyl CoA + Glycine

↓ ALA Synthase (mt), PLP ($\rightarrow$ COASH, CO_2)

2x δ Aminolevulenic acid (δALA)

↓ ALA dehydratase (Zn) ($\rightarrow$ $2H_2O$)

4x Porphobilinogen (PBG)

↓ Uroporphyrinogen I synthase OR HMB synthase OR PBG deaminase ($\rightarrow$ $4NH_3$)

(Linear tetrapyrrole)

Hydroxymethylbilane (HMB)

↓ Cyclization — Uroporphyrinogen III cosynthase

A P
A A
P P
P A

Uroporphyrinogen III

↓ Uroporphyrinogen decarboxylase ($\rightarrow$ $4CO_2$)

M P
M M
P P
P M

Coproporphyrinogen III

↓ Coproporphyrinogen oxidase (mt) ($\rightarrow$ $2CO_2$)

M V
M M
P V
P M

Protoporphyrinogen IX

↓ Protoporphyrinogen oxidase (mt)

Protoporphyrin

↓ Ferrochelatase (mt) (Fe^{2+} $\rightarrow$)

Heme

ALA synthase 1	ALA synthase 2
Expressed in All cells	Expressed in erythrocyte precursor cells
Rate limiting step in liver	
Heme inhibit ALAS 1 by feedback regulation	No feedback Inhibition
Induced by the Drugs	Not induced by Drugs

Regulation

① ALA synthase is the rate limiting and regulatory step in Heme synthesis.

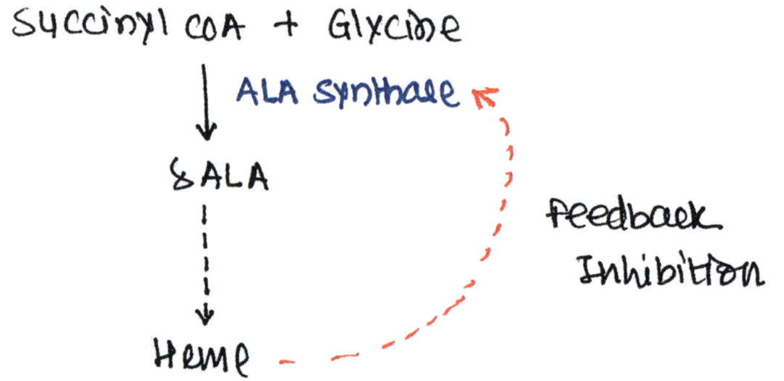

ALA synthase is also inhibited by Heme by repression and derepression at genetic Level.

② Transport of ALA from mitochondria to cytosol is also a regulatory step.

PORPHYRIAS

Inborn errors in Heme synthesis pathway

Two types

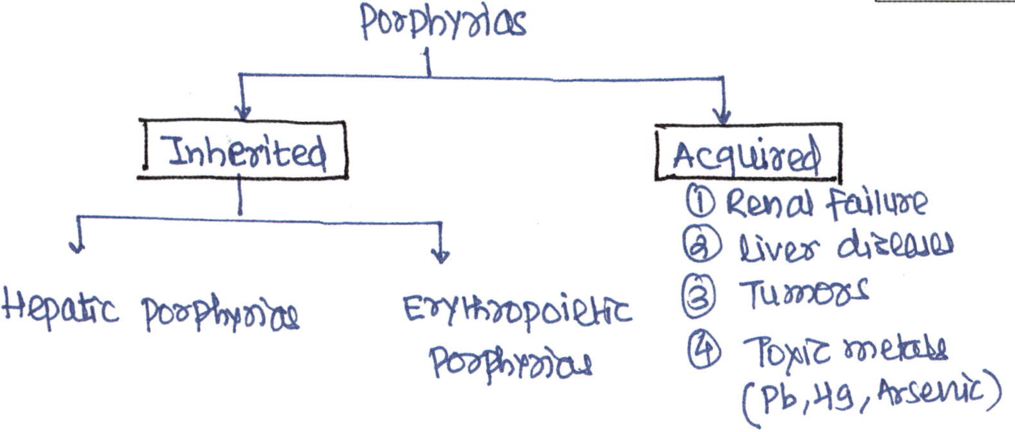

Based on the major site where enzyme deficiency is manifested — Hepatic / Erythropoietic

Hepatic Porphyrias

	Type of Porphyria	Enzyme defect	Clinical features
1.	Acute Intermittent porphyria	Uroporphyrinogen I synthase	Abdominal pain Neuropsychiatric symptoms.
2.	Porphyria cutanea tarda	Uroporphyrinogen decarboxylase	Photosensitivity
3.	Hereditary coproporphyria	coproporphyrinogen oxidase	Abd. pain Neuropsychiatric symptoms Photosensitivity
4	Variegate porphyria	Protoporphyrinogen oxidase	Abd. pain Neuropsychiatric symptoms Photosensitivity

Erythropoietic Porphyrias

	Type of Porphyria	Enzyme defect	Clinical features
1.	Congenital erythropoietic porphyria	Uroporphyrinogen III cosynthase	Photosensitivity ↑ Hemolysis
2.	Protoporphyria	Ferrochelatase	Photosensitivity

All porphyrias are autosomal dominant
EXCEPT :- X-linked sideroblastic anemia MCQ
Congenital erythropoietic porphyria

1. X-linked sideroblastic anemia

- defect in ALA synthase
- Anemia

2. Plumboporphyria

- also known as Doss porphyria
- due to lead poisoning - defect in ALA dehydratase

3. Acute intermittent porphyria

- defect in HMB synthase | PBG deaminase | Uroporphyrinogen I synthase
- Intermittent symptoms
- Neuropsychiatric manifestation
- Abdominal pain
- Autosomal dominant

(Neuropsychiatric manifestation and abdominal pain:) due to ↑ accumulation of PBG, ALA and HMB
↓ Activity of Try pyrrolase which causes ↑ accumulation of Try and serotonin.

Lab Δ :- ↑ porphobilinogen (PBG) and ALA in urine

T/t :- Hematin/Heme
Avoid Phenobarbitone, barbiturates, Anticonvulsants
[Induces cyt P450 - Aggrevates symptoms]

4. Congenital erythropoietic porphyria

Also known as Gunther's disease
Autosomal recessive
Defect :- Uroporphyrinogen III cosynthase
Photosensitivity and blistering skin lesions

↓ UPG III cosynthase ⟶ Accumulation of porphyrins ⟶ (UV light, 400 nm) excited and forms ROS
↓
Lysosomal damage
Release of enzymes
↓
Skin lesions

patient appear like Vampires or Werewolf – due to repeated scarring and facial hairs.

Erythrodontia – fluorescent teeth under UV-rays MCQ

passage of red urine – causes staining of diaper MCQ

5. Porphyria cutanea tarda

due to defect in Uroporphyrinogen decarboxylase

Autosomal dominant

C/F – Bullous skin lesion's – due to porphyrins

Porphyrins $\xrightarrow[\text{400 nm}]{\text{UV light}}$ ROS ⟶ Lysosomal damage ⟶ Release of enzymes ⟶ skin Lesions

Most common porphyria

6. Hereditary porphyria → AD

defective enzyme → Coproporphyrinogen oxidase

Neuropsychiatric manifestations, Abdominal pain, skin lesions

7. Variegate porphyria → AD

defective enzyme – protoporphyrinogen oxidase

Neuropsychiatric symptoms, Abdominal pain, skin lesions.

8. Protoporphyria

defect in ferrochelatase OR Heme synthase

photosensitivity

T/t of porphyrias – ① Hematin

② Anti-oxidant vitamins – Vit A
Vit C
Vit E

③ Use of sunscreen

④ Avoid Cyt P450 inducer drugs

Clinical case-1

A 9-year-old boy was brought to the hospital with chief complaints of photosensitivity and spontaneous development of small blistering skin lesions on exposure to sun. On examination there were many scars on the body, representing the healed lesions. The teeth were Red on wood lamp examination and urine showed Red fluoroscence.

Diagnosis:- Congenital erythropoietic porphyria

Enzyme deficient:- Uroporphyrinogen III cosynthase

Clinical case-2

A mentally retarded boy was brought to the psychiatry department with chief complaint of repeated episodes of abdominal pain. Urine examination showed elevated levels of PBG and ALA.

Diagnosis:- Acute intermittent porphyria

Enzyme deficient:- Uroporphyrinogen I synthase
OR PBG deaminase
OR HMB synthase

Clinical case-3

An autosomal dominant condition, causing bullous skin lesions on sun exposed part of the body. The condition is more common in patient with HCV or HIV infection. Laboratory investigation - ↓ Activity of Uroporphyrinogen decarboxylase.

Mention the defect → Porphyria cutanea tarda

Enzyme deficient → Uroporphyrinogen decarboxylase

C/F → Bullous skin lesions (photosensitivity)

HEME CATABOLISM/BILIRUBIN METABOLISM

LAQ :- Write in detail about synthesis, uptake and conjugation of bilirubin. Add a note on Jaundice.

Site :- Macrophages of Reticuloendothelial system (liver, spleen, Bone marrow)

Hb (Hemoglobin)
→ Globin → Amino acids
↓
Heme
↓ NADPH, O_2 — Heme oxygenase (Microsome) → Fe^{3+}, CO

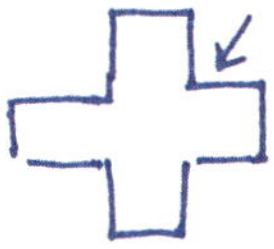

O_2 ↓ α-methenyl bridge

↓
Biliverdin (green)
↓ Biliverdin Reductase — methenyl bridge → Methylene bridge [III & IV Pyrrole]
↓
Bilirubin (yellow) [Unconjugated bilirubin]

Transport

↓ + Albumin
Albumin - Bilirubin complex
↓
Liver

Albumin
- High affinity binding site ~25 mg
- Low affinity site → easily detached and diffuses into tissues in diseases

Metabolism of Bilirubin

1. Uptake of bilirubin in Hepatocytes
2. Conjugation
3. Secretion into bile canaliculi

1. Uptake of Bilirubin in Hepatocyte → facilitated diffusion

endothelium

Hepatocyte

Alb

Alb-Bil → Bil → Bilirubin

Ligandin Protein] — Prevent efflux of bilirubin to blood
Keeps bilirubin in solubilized state

Blood

facilitated diffusion] large capacity
Not a rate limiting step even in pathological state

2. Conjugation

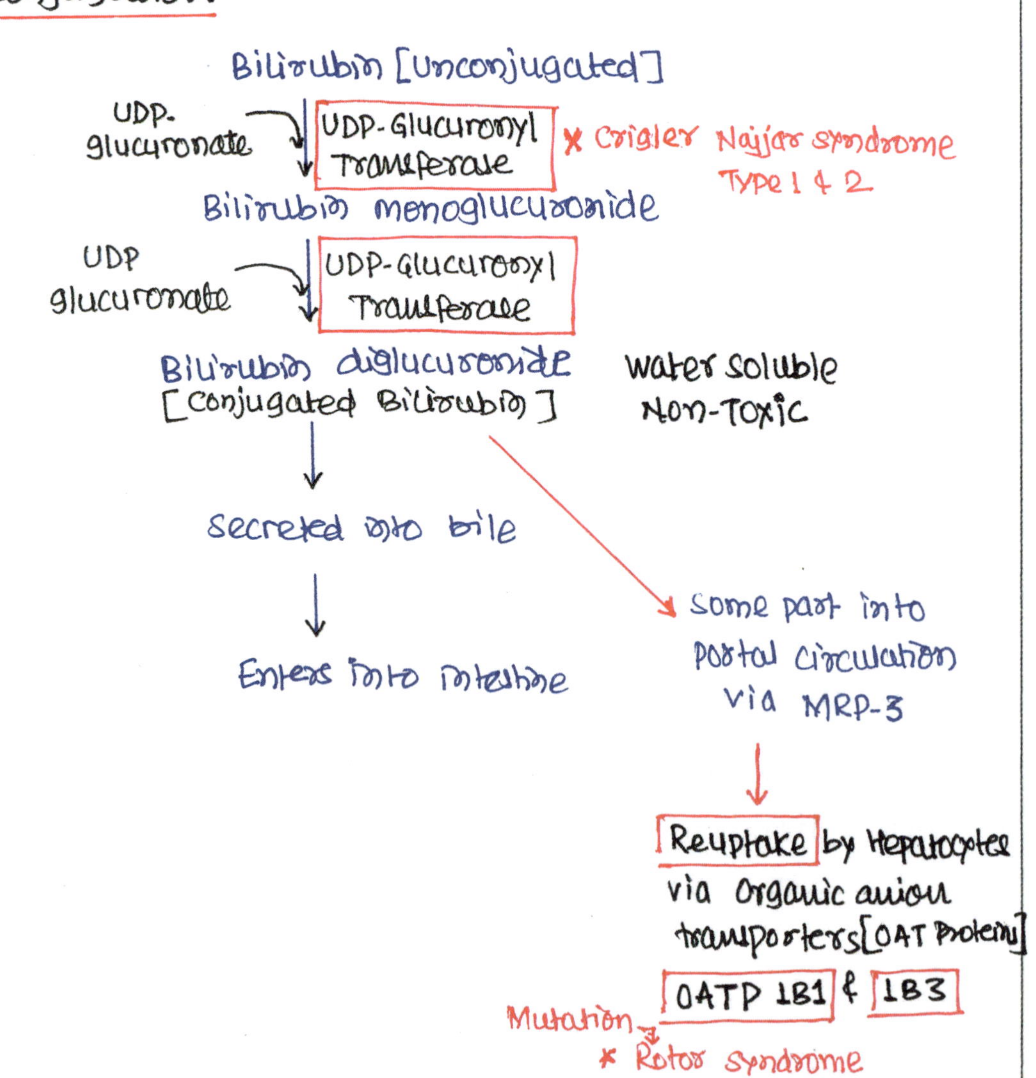

3. Secretion into Bile Canaliculi

Rate limiting step

Active transport → By MRP 2 [Multidrug Resistance like Protein 2]

Mutation :- Dubin Johnson Syndrome

Conjugated bilirubin
↓
Bile
↓
Ileum, colon
↓ Colonic Bacteria
Urobilinogen — Enterohepatic circulation
↓ → Kidneys → Urobilin → Excreted in Urine, Yellow color
Stercobilinogen
↓
Feces

1 gm Hb ⟶ 35 mg bilirubin

6 gm degraded per day

Congenital Hyperbilirubinemia

- ↑ Unconjugated bilirubin (↑ Indirect bilirubin)
  - **Crigler Najjar syndrome**
    - Type 1: Complete deficiency of UDP-Glucuronyl Transferase
    - Type 2: Partial deficiency UGT
  - **Gilbert ds**
    - defect in uptake of bilirubin in liver
- ↑ Conjugated Bilirubin (↑ Direct bilirubin)
  - **Dubin Johnson Syndrome**
    - Mutation-MRP2
    - defect in excretion of conjugated bilirubin
    - Black colored liver
  - **Rotor Syndrome**
    - Mutation OAT 1B1, 1B3
    - defect in excretion
    - No staining of liver

Types of Hb

Imp. BAQ

Type of Hb		Component subunit	
Adult Hb	HbA	$\alpha_2\beta_2$	The major Hb in humans (90%)
	HbA_2	$\alpha_2\delta_2$	Minor component in Adult (2-5%)
Fetal Hb	HbF	$\alpha_2\gamma_2$	< 2%
Glycated Hb →	HbA_{1c}	$\alpha_2\beta_2$-Glucose	< 5%, Increases in DM

Hemoglobin derivatives

Normal → OxyHb, deoxyHb, $HbCO_2$

Abnormal → MetHb, HbCO, SulfHb, Cyanhemoglobin

1. Oxyhemoglobin (OxyHb) → Hb with O_2
2. Deoxyhemoglobin (deoxyHb) → Hb without O_2
3. Carbaminohemoglobin ($HbCO_2$) → CO_2 is noncovalently bound to globin chain of Hb. $HbCO_2$ transport CO_2 in blood.
4. Methemoglobin (MetHb) → contains Fe^{3+} instead of Fe^{2+} in heme group
   unable to transport oxygen
5. Carbonylhemoglobin (HbCO) → CO bind to Fe^{2+} in heme in case of CO poisoning or smoking. carbon monoxide has higher affinity to Fe^{2+} than O_2
6. Sulfhemoglobin – poisoning by H_2S
   Cyanhemoglobin - cyanide poisoning

Sickle Cell Anemia

Autosomal recessive disorder

It is an example of Partially acceptable missense point mutation

Glutamate is replaced by valine at 6th position of beta chain

This substitution causes localized stickiness on the surface of deoxyHbS.

This change causes polymerisation of HbS under deoxygenated state and results in formation of fiber which distort the shape of RBC's into sickled shape.

These distorted RBCs occlude the microvasculature and causes intense pain in the affected areas

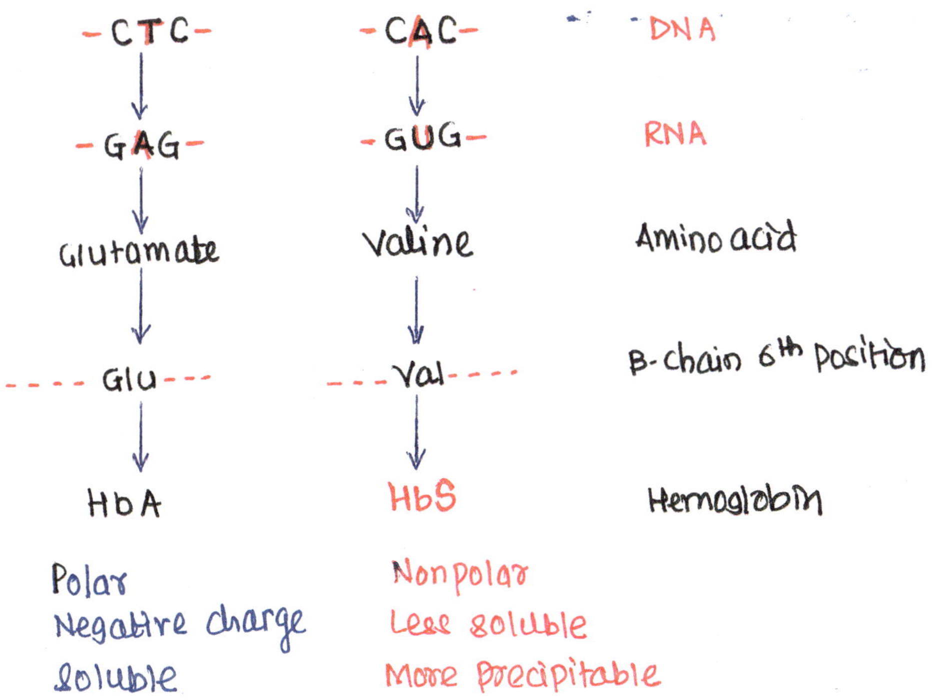

Molecular basis of sickle cell disease

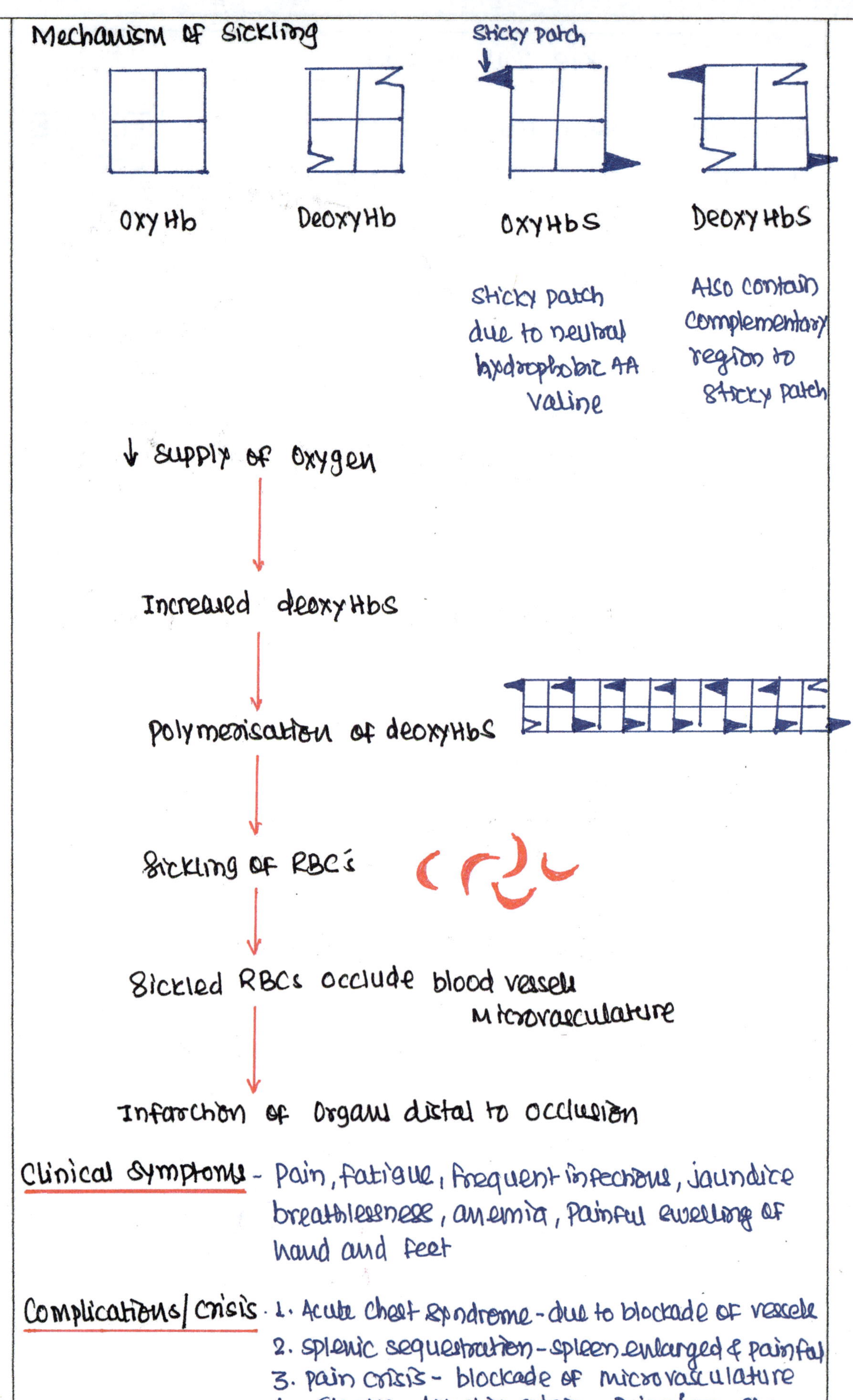
Mechanism of Sickling
Sticky Patch
OxyHb
DeoxyHb
OxyHbS
DeoxyHbS
Sticky patch due to neutral hydrophobic AA valine
Also contain complementary region to sticky patch
↓ supply of oxygen
Increased deoxyHbS
Polymerisation of deoxyHbS
Sickling of RBC's
Sickled RBCs occlude blood vessels microvasculature
Infarction of organs distal to occlusion
Clinical symptoms - Pain, fatigue, frequent infections, jaundice breathlessness, anemia, Painful swelling of hand and feet
Complications/Crisis - 1. Acute Chest Syndrome - due to blockade of vessels
2. splenic sequestration - spleen enlarged & painful
3. pain crisis - blockade of microvasculature
4. Stroke, Aplastic crisis, priapism etc.

Lab Diagnosis

1. **Sickling test** :- Screening test for sickle cell anemia
   Prepare peripheral smear by adding reducing agent
   sodium metabisulfite OR
   sodium dithionite
   Observe the sickled RBCs under microscope.

2. **Hb electrophoresis**

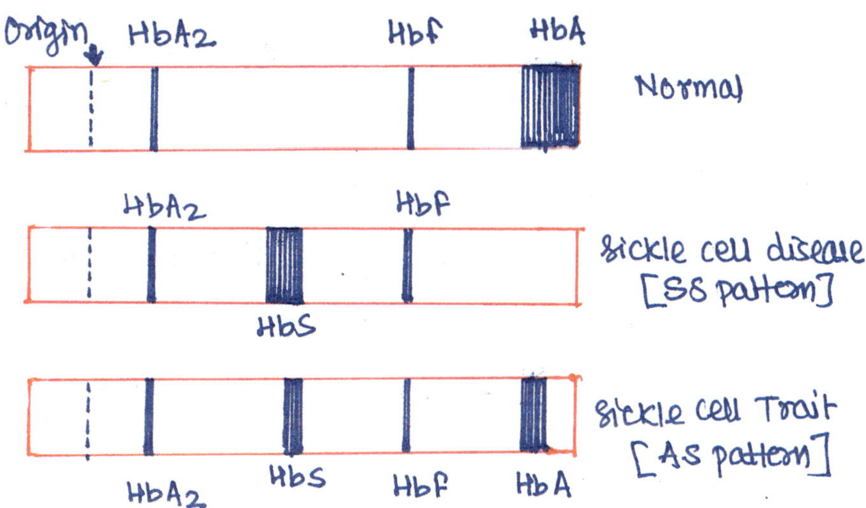

HbS moves slowly in electrophoresis, HbA moves fast because it carries more negative charge, due to presence of glutamic acid.

3. **HPLC** - High performance liquid chromatography
4. **Capillary electrophoresis**

Treatment

1. Hydroxyurea
2. Blood transfusion
3. Stem cell transplant therapy
4. Gene therapy

Inheritance

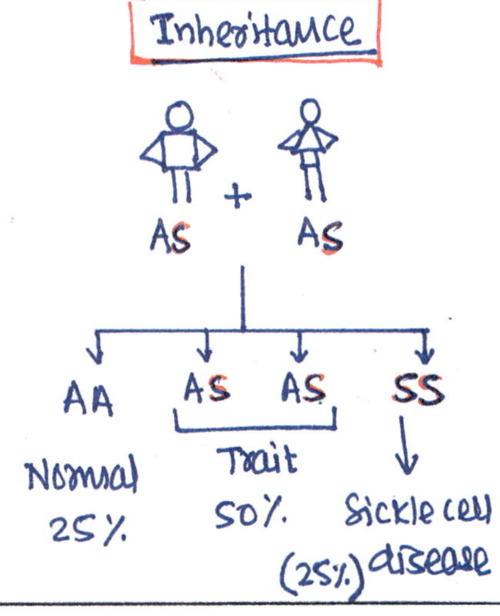

THALASSEMIA

Inherited disorder with decrease synthesis or total absence of α- globin chain — Alpha thalassemia
β- globin chain — Beta thalassemia

Alpha thalassemia

Two genes on chromosome 16 code for α-globin
Two copies of chromosome 16 in each diploid cell makes 4, α-globin allele.

↓

Deletion of one or more α-globin allele

↓

↓ quantity of α-globin production

	Types	Number of Missing genes	Clinical Symptoms
1.	Silent carrier	① ~~α_1~~ - α_2, α_3 - α_4. Remaining 3 functional allele synthesize adequate α-globin	No symptoms
2.	α-Thalassemia Trait	② ~~α_1~~ - ~~α_2~~, α_3 - α_4 OR ~~α_1~~ - α_2, ~~α_3~~ - α_4. decreased α-globin production. Normal β-globin production	Mild Anemia
3	Hemoglobin H disease	③ ~~α_1~~ - ~~α_2~~, ~~α_3~~ - α_4. decreased α-globin. Excess β-globin bind to form HbH (β_4)	Moderate anemia c̄ microcytic hypochromic RBCs.
4	Hydrops fetalis	④ ~~α_1~~ - ~~α_2~~, ~~α_3~~ - ~~α_4~~. No α-globin. Excess γ-globin bind to each other to form Hb Barts (γ_4), prevent release of oxygen to tissues	Fetal death occurs at birth

Beta Thalassemia

decrease synthesis or total absence of β-globin chain

2 Types

β-Thalassemia major

defect in both gene (Homozygous)

few or No β chains

↓↓ HbA or Absent

↑ HbF

↑ HbA_2

β-Thalassemia Minor

defect in single gene (Heterozygous)

↓ HbA synthesis

Mild anemia, Jaundice, splenomegaly.

↑HbF, ↑HbA_2

C/F :- Anemia

Jaundice – due to hemolysis of RBC's

Hepatomegaly – Hematopoiesis in liver

Splenomegaly

Frontal bossing
Maxillary overgrowth
Hair-on-end appearance on X-ray skull
} Hematopoiesis in bones causes skeletal deformities

Treatment :- Supportive care

Frequent blood transfusion

Iron chelating agents.

Stem cell transplant.

Lab diagnosis :- Hb electrophoresis and HPLC

DNA Analysis for α and β globin mutations

↓ or Absent HbA

Increased HbF & HbA_2

Clinical case-1

A 4-Year-Old boy was brought to the pediatric department with chief complaints of fatigue, breathlessness and pain in the extremities. On examination he was found to have pallor, jaundice and splenomegaly. Hb was 7 gm%. and peripheral blood smear examination revealed presence of sickle shaped RBC's.

Δ Sickle cell anemia

Screening test :- Demonstration of sickling by using sodium dithionate under microscopy.

Confirmatory Δ :- Hb electrophoresis

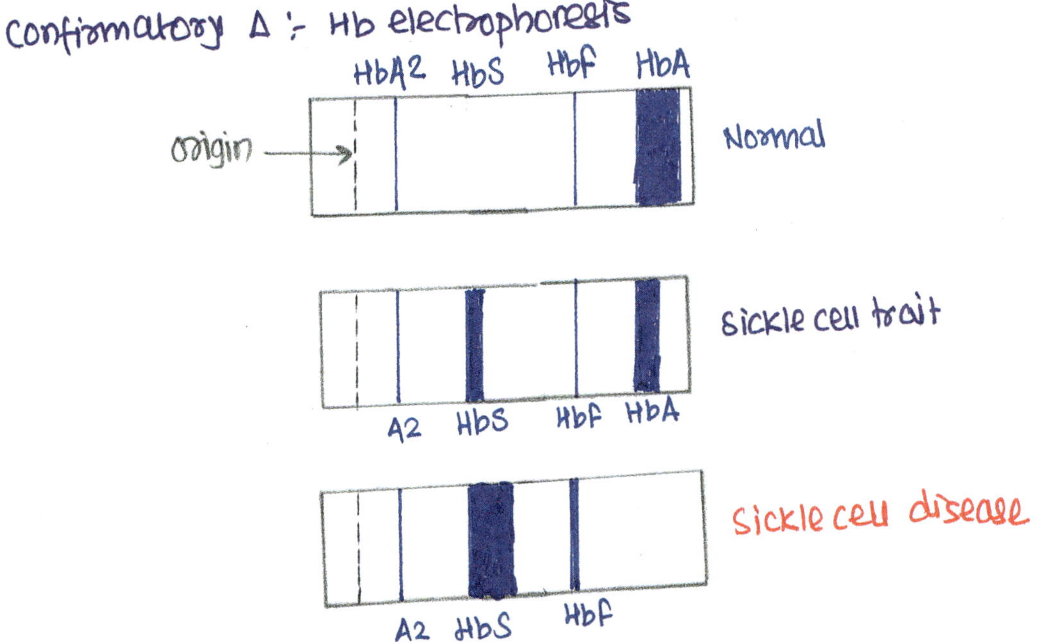

Clinical case-2

A 3-year-old girl child was brought to the hospital with hepatosplenomegaly. On examination she had pallor and icterus. Laboratory investigation showed decrease HbA to a critically low level and increased in Hbf and HbA2. She had characteristic facial appearance - maxillary overgrowth, depressed nasal bridge, small lower jaw etc., x-ray skull showed - "Hair on end" appearance.

What is your probable diagnosis :- β-Thalassemia major

VITAMINS

MCI COMPETENCY BI 6.5

Describe the biochemical role of vitamins in the body and explain the manifestations of their deficiency.

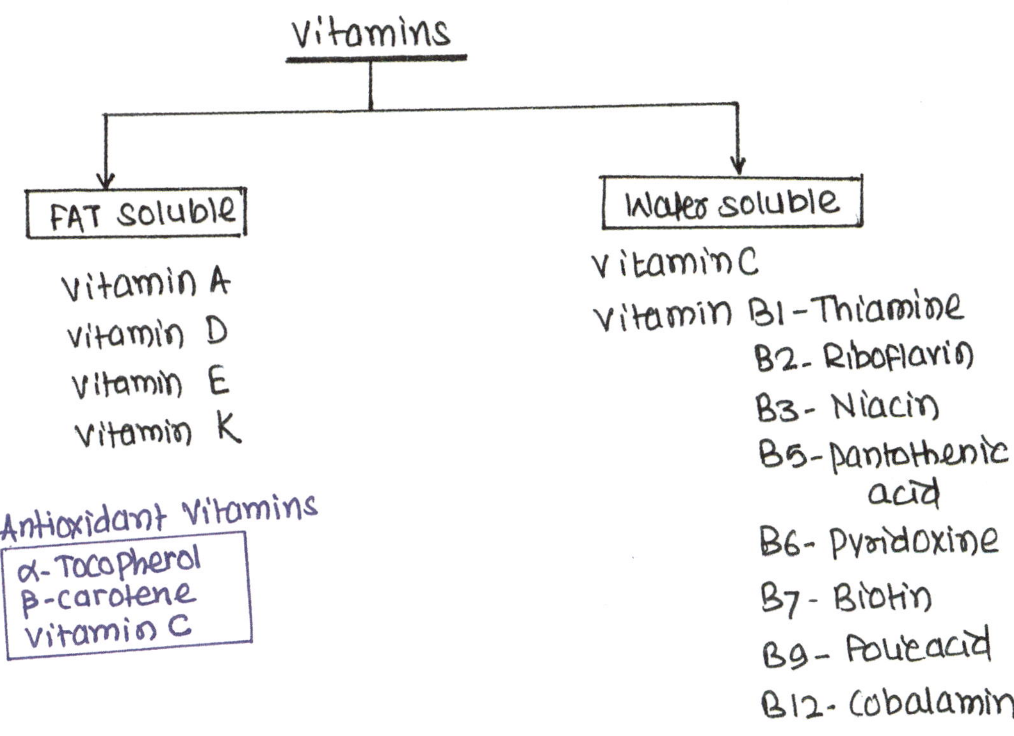

Clinical cases

1. Bitots spot, Night blindness, xerophthalmia
2. Rickets, osteomalacia
3. Bleeding disorder – Vit K deficiency
4. Scurvy
5. Beriberi
6. Pellagra
7. Megaloblastic anemia
8. Ariboflavinosis
9. Wernicke- Korsakoff syndrome

* Endogenously produced vitamins

- Vitamin D
- Niacin
- Biotin
- Vitamin K2
- pantothenic acid

VITAMIN-A
LAQ CASE

LAQ:- Write about RDA, dietary sources, functions and deficiency manifestations of Vit A

Vitamers of Vit A

Compound having characteristics of Vit A/ Retinol

Retinal [Aldehyde]
Retinol [Alcohol] } Retinoids.
Retinoic acid [Acid]

Provitamin – β-carotene – plant sources

Digestion, Absorption & storage

β-carotene
↓ Dioxygenase
Retinal
↓ Retinal reductase
Retinol

Retinyl ester [Animal sources]
↓ Intestinal brush border Hydrolases
Retinol + FFA

Retinol (from both pathways) → esterified in intestinal mucosal cells → Retinyl esters
↓ Incorporated into
Chylomicrons
↓ via lymph
Liver: RE → Retinol → Blood (RBP + Retinol)

Liver: Stored as Retinyl ester in the liver [Retinyl palmitate]

Blood: In blood, transported c̄ RBP
↓
Target tissue:
Retinol
↓
Retinoic acid
↓ Acts on
Intranuclear receptor
↓
Protein synthesis

Biochemical Functions

1. Vision : Walds visual cycle

Rhodopsin
[11-cis retinal + opsin]

Light →

Opsin

All Trans retinal → (partial conversion) → 11-cis retinal

All Trans retinal → (Retinol DH) → All Trans retinol

All Trans retinol → (Isomerase) → 11-cis retinol

11-cis retinol → (Retinol DH) → 11-cis retinal

Liver

11-cis retinal + Opsin → Rhodopsin

Rhodopsin —(Light)→ prelumirhodopsin → Lumirhodopsin → Metarhodopsin I

→ Metarhodopsin II → All trans retinal + opsin

Metarhodopsin II

↓

Activates Transducin

↓ (+)

Phosphodiesterase

↓

↓ cGMP levels

↓

Closure of Na^+ channels in Rods

↓

Hyperpolarisation

↓

Excitation of visual cortex

Rods - Dim light vision

Cones - Bright light vision

color vision

Porphyropsin (Red)

Iodopsin (Green)

Cyanopsin (Blue)

2. Protein synthesis – cell growth and differentiation.
3. Prevent keratinisation of skin & mucus membrane
4. Glycoprotein synthesis, mucus formation
5. Cholesterol synthesis
6. Antioxidant – prevent CAD, cataract, Atherosclerosis
7. Maintenance of germinal epithelium – Reproduction

Sources:- Cod or shark liver oil
Meat, dairy product, eggs.

β-carotene :- carrots, spinach, papaya, Mango etc.

RDA :- 1 Retinol equivalent = 1 μg of Retinol

Male → 600 μg
Female → 600 μg
Pregnancy 800 μg
Lactation 950 μg

Bitot's spot
Corneal ulcer

Deficiency manifestations.

1. Vision impairment (WHO)

XN – Night blindness → 1st symptom of vit A deficiency
X1A – conjunctival xerosis
X1B – Bitot's spot [Δ white, opaque, plaques on conjunctiva]
X2 – Corneal xerosis
X3A – Corneal ulcer [< 1/3 corneal involvement]
X3B – Corneal ulcer – keratomalacia [> 1/3 cornea involvement]
XS – Corneal scarring
XF – Xerophthalmic fundus

2. Sterility
3. ↑↑ keratinization → Horny eruptions on skin and mucous membrane (Phrynoderma)
4. Growth impairement

Treatment ① 2 Lakh IU of vit A
② Increase dietary sources

Toxicity :- CNS : raised intracranial pressure [pseudotumor cerebri]
Symptoms ⇒ Headache, nausea, vomiting, blurred vision, Ataxia
Liver : Hepatomegaly and hyperlipidemia
Bones : Thickening of long bones, soft tissue calcification
Exfoliative dermatitis, Alopecia, Teratogenic vitamin

Clinical case-1

An 8-year-old child was brought to the clinic with chief complaints of difficult to see in the dim light and triangular white spot on conjunctiva. In the dietary history he do not take any green leafy vegetables and do not consume fish. Which vitamin is deficient in this child?

- Vitamin A deficiency

Clinical case-2

A software engineer from Banglore city treated himself for Acne and consumed retinoid medication Isotretinoin. He develops nausea, vomiting, headache, dizziness, irritability, blurred vision and lack of coordination. The intracranial pressure was raised. What is this condition known as?

- Pseudotumor cerebri -(Intracranial Hypertension)

Toxic manifestation of vitamin A due to overdose

Clinical case-3

A 9-year-old child develops bilateral hyperkeratotic skin lesions on elbows, knees, buttocks and extensor aspect of arms. What is your probable diagnosis?

- Phrynoderma OR Toad skin

Follicular hyperkeratosis due multinutrient deficiency. More commonly due to vitamin A deficiency OR due to deficiency of essential fatty acids.

VITAMIN-D

LAQ:- Describe the RDA, dietary sources, functions and deficiency manifestations of vit-D

Two forms / Provitamins
① Ergocalciferol (Plants) - vit D2
② Cholecalciferol (Animals) - vit D3

Vit D [calcitriol] synthesis

7-dehydrocholesterol (epidermis of skin)

↓ UV-B rays, sunlight

Cholecalciferol

↓ NADPH, Cyt P450, O_2 — 25 hydroxylase (Liver)

25-OH-cholecalciferol [Storage & transport form]

↓ 1α hydroxylase (kidneys)

1,25 dihydroxy cholecalciferol [Calcitriol]

Active form of vit-D

Sources ① On exposure to sunlight [UV-B rays]
② Fish, Milk, Egg yolk

RDA :- Children → 400 IU
Adult → 200 IU
Pregnancy & lactation → 400 IU
Old age (>60 yrs) → 600 IU

Why vitamin D is also known as Hormone ?

- Synthesis occurs in the skin [epidermis by UV-B rays]
- Has target tissues ① Bone ② Intestine ③ Kidneys
- Intracellular receptor for its action [Group I Hormones]
- Acts like steroid Hormones
- Regulate the calcium level

Biochemical functions :-

① Regulate the calcium level ē PTH and calcitonin

Three target tissues

① **Intestine**

Vit D
↓
Intracellular HR complex
↓
Binds ē vit D responsive element (DRE) of DNA
↓
Transcription
Translation
↓
Synthesis of calcium binding protein
↓
↑ Intestinal absorption of calcium & Ⓟ

② **Kidney**

Act on Renal tubular cell
↓
↑ Reabsorption of calcium and Ⓟ

③ **Bone**

Formation of bone
Osteoblastic activity
Bone mineralisation
- deposition of Ca^{2+} & Ⓟ in Bone.

Sr. calcium 9 – 11 mg/dl

Vit D & PTH ↑ calcium level

Calcitonin ↓ calcium level

② Vit D is helpful in t/t of psoriasis
③ Immunomodulatory function
④ Prevent from coronary artery diseases
⑤ Useful in some cancers.

Deficiency manifestation :- Rickets in children

① Craniotabes :- soft and pliable skull bones (< 1 yr old child)
② Delayed closure of fontanelles
③ Frontal bossing (Prominent forehead)
④ Widening of wrist (Radius & Ulna)
⑤ Bow legs (when child starts walking due to weight)
⑥ Rachitic rosary (Prominent costochondral junction)
⑦ Harrison's sulcus (transverse groove along the insertion of diaphragm)

Osteomalacia in Adults

① Soft bones affects back, Pelvis, hips, Legs and ribs.
② Easy fractures and Pain
③ Osteoporosis
④ Decreased calcium level
⑤ Increase ALP activity

Renal Rickets:- defect in activation of vit D in kidneys.

Vit-D Resistant Ricket:- defect in renal tubular absorption Associated with fanconi anemia.

Laboratory Diagnosis: ① Serum Vit-D level ↓

<30 ng/ml ⟶ Insufficient
<20 ng/ml ⟶ Deficient
<10 ng/ml ⟶ Severe deficiency
>150 ng/ml ⟶ Toxicity

② ↑ Activity of Alkaline phosphatase

Treatment:

High single dose of vit D followed by maintenance doses
Increase exposure to sunlight
↓
Increase outdoor activities of child
Calcium supplementation

Clinical case

A 3-year-old child was brought to the pediatric OPD with chief complaints of bow legs. On observation- prominent forehead, beaded appearance of chest was noted. Serum Alkaline phosphatase levels were high.
What is your probable diagnosis?

Δ Rickets - vit D deficiency

Clinical case- 2

A 50-year-old female came to OPD with H/O frequent fractures. Lab investigation showed ↑ ALP and ↓ levels of vitamin D, calcium & Ⓟ. Δ :- Osteomalacia (Vit D ↓ in Adult)

VITAMIN-E

Fat soluble
Antisterility vitamin
Vitamin-in-search of disease

RDA : 10 mg/day – Males
8 mg/day – females

Includes
- Tocopherols
- Tocotrienols

α-Tocopherol [Most active]
6-OH Chromane ring with isoprenoid side chain
Antioxidant property.

Sources

Vegetable oils – Sunflower, Safflower, Soyabean, peanut, Wheat germ

Deficiency manifestations

Mild hemolytic anemia
Neurological – spinocerebellar ataxia, Retinopathy, Neuropathy, Myopathy

Biochemical functions

1. Membrane antioxidant
   prevents lipid peroxidation

   Vit E + Free radicals ⟶ Vit E Quinone form

   prevent PUFA from free radicals
2. Prevents hemolysis in RBC's
   Maintain integrity of RBC membrane
3. Maintains germinal epithelium of gonads, prevents sterility
4. Prevents oxidised-LDL formation
   prevents coronary artery diseases
   Helpful in Myocardial infarction
5. Act synergistically with selenium.
   selenium is important cofactor for glutathione peroxidase ↓ scavenges free radicals.
6. Immune response
7. Prevents oxidation of vit A
8. prevent cataract formation (Along with vit A & vit C)
9. Heme synthesis [Activates ALA synthase & dehydratase]
10. Nucleic acid synthesis
11. Proper storage of creatine in muscle.

VITAMIN K

K for Koagulation in german

LAQ: Write about RDA, dietary sources, Biochemical functions and deficiency manifestations of Vit K.

Forms: ③ K_1, K2, K3 K_1K_2 - Naphthoquinone derivatives

K1 → Phylloquinone [20C isoprenoid side chain] - Green leafy vegetables

K2 → Menaquinone [30C isoprenoid side chain] - Intestinal bacteria, Animal sources

K3 → Menadione - Synthetic, Water soluble; Given to newborn 1 mg IM (Prophylaxis)

Biochemical functions

① Post-translational modification of clotting factors II, VII, IX and X [γ-carboxylation] in ER

Glutamate: ~N(H) - CH - C(=O)~ with side chain CH_2 - CH_2 - COOH

II, VII, IX, X — Inactive zymogens

→ Vit K →

γ-carboxy Glutamate: ~N(H) - CH - C(=O)~ with side chain CH_2 - CH(COO^-)(COO^-)

II, VII, IX, X — Active factors

Vit K - converts glutamate into γ-carboxyglutamate which carries extra negative charge, This negatively charged clotting factor attached with positively charged calcium ions.

Active clotting factor → Ca^{2+}-Prothrombin complex → Attach to PL on Platelet Membrane → Prothrombin → Thrombin → Coagulation

II, VII, IX, X

MCQ → Warfarin, Dicumarol] Competitive inhibitor of vit-K epoxide Reductase

(2) Activation of osteocalcin – Bone formation

(3) Activation of Protein C & S – Anticoagulant

RDA :- 70–140 μg/day

Sources :- Green leafy vegetable
Intestinal bacteria
Non-vegetable sources

Deficiency manifestations

Hemorrhagic manifestations
Ecchymotic patches all over the body
Red spots, Rash

Causes :- Poor placental transport
Low hepatic stores
Sterile intestine
Immature hepatocytes
Milk - Poor source
Prematurity

Prophylaxis - 1mg Menadione is given IM at birth

Lab Δ :- ↑ Prothrombin time

Toxicity :- Hemolysis, Kernicterus

ASCORBIC ACID [VITAMIN C]

LAQ:- Write about RDA, sources, biochemical functions and deficiency manifestations of vitamin C

Active form - L-Ascorbic acid

Dehydroascorbate → oxalic acid → Excreted

Sources: Indian Gooseberry, citrus fruits (lemon, orange etc) Guava

Synthesized in lower animals via Uronic acid pathway

Human and other higher primates cannot synthesize vit C because they lack 'L-Gulonolactone oxidase'.

RDA: 60-75 mg/day

Biochemical functions

① Hydroxylation of proline and Lysine ↓ synthesis of **Collagen**

Proline —Prolyl hydroxylase→ Hydroxyproline [to stabilize collagen]

Vit-C

Lysine —Lysyl hydroxylase→ Hydroxylysine [to give structural strength cross-linking]

② Antioxidant - free radical scavenger

③ Prevention of cataract, Atherosclerosis and coronary artery diseases.

④ Oxidation-Reduction reaction - cellular respiration through cyt. oxidase

⑤ Aromatic amino acid metabolism

Tryptophan ⟶ 5 Hydroxytryptophan

P-OH phenylpyruvate ⟶ Homogentisate

⑥ Immunological function

⑦ Iron absorption [Fe^{3+} → Fe^{2+}]

⑧ Met Hb ⟶ Hb

⑨ Bone formation

⑩ Steroid hormone synthesis

⑪ Spares vit A and vit E in antioxidant function

It reduces oxidized vit E to regenerate functional vit E

Deficiency Manifestations

SCURVY

Swollen and painful gums, bleeding gums – defective cross linking & strengthning of collagen fibre, fragile blood vessels

Petechiae
Ecchymoses } bleeding tendency through fragile vessels
Hematomas

Delayed wound healing

Anemia

Scorbutic rosary – beaded appearance of chest due to prominent costochondral junction

Depression of sternum

Joint effusion – pseudoparalysis

Barlow's syndrome – Infantile scurvy [6 months to 1 year]

Treatment

High doses of Ascorbic acid

Diet rich in fruits and vegetables rich in vitamin C

Clinical case

An 11-year-old child was brought to the hospital with chief complaints of painful bleeding gums and delayed wound healing. Red hemorrhagic spots were noted all over the body. What is your probable diagnosis?

Δ Scurvy – due to deficiency of vitamin C

A 55-year-old lady came to hospital with chief complaints of joint pain and swollen joints. She also gave h/o delayed wound healing. She noticed bleeding gum while brushing teeth. What is your probable diagnosis?

Δ Scurvy – due to vitamin C deficiency

B1 - THIAMINE

Anti-Beriberi vitamin
Anti-Neuritic vitamin

Active form:- TPP [Thiamine pyrophosphate]

Chemistry: pyrimidine – Thiazole

Thiamine —Ⓟ—Ⓟ

Biochemical functions

① Energy generation
② Transmission of Nerve impulse
③ Cleavage of C-C bond
④ Oxidative decarboxylation [α-keto acids] PDH, αKGDH, BCKADH
⑤ Transketolase reaction [HMP shunt]

Reactions: Oxidative decarboxylation

① pyruvate dehydrogenase
② α ketoglutarate dehydrogenase
③ Branched chain keto acid DH

① Pyruvate $\xrightarrow[\text{PDH}]{\text{NAD}^+ \;\; \text{TPP} \;\; \text{NADH+H}^+}$ Acetyl CoA (+ CO_2)

② α ketoglutarate $\xrightarrow[\text{αKGDH}]{\text{NAD}^+ \;\; \text{TPP} \;\; \text{NADH+H}^+}$ Succinyl CoA (+ CO_2)

③ Branched chain α keto acids $\xrightarrow[\text{α keto acid DH}]{\text{NAD}^+ \;\; \text{TPP} \;\; \text{NADH+H}^+}$ Acyl CoA (+ CO_2)

Sources:- Rice - Present in outer aleurone layer
polishing removes 80% of Thiamine

Whole wheat

Yeast, liver, kidneys etc.

RDA:- 0.5 mg/1000 cal intake

1 to 1.5 mg/day

Deficiency Manifestations: BERI BERI

Causes of Thiamine deficiency

① Nutritional deficiency
due to intake of polished rice as a staple food

② Thiaminase - Present in tea & coffee

③ Chronic Alcoholic → ↓ Absorption and conversion of thiamine to TPP

④ Hyperemesis gravidarum → causes deficiency in infant.

⑤ Diuretic therapy → loss in urine

BERI-BERI
Weakness

Dry Beri-beri
- Weakness
- Peripheral neuropathy

Wet Beri-beri
- CVS- Cardiomyopathy
  Dilatation of Heart
- Fluid accumulation in pericardial space, ↑ BP
- Edema
- High output cardiac failure

Infantile Beri-beri
- Involves CVS
- Acute fulminant form
  ↓
  Shoshin Beriberi

4. Wernicke - Korsakoff syndrome

In chronic alcoholic due to ↓ Absorption of Thiamine and defective conversion of Thiamine → TPP

Neurological manifestations are due to ↓ ATP synthesis

C/F:-		Mnemonic
Global confusion Ophthalmoplegia Ataxia	Wernicke's encephalopathy	GOA
Memory loss Psychosis	Korsakoff psychosis	MP

t/t:- 100 mg/day for 7 days
followed by 10 mg/day maintenance dose.

Clinical case-1

A 30-year-old male chronic alcoholic was brought to the hospital with high blood pressure, high output cardiac failure and pericardial effusion. Investigations revealed dilated cardiomyopathy and ↓ RBC transketolase activity. Which vitamin deficiency can cause above manifestations?

- Vitamin B1 - Thiamine deficiency – Wet Beriberi

Clinical case-2

A patient on polished rice as staple diet develops weakness and loss of sensation in the legs. He also has calf muscle tenderness. Erythrocyte transketolase activity was found low. Mention the diagnosis?

Δ Dry Beriberi – Thiamine deficiency

Clinical case-3

A chronic alcoholic was brought to the emergency dept. in confused state, lack of coordination and paralysis of ocular muscles. His friends gave H/o memory loss and psychotic episodes. Mention the probable diagnosis?

Δ Wernicke's - Korsakoff syndrome

What is biochemical basis of ↑ pyruvate & neurological signs?

$$\text{Pyruvate} \xrightarrow[\text{TPP}]{\text{PDH Complex}} \text{Acetyl CoA}$$

- block in pyruvate dehydrogenase step ↑ pyruvate
- Thiamine pyrophosphate – (P)ylation of chloride channels and hence activation in the nerve membrane.

TPP have role in nerve conduction.

Also ↓ ATP production in TPP deficiency causes neurological symptoms

In Alcoholic ↓ Absorption of thiamine

↓ conversion of thiamine ⟶ TPP

B2 - RIBOFLAVIN

- Warburg Yellow enzyme of cellular respiration

Chemistry :

Isoalloxazine - N - Ribitol

Active form :- FMN
FAD

Biochemical functions

Catalyzes following reactions

FMN :- L-Aminoacid oxidase
Monoamine oxidase
Complex I of ETC - NADH DH

FAD :- Complex II of ETC - Succinate DH
Acyl CoA DH
D-Aminoacid oxidase
Xanthine oxidase
Glutathione reductase
PDH complex
α KG DH complex
Branched chain ketoacid DH

} MCQ required 5 Coenzymes TPP, CoA, Lipoic acid, FAD, NAD

Sources :- Liver, Kidney, dairy products, outer layer of cereals.

RDA :- 1.5 mg/day

Deficiency manifestations - **Ariboflavinosis**

Glossitis (Magenta tongue)
Cheilosis (swelling & fissuring of lips)
Angular stomatitis (Angle of mouth)
Seborrheic dermatitis
Corneal vascularization

Clinical case

A child was brought to the clinic with chief complaints of glossitis, cheilosis, angular stomatitis, pink eye and seborrheic dermatitis. The RBC glutathione reductase activity was low. Mention the probable cause?

A. Ariboflavinosis - due to vitamin B2 (Riboflavin) deficiency

B3 - NIACIN - Nicotinic Acid

Pellagra preventive factor of Goldburger

- Not a true vitamin, partial synthesis occurs in the body

60 mg Tryptophan $\xrightarrow[\text{B2, B6, Iron}]{}$ Niacin 1 mg

Chemistry

Active forms - NAD^+
$NADP^+$

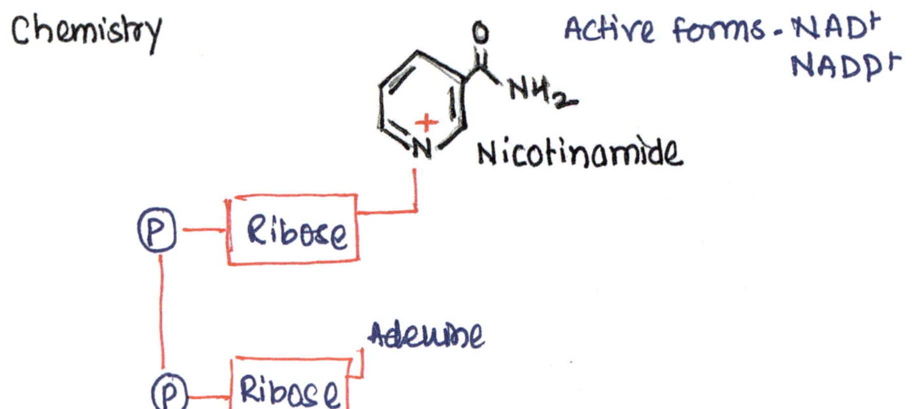

Nicotinamide Adenine dinucleotide [NAD^+]

Biochemical Functions

Oxidation - Reduction reaction

NAD^+ / $NADP^+$ ⟶ $NADH + H^+$ / $NADPH + H^+$ $\xrightarrow{ETC}$ 2.5 ATP

Important Reactions

NAD^+ → Glyceraldehyde 3 Ⓟ DH
- Lactate DH
- Pyruvate DH
- Isocitrate DH
- α KG DH
- Malate DH
- β-OH-Acyl CoA DH
- Branched chain ketoacid DH

$NADP^+$ → Glucose 6 Ⓟ DH
- Phosphogluconate DH
- Malic enzyme

NADPH → Fatty acid synthesis
- Cholesterol synthesis
- Vit D synthesis
- steroid Ⓗ synthesis

Sources :- Liver, kidneys, fish, cereals etc

RDA : 20 mg of Niacin equivalents/ day

Deficiency manifestation : pellagra - Diarrhea
Dementia
Dermatitis

PELLAGRA

Clinical features → Diarrhea
Dementia – loss of thinking, remembering & reasoning

4 D's

Dermatitis
on sun exposed areas of the body
Typically around the neck – Casal necklace appearance
and hand & feet – Glove & stocking appearance.

Death if untreated

Causes of Niacin deficiency

1. ↓ intake – dietary deficiency.
2. Deficiency of B6 (PLP) & B2,
3. Hartnups ds → defective absorption of Tryptophan
4. Carcinoid syndrome :- Try $\xrightarrow{60\%}$ serotonin
5. Jowar and Maize as a staple food

Jowar → High leucine content ↓ ⊖ QPRT Quinolinate Phosphoribosyl Transferase

Try ✗→ NAD

Maize → Tryptophan is present in bound form
Not available for body to utilize

Clinical case

A patient develops necklace shaped rash around neck. He has diarrhea and also suffered from memory loss. He is continuously taking Jowar and Maize as staple diet.
Mention the diagnosis ?

A. Pellagra – due to Niacin deficiency

Pantothenic acid - B5

coenzyme :- Coenzyme-A

RDA :- 10 mg/day

Sources :- Eggs, liver, milk, Yeast, meat, Plant sources

Functions :- Serves as a carrier of active acyl groups

Coenzyme for :- Pyruvate DH complex
α-KG DH
FAS complex

for synthesis of succinyl CoA, Acyl CoA, Propionyl CoA, Malonyl CoA etc.

Deficiency manifestation :- Burning feet syndrome
pain, numbness and feeling of hotness in the legs.

B6 - PYRIDOXINE

1 Coenzyme - PLP - Pyridoxal Phosphate

RDA 2 mg/day

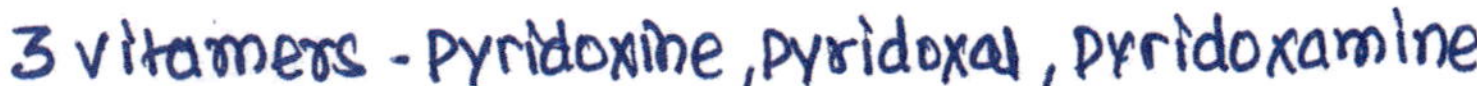

3 vitamers - Pyridoxine, Pyridoxal, Pyridoxamine

4 Drug interactions

5 Deficiency manifestations

6 Functions

Chemistry

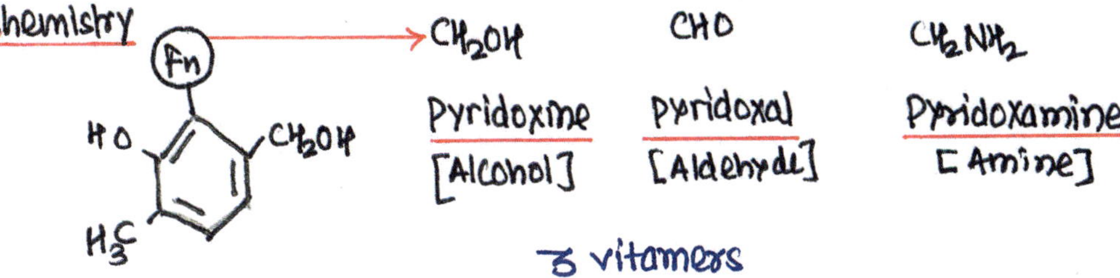

3 vitamers

Sources : Fish, meat, Milk, Wheat, Rice etc.

Biochemical functions
- ① Energy releasing
- ② Hematopoiesis

1. Transamination

Aspartate + α ketoglutarate —AST/SGOT, PLP→ Glutamate + oxaloacetate

Alanine + α Ketoglutarate —PLP, ALT/SGPT→ Glutamate + pyruvate

2. Decarboxylation

5-OH-Tryptophan —PLP (CO_2)→ Serotonin

DOPA —PLP (CO_2)→ Dopamine

Glutamate —PLP (CO_2)→ GABA (γ-Amino butyrate)

3. Transsulfuration

Homocysteine + serine —PLP→ Cysteine + Homoserine

4. Heme synthesis : 1st step in Heme synthesis

Succinyl CoA + Glycine —ALA Synthase, PLP→ δ ALA

5. Tryptophan metabolism

3-OH-Kynurenine —B6 deficiency→ Xanthurenate excretion in Urine.

↓ PLP | Kynureninase

3-OH-Anthranilate

↓

↓

NAD+

6. Glycogen phosphorylase

80% of PLP activity in muscle

Glycogen

↓ PLP | Glycogen Phosphorylase

Glucose

Deficiency manifestations

1. peripheral neuropathy
2. convulsions
3. pellagra like symptoms
4. Excretion of xanthurenate in urine
5. sideroblastic anemia [Hypochromic microcytic]

Drugs causing B6 deficiency

1. INH
2. Penicillamine
3. cycloserine
4. L-DOPA

B6 is used in treatment of

Homocysteinemia, Anemia, Retinal atrophy } In therapeutic doses

Premenstrual syndrome, Schizophrenia, carpel tunnel syndrome } In High doses

B6 + Doxylamine → t/t of Morning sickness

Clinical case

A patient was on Isoniazide therapy for tuberculosis develops neuropathy, convulsions and rash around neck & face. xanthurenate excretion was detected in urine sample. He also develops microcytic anemia. Mention the deficient vitamin?

Ans → Vitamin B6 deficiency (PLP)

B7 - BIOTIN , Vitamin H / Coenzyme R

Anti-eggwhite Injury factor

Chemistry

Active form: Biotin or carboxybiotin

Sources: Egg yolk
Meat, fish, Milk
Tomatoes, Grains
Also synthesized by intestinal flora

RDA: 30 to 100 µg/day

Biochemical function - Carboxylation reactions.

① Pyruvate $\xrightarrow[\text{ATP, Biotin, } CO_2]{\text{Pyruvate carboxylase}}$ Oxaloacetate [Gluconeogenesis]

② Acetyl CoA $\xrightarrow[\text{ATP, Biotin, } CO_2]{\text{Acetyl CoA Carboxylase}}$ Malonyl CoA [FA synthesis]

③ Propionyl CoA $\xrightarrow[\text{ATP, Biotin, } CO_2]{\text{Propionyl CoA Carboxylase}}$ Methyl Malonyl CoA

Viva **Biotin independent carboxylation Reactions**

① Carbamoyl Ⓟ synthetase 1 & 2
② AIR carboxylase — purine nucleotide synthesis
③ Malic enzyme
④ γ- carboxylation (Vit K)

Deficiency manifestations

- Rare
- due to prolonged Antibiotic therapy

① Neurological manifestations — Hallucination, Depression

② Skin Rash

③ Hair loss

Antidote

Avidin : More affinity towards Biotin than Ag-Ab complex

Streptavidin - A form of Avidin

Used in - ELISA
DNA labelling
DNA probes

Clinical case

An 8-Year-old child was brought to the pediatric clinic with dermatitis and Alopecia (Hair loss). His mother gave him raw eggs daily to increase the weight gain. Mention the probable deficiency.

A Biotin deficiency

- Avidin present in the egg white tightly binds with biotin. hence biotin will not be available for intestinal absorption. causing biotin deficiency.

B9 - FOLIC ACID

Water soluble, Hematopoietic vitamin

Chemistry

Pteridine + PABA — Glutamate

Pteroic acid

Folic Acid [Folium - leaf of green vegetable]
Imp. source

Active form

Folic Acid
↓ Dihydrofolate Reductase
Dihydrofolate
↓ Dihydrofolate Reductase
Tetrahydrofolate
[Active coenzyme]

Biochemical Functions

One carbon metabolism

The organic compounds which contain one transferable carbon are called one-carbon compounds

- $-CH_3$ Methyl
- $-CH_2-$ Methylene
- $=CH-$ Methenyl
- $-CHO$ Formyl
- $-CH_2OH$ Hydroxymethyl
- $-CH=NH$ Formimino

① Homocysteine —Methyl THF→ Methionine

② dUMP —Methylene THF (Thymidylate synthase)→ dTMP

③ Glycinamide Ribosyl 5(P) —Methenyl THF→ Formylglycinamide Ribosyl 5(P) [C-8 Purine]

Aminoimidazole Carboxamide Ribosyl 5(P) —Formyl THF→ Formamino imidazole carboxamide Ribosyl 5(P) [C-2 Purine]

④ Glycine —Methylene THF→ Serine

⑤ Histidine → FIGLU → L-Glutamate

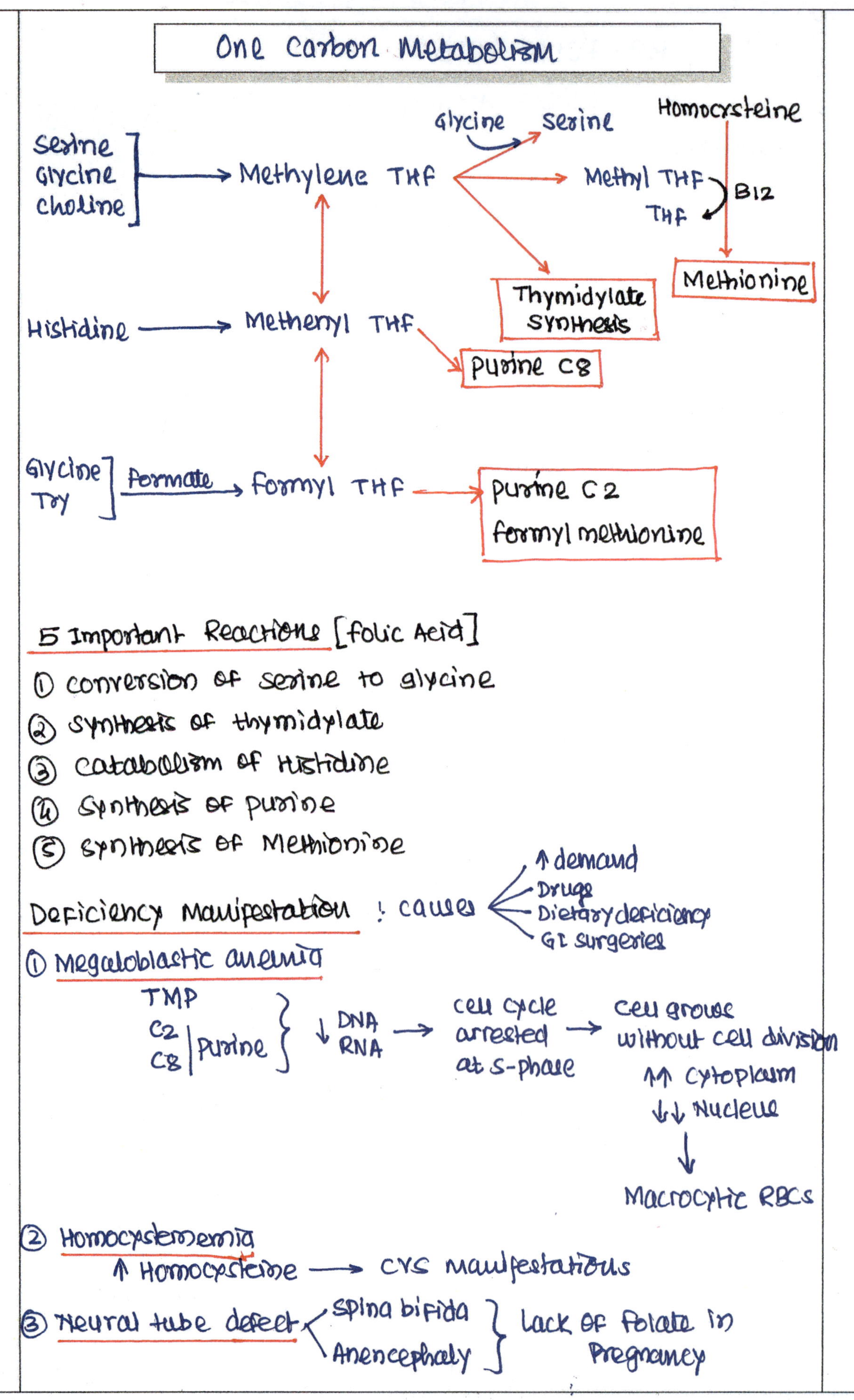
One carbon Metabolism
Serine
Glycine
Choline
Methylene THF
Glycine
Serine
Homocysteine
Methyl THF
B12
THF
Methionine
Thymidylate synthesis
Histidine
Methenyl THF
purine C8
Glycine
Try
Formate
formyl THF
purine C2
formyl methionine
5 Important Reactions [folic Acid]
① conversion of serine to glycine
② synthesis of thymidylate
③ catabolism of Histidine
④ Synthesis of purine
⑤ synthesis of Methionine
Deficiency Manifestation
: causes
↑ demand
Drugs
Dietary deficiency
GI surgeries
① Megaloblastic anemia
TMP
C2
C8
Purine
↓ DNA
RNA
cell cycle arrested at S-phase
cell grows without cell division
↑↑ cytoplasm
↓↓ Nucleus
Macrocytic RBCs
② Homocysteinemia
↑ Homocysteine
CVS manifestations
③ Neural tube defect
Spina bifida
Anencephaly
lack of Folate in Pregnancy

RDA: Adult = 200 µg/day
Pregnancy = 400 µg/day

Dietary sources:- Vegetables [spinach, beans]
Orange, papaya, Beet

Antagonist

① Sulphonamide

Antibacterial
Structurally similar to PABA - Competitive inhibition of folic acid synthesis
↓
Inhibit growth of bacteria

② Aminopterin & Amethopterin [Methotrexate]

Competitive inhibitor of

⊖ Dihydrofolate Reductase → ↓THF → ↓↓ cell division
↓
Death of cancer cells.

Lab Δ Megaloblastic Anemia

Peripheral smear: Macrocytosis, Reticulocytosis
Histidine load test and excretion of FIGLU in urine
Serum folate level and RBC folate level.

Block occurs in folate deficiency

Histidine → formiminoglutamate [FIGLU] → L-Glutamate (THF → N5-formimino THF)

[FIGLU] ↓ Excreted in urine in folic acid deficiency

In pregnancy - folic acid c̄ elemental iron is given
500 mcg 100 mg

B12- COBALAMIN

Co Central cobalt corrin ring
B12
Active - Adenosyl B12
M Methyl B12
Intrinsic factor
Neurological
Extrinsic factor of Castle

M | Methyl Malonyl CoA mutase
M |
S | Methionine synthase
S | Subacute combined degeneration
C |
a
Methyl malonate in urine

Digestion, Absorption and Transport

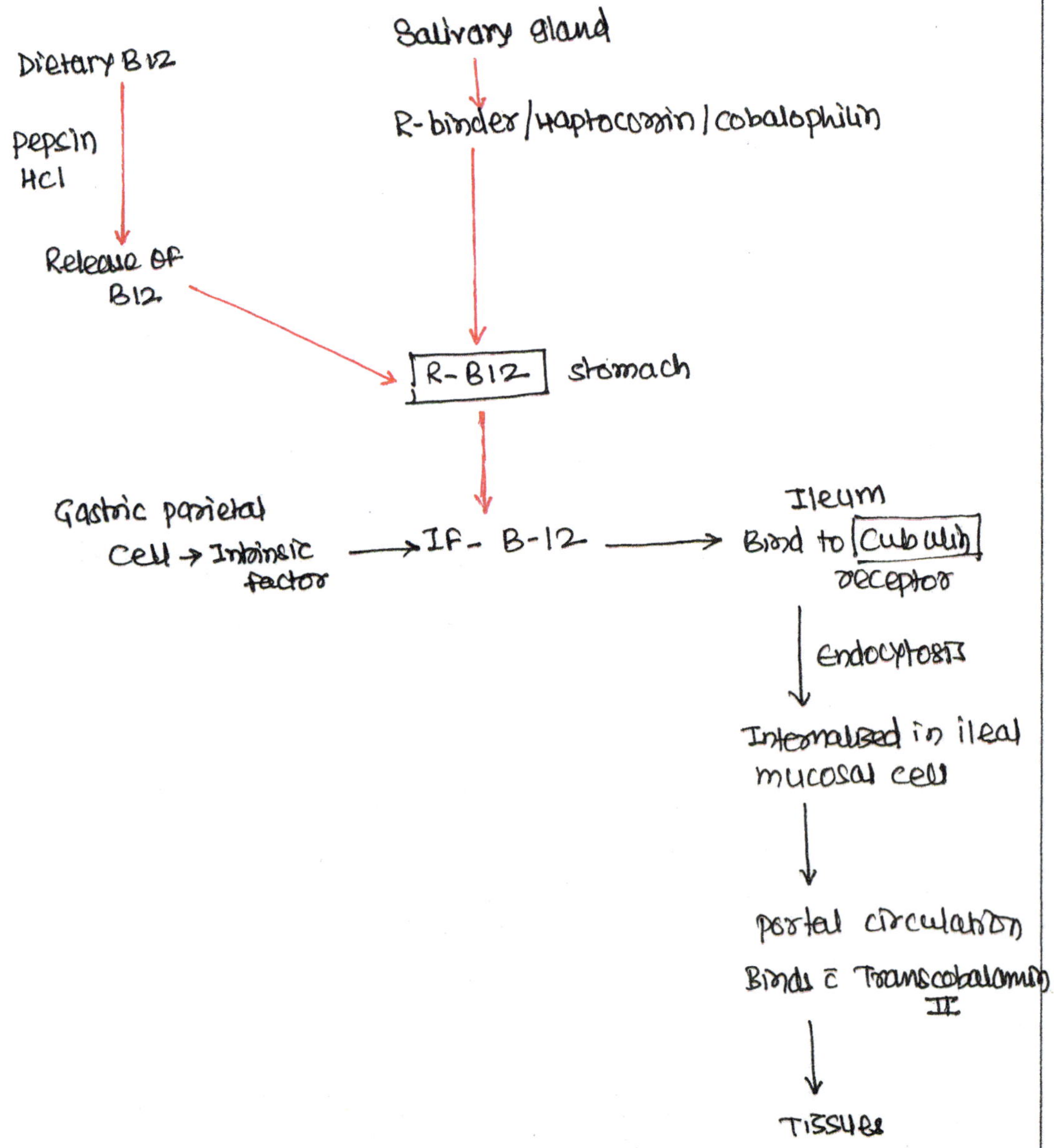

Dietary B12 → (Pepsin, HCl) → Release of B12 → R-B12 (stomach)

Salivary gland → R-binder/Haptocorrin/cobalophilin → R-B12 (stomach)

R-B12 → IF- B-12

Gastric parietal cell → Intrinsic factor → IF- B-12 → Ileum: Bind to Cubulin receptor

↓ Endocytosis

Internalised in ileal mucosal cell

↓

Portal circulation
Binds c̄ Transcobalamin II

↓

Tissues

Sources :- produced by microorganisms in the intestine

present in food of Animal origin (Meat, Fish etc)

RDA - 1 μg

Biochemical functions ① Conversion of Homocysteine to Methionine.

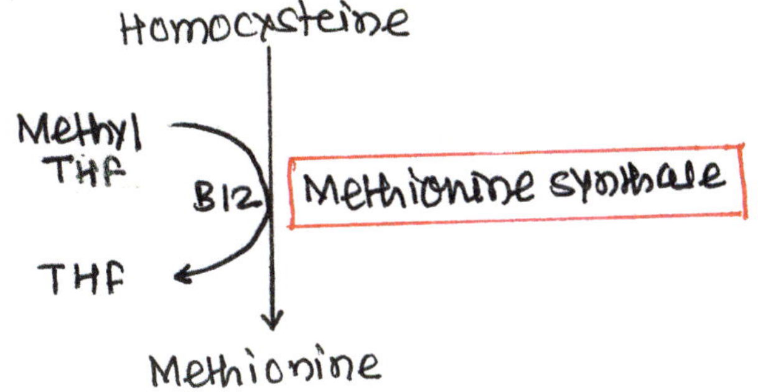

② Methyl malonyl COA Mutase

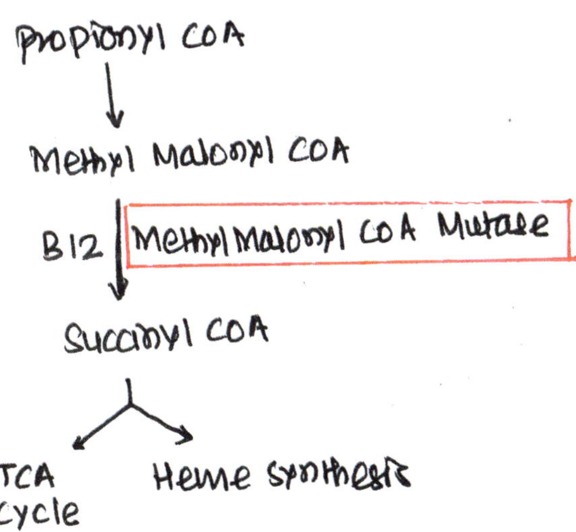

Deficiency Manifestations

Causes ① Nutritional deficiency → ↓ intake, strict vegetarian

② Malabsorption syndrome

③ ↑ demand (Pregnancy)

④ pernicious anemia – Autoantibodies against gastric Parietal cells

↓

Deficiency of Intrinsic factor

↓

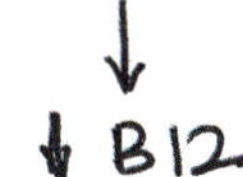

① Megaloblastic Anemia

B12 OR B9 deficiency → ↓ DNA & RNA | Cell cycle arrest in S phase

↓

Cell grows continuously without cell division

Cytoplasm ↑↑

Nucleus ↓

↓

Macrocytic RBCs

② Sub-acute combined degeneration of spinal cord

due to defective myelination

① Methyl malonyl CoA competes with Malonyl CoA

↓

↓ FA synthesis

↓

↓ Myelin synthesis

② Methyl malonyl CoA incorporated into sphingolipids/Glycolipids

↓

Defective Myelin synthesis

↓

Degeneration

③ Excretion of Methyl Malonate in urine

Lab. Diagnosis:-

① Serum Cobalamin
② Serum Methyl malonic acid
③ Serum Homocysteine
④ Megaloblast in peripheral smear
⑤ Bone marrow Aspiration

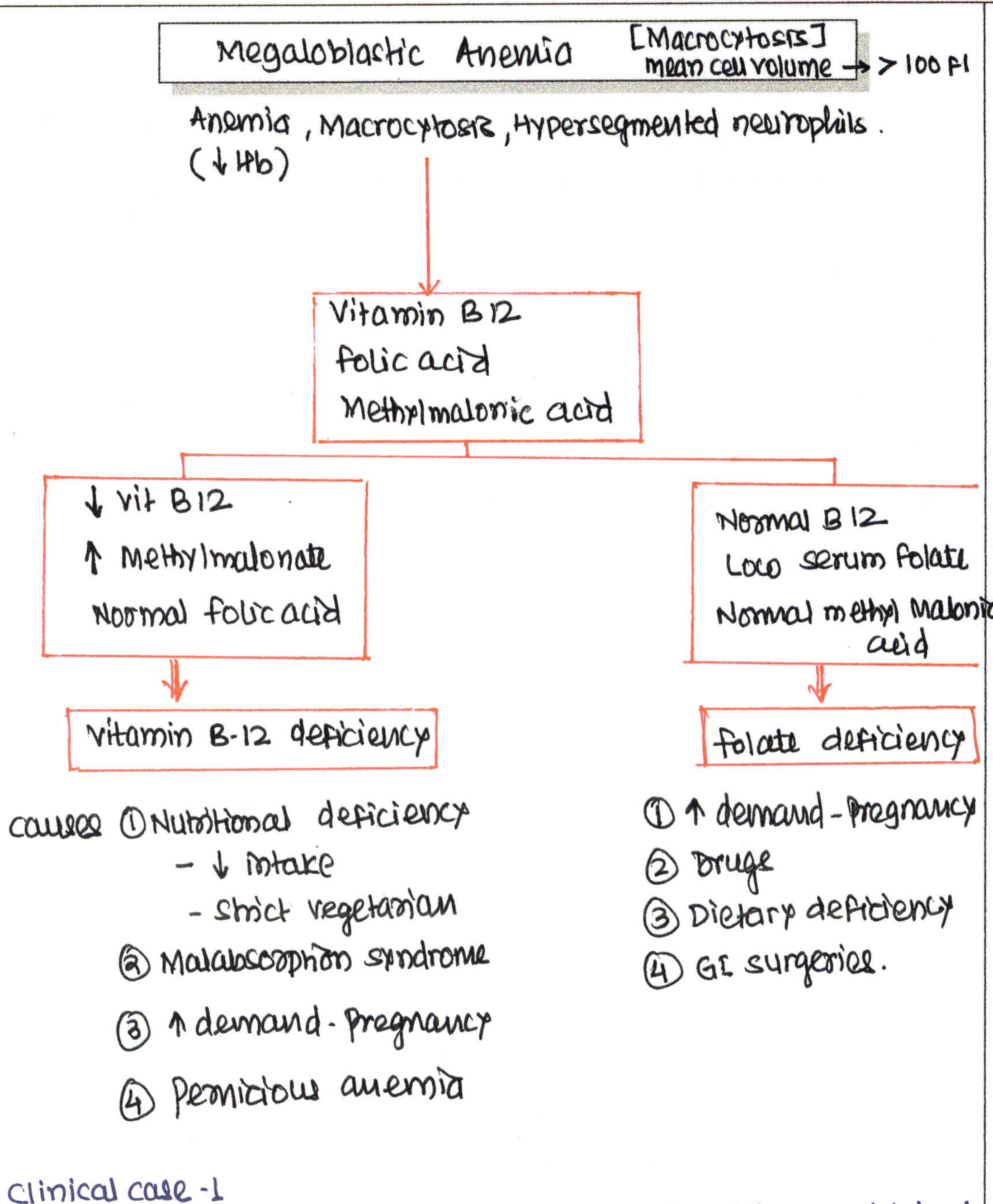

Clinical case - 1

A pregnant women was came to OPD with chief complaints of fatigue, weakness and breathlessness. Laboratory investigation showed Macrocytic RBC's, hypersegmented neutrophils and decreased Hb levels. Histidine load test shows increased excretion of FIGLU in urine. What is your diagnosis & what is risk to fetus?

Δ Megaloblastic anemia due to folate deficiency
Risk of - Neural tube defect [spina bifida & anencephaly] in fetus

Clinical case - 2

A 30 year old patient was brought to the hospital with tingling sensation in the lower limb. His position & vibration sense were reduced. PS-showed macrocytosis, hypersegmented neutrophils and Methylmalonate in urine. What is your probable diagnosis?

Δ Vitamin B-12 deficiency – Megaloblastic anemia c̄ subacute combined degeneration sp. cord.

MINERALS

MCI COMPETENCY BI 6.9-6.10

BI6.9 → Functions, Metabolism & Homeostasis of various minerals.

BI6.10 → Disorders related with Minerals.

MCQs * Chromium - Glucose tolerance factor potentiates action of Insulin

LAQs

1. Write about RDA, sources, functions, deficiency manifestations and homeostasis of Calcium.
2. Write about RDA, sources, absorption, transport, storage, functions and deficiency manifestations of Iron.

SAQ

1. Copper
2. Zinc
3. Phosphorus
4. Na^+ & K^+

Ultratrace elements

Daily requirement < 1 mg/day

Selenium
Iodine
MCQ * Molybdenum → xanthine oxidase

Clinical cases

1. Iron deficiency Anemia → ↓Hb, ↓Ferritin, ↑TIBC, Apathy
2. Tetany - Carpopedal spasm
3. Acrodermatitis enteropathica - skin rash
4. Wilson's disease - Hepatolenticular degeneration
5. Fluorosis - ↑ Fluorine → Dental & skeletal fluorosis
6. Iodine deficiency goitre → Enlargement of thyroid gland, swelling in the neck

Major Elements

Requirement is >100 mg/day

1. Calcium
2. Phosphorus
3. Na^+
4. K^+
5. Cl^-
6. Magnesium

Trace elements

Daily requirement is < 100 mg

1. Iron
2. Zinc
3. Copper
4. Iodine
5. Fluorine
6. Selenium
7. Molybdenum

CALCIUM

Important LAQ:- Write about RDA, Sources, functions, deficiency manifestations and homeostasis of calcium

Total body content: 1 to 1.5 kg
- 99% Bones & teeth
- 1% ECF

Sources: Milk and Milk products
Meat, Eggs, Fish
vegetables
Cereals and Pulses

RDA: Adult 800 mg
Children 1500 mg
Pregnancy, Lactation, After menopause 1500 mg

Functions:- Mnemonic -[BCCI MENS Have Heart]

1. Bones and teeth — formation of hydroxyapatite
   Mineralisation of Bone
2. Muscle contraction - Increases Actin & Myosin binding
   Excitation contraction coupling
3. Coagulation - factor IV (calcium) + Prothrombin
   ↓
   Thrombin
4. Integrity of cell membrane and permeability
5. Second messenger - IP_3-DAG Pathway
6. Enzyme activation - ATPase, Adenylate cyclase, Pyruvate kinase, PDH, Pyruvate carboxylase, Gly. synthase etc.
7. Nerve impulse transmission
8. Secretory processes - Endocytosis, Exocytosis
9. Release of Hormones - Insulin, vasopressin, PTH
10. Heart - prolongs systole

Calcium Homeostasis

Normal level Sr. Calcium: 9 to 11 mg/dl
- 50% Ionised
- 40% Protein bound
- 10% c̄ Bicarbonate & phosphate

Regulation of Sr. calcium

Vitamin D	PTH	Calcitonin
↑ absorption of calcium from intestine ↑ Bone mineralization ↑ Reabsorption of Ca^{++} & (P) from renal tubules.	Demineralization ↑ Absorption of calcium through Vit-D ↑ Reabsorption of Ca^{++} from renal tubules	Mineralization of bones ↓ Reabsorption of calcium from renal tubule ↓ Intestinal absorption

Absorption :- Against concentration gradient - from 1st & 2nd part of duodenum

factors ↑ Absorption of calcium	factor ↓ Absorption of calcium
Vit D, PTH, Lysine, Arginine Low pH, Lactose	Phytates, Oxalates, Phosphate High pH, Dietary fibres, free FA

Clinical Significance

Hypocalcemia ↓ 9 - 11 mg/dl **Hypercalcemia**

Hypocalcemia

symptoms - convulsion
Arrythmia
Tetany
spasm
stridor

causes - Rickets
Osteomalacia
Osteoporosis
↓ PTH
Medullary carcinoma of thyroid

Hypercalcemia

Pain in abdomen & bones
Nausea, weakness
confusion, Fatigue

causes - PTH Tumor
Bone tumor or Metastasis
Multiple myeloma
Pagets ds of bone

T/t of Tetany
Inj. calcium gluconate 10%, 10 ml IV followed by maintenance drip

Tetany :- When Sr. calcium < 7 mg/dl
Carpopedal spasm - flexion of wrist & MCP joint
Extension of interphalangeal joint
Chvostek's sign - facial twitching on tapping of facial nerve
Trousseau sign - Carpopedal spasm on application of BP cuff

Phosphorus

LAQ/SAQ

Que: Write about RDA, sources, Biochemical functions and deficiency manifestations of phosphorus.

Total body content - 1 kg
- 80% in bones and teeth
- 10% Muscles and RBCs
- 10% with other compounds

RDA: 800 mg/day

Sources: Milk, meat, Eggs, cereals, vegetables etc

Biochemical functions

1. formation of bones and teeths
2. synthesis of nucleoproteins - DNA & RNA
   formation of 3'-5' phosphodiester linkage
   component of nucleotides.
3. phospholipid and phosphoprotein synthesis
4. Regulation of enzyme activity by phosphorylation
5. Synthesis of high energy compounds - ATP, Creatine Ⓟ
6. Major intracellular buffer, important in acid-base balance
7. synthesis of various coenzymes - PLP, TPP, NADP etc.

Serum level: 3 to 4 mg/dl
Blood level 30 to 40 mg/dl

Deficiency manifestations (Rare)

↓ Ⓝ Serum level 3 to 4 mg/dl ↑

Hypophosphatemia
- Vitamin D deficiency
- Hyperparathyroidism
- Malnutrition
- Malabsorption

Hyperphosphatemia
- Vitamin D toxicity
- Hypoparathyroidism
- Cancer Chemotherapy
- Secondaries in bone

Magnesium

Que: Enumerate important functions of magnesium in our body.

RDA :- 350 mg/day Males
300 mg/day females

sources :- vegetables, milk, meat, cereals

functions :-

1. Bone and teeth formation
2. Activation of enzymes - Glucokinase
   Hexokinase
   Phosphofructokinase, Aminoacyl tRNA synthetase
   Adenylate cyclase, RNA Polymerase
3. Neuromuscular function.

Deficiency manifestations - Neuromuscular irritability
convulsions
spasm

Sulfur

Que: Enumerate important Biochemical functions of Sulfur

Functions

1. Important constituent of sulfur containing vitamins
   cysteine, cystine and Methionine

   Important in protein synthesis and orientation of aminoacids in protein structure (disulfide bond)
2. constituent of vitamins - Thiamine, Biotin, Lipoic acid & co A
3. Important for synthesis of - Glutathione
   Heparin, CS, DS, HS (GAGs)
   Taurocholate
   PAPS
4. Help in detoxification reactions.
5. Important in transmethylation reactions.

Sodium [Na^+]

Que: Write about RDA, sources, functions and disorders of Na^+.

Que: What is normal serum Na^+ level & Add a note on Hyponatremia.

Que: Enumerate causes of Hyponatremia + Hypernatremia.

RDA: 5 gm/day

For Hypertensive family < 1 gm/day

Sources: Table salt, cereals, legumes, Eggs, Milk, Meat etc.

Functions:-

① Maintenance of plasma volume

② Acid-base balance

③ Na^+-K^+ pump - for exchange of $3Na^+$ and $2K^+$

④ Absorption of Glucose, Galactose and AAs from intestine

⑤ Required for Nerve muscle conduction

Distribution

Total body content is 4000 mEq

75% Exchangeable form

ECF - Plasma : 135-145 mEq/L

ICF :- 4-10 mEq/L

25% Non exchangeable form : In tissues

Regulation:- ① Aldosterone ② ANF

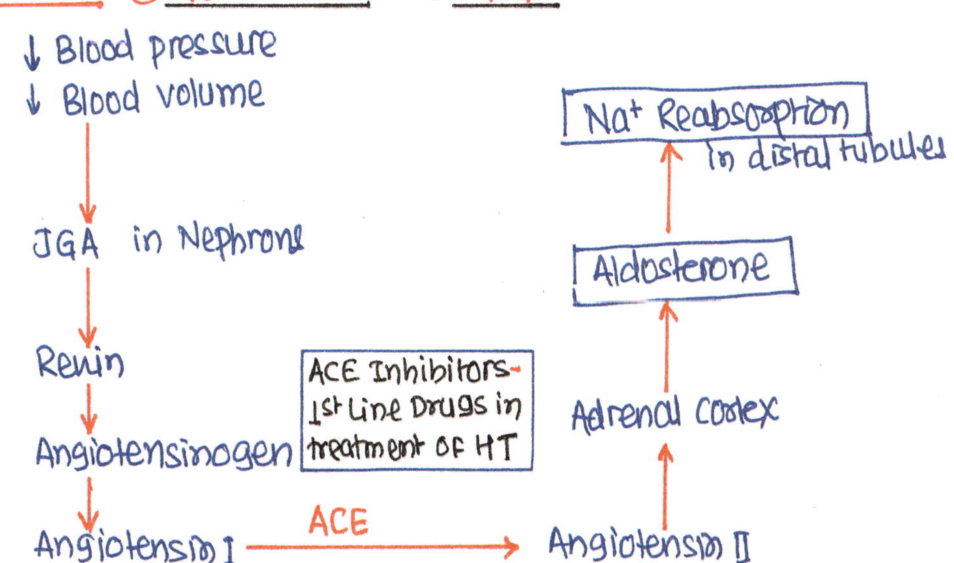

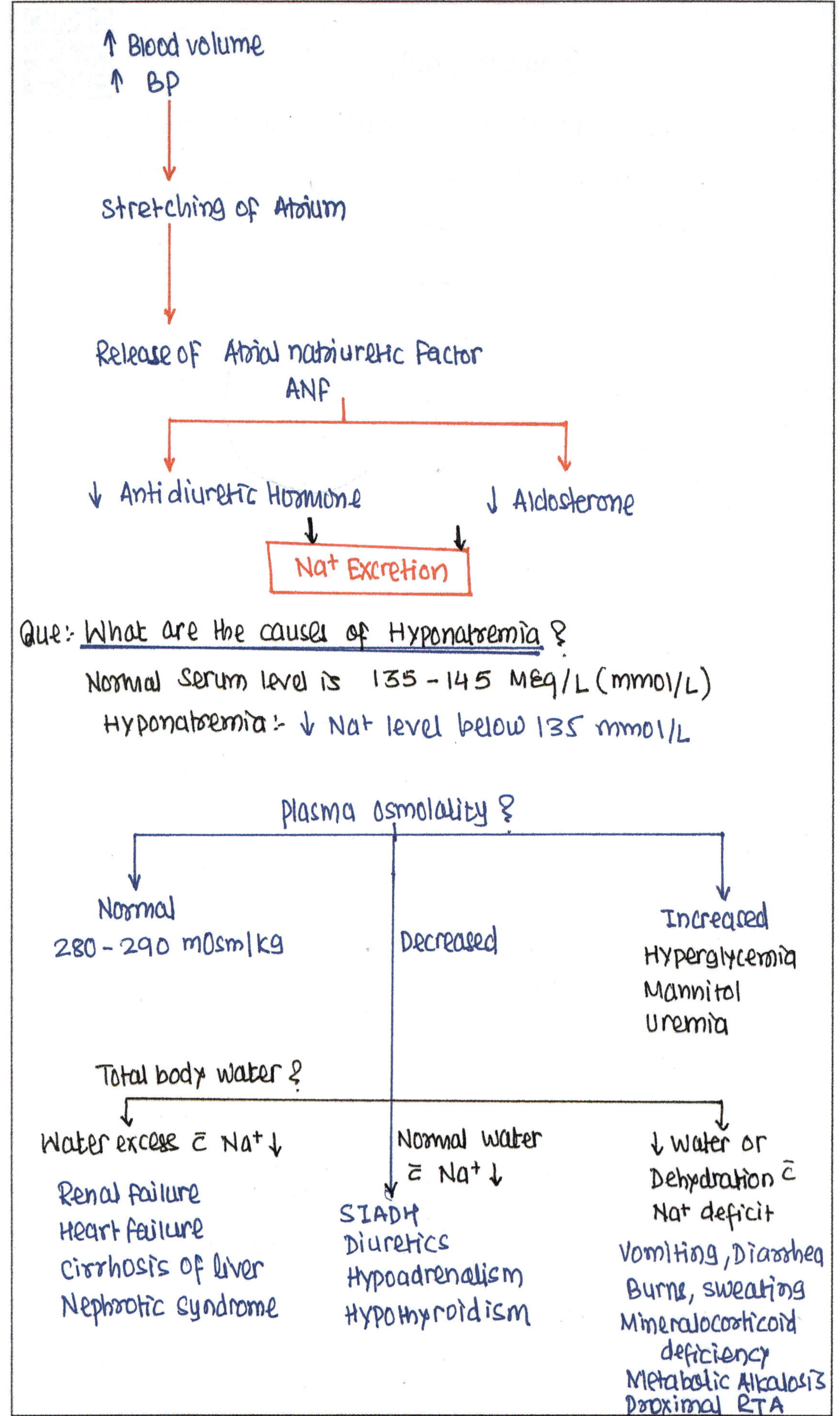
↑ Blood volume
↑ BP
Stretching of Atrium
Release of Atrial natriuretic Factor
ANF
↓ Antidiuretic Hormone
↓ Aldosterone
Na+ Excretion
Que:- What are the causes of Hyponatremia?
Normal Serum level is 135 - 145 MEq/L (mmol/L)
Hyponatremia:- ↓ Na+ level below 135 mmol/L
Plasma osmolality?
Normal
280 - 290 mOsm/kg
Decreased
Increased
Hyperglycemia
Mannitol
Uremia
Total body water?
Water excess c̄ Na+ ↓
Renal failure
Heart failure
Cirrhosis of liver
Nephrotic Syndrome
Normal water
c̄ Na+ ↓
SIADH
Diuretics
Hypoadrenalism
Hypothyroidism
↓ water or
Dehydration c̄
Na+ deficit
Vomiting, Diarrhea
Burns, sweating
Mineralocorticoid
deficiency
Metabolic Alkalosis
Proximal RTA

Que: What are the causes of Hypernatremia?

↑ serum Na^+ more than 150 mmol/L

Total body water?

Water excess $\bar{c}$ Large Na^+ excess	Normal water Only Na^+ excess	Less water Dehydration $\bar{c}$ Na^+ excess
Hyperaldosteronism Cushing syndrome Excess IV Fluids	Diabetes insipidus	↓ Water intake $\bar{c}$ diuretics diarrhea sweating

Potassium

Que:- Write about RDA, sources, functions and clinical disorders related with potassium.

RDA :- 3 to 4 gm/day

Sources :- Bananas, orange, Apple, Pineapple, tender coconut vegetables, Legumes, nuts, Beans etc.

Functions

1. It is a major intracellular cation and regulates intracellular volume
2. Required for generation of resting membrane potential, depolarisation and contraction of Heart
3. Nerve transmission
4. component of Na^+-K^+ pump
5. secretion of HCl in gastric juice via H^+-K^+ exchanger
6. Activates enzymes - Pyruvate kinase

Regulation :-

① Aldosterone and corticosteroids : ↑ excretion of K^+

② K^+ Redistribution : Insulin : ↑ cellular uptake
H^+ concentration :- Acidosis & Alkalosis

Normal level of serum K^+ level :- 3.5 to 5.0 mmol/L
Intracellular concentration :- 160 mmol/L

Hypokalemia

serum level < 3.5 mmol/L

K^+ Redistribution	True K^+ deficit
Insulin therapy	Renal tubular acidosis (RTA)
Alkalosis	Acute tubular necrosis
↑ WBC	Diarrhea
Hypothermia	Vomiting
Periodic paralysis	Fistula
	Excessive sweating
	Cushing syndrome
	Mineralocorticoid excess

Hyperkalemia

Serum $K^+ > 5.5$ mmol/L

Pseudohyperkalemia
- Hemolysis
- Thrombocytosis
- Leukocytosis

K^+- Redistribution
- Metabolic acidosis
- Insulin deficiency
- Dehydration
- Massive tissue hypoxia

K^+ Retention
- Hemolysis
- Tissue necrosis
- Massive blood transfusion
- Addison's disease
- Obstructive nephropathy
- Sickle cell disease
- SLE - systemic lupus erythematosus
- K^+- sparing diuretics
- Renal transplant

IRON

Important LAQ :- Write about RDA, sources, deficiency manifestations, biochemical functions of Iron

SAQ :- Digestion, absorption & transport of Iron

Trace element

Total body content : 3 to 5 gm (75% → Hb)

Proteins containing Iron – Hemoglobin
Myoglobin
Cytochromes
Iron-sulfur proteins
Cyt P450
Enzymes – Catalase
Xanthine oxidase
Tryptophan pyrrolase
NOS
Aconitase

Functions

1. Transport of O_2 to tissues and CO_2 to lungs
2. Important constituent of Myoglobin in muscle
3. Electron transport chain (Cytochrome & Fe-S proteins)
4. Detoxification (Cyt P450)
5. Antioxidant
6. Purine degradation (xanthine oxidase)
7. Tyrosine metabolism

RDA :- Adult male – 10 mg
Female – 20 mg
Pregnancy and lactation – 40 mg

Sources :- Green leafy vegetables (20 mg/100 gm)
Pulses (10 mg/100 gm)
Cereals (5 mg/100 gm)
Liver (5 mg/100 gm)
Meat (2 mg /100 gm)
Jaggery, Dry fruits, Beetroots etc

Absorption, Transport and Storage

Factors ↑↑ Iron Absorption

Vitamin C, Gastric HCl
Small peptides, Cysteine

Factors ↓↓ Iron Absorption

Phytate, Oxalate, Phosphate
Lead

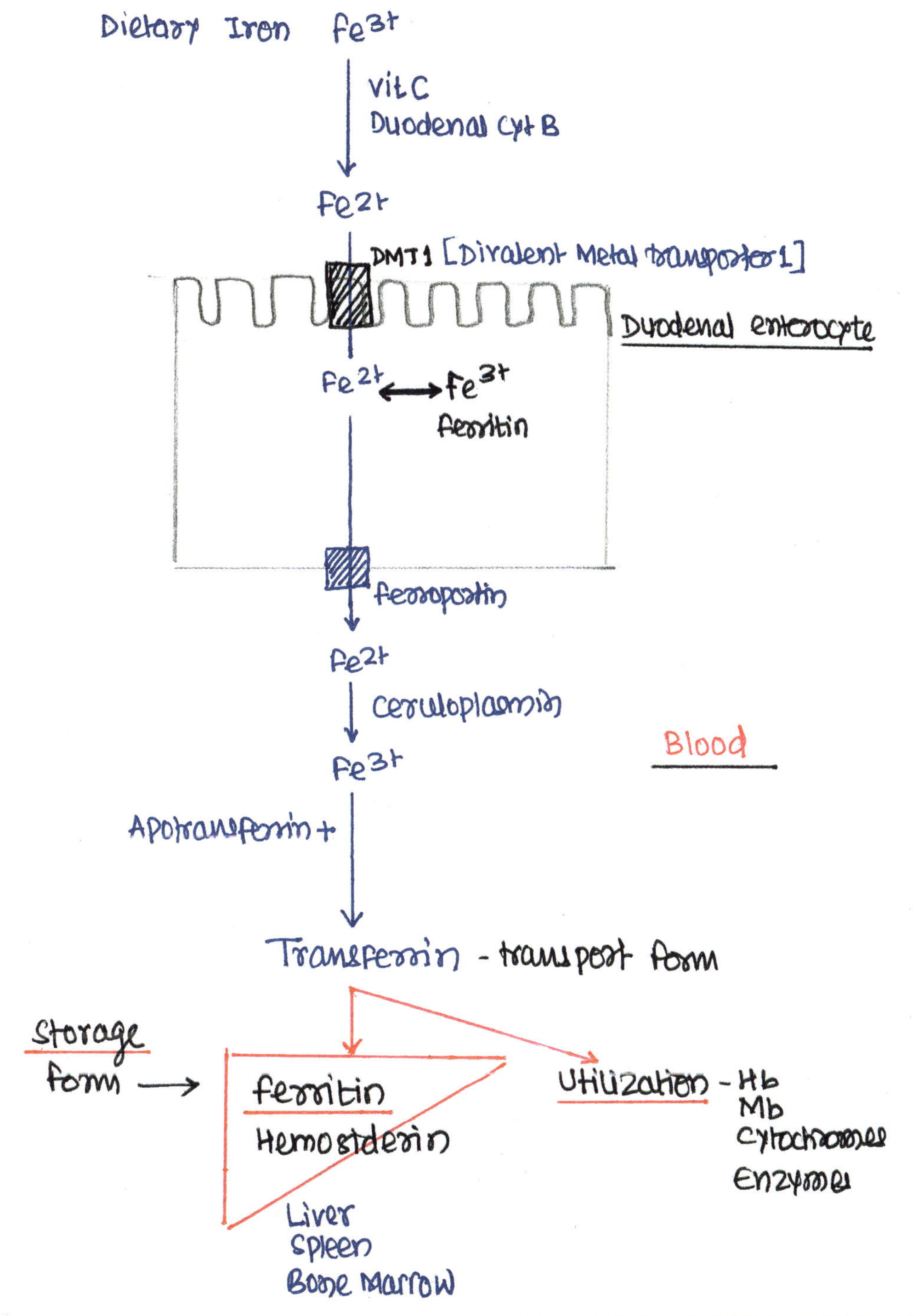

Hepcidin - Regulate Iron level → ↓ intestinal absorption

blocks the ferroportin

peptide produced by liver, causes internalization and degradation of cellular Iron exporter [DMT-1 Ferroportin]

Iron is the one way element

- Utilized and reutilized again & again
- < 1 mg/day excreted in → Bile, sweat, Hair loss
- At mucosal level - Regulation of Iron level known as Mucosal block theory
- Almost No Iron excreted through Urine. feces contains unabsorbed & trapped Iron in intestinal cell that are desquamated

Clinical Significance - Iron Deficiency Anemia

Causes :- ↑ demand
↓ Intake
Chronic blood loss
Lead
Repeated pregnancies

Symptoms ① Anemia - Microcytic hypochromic anemia
RBC have less Hb than Normal
Small sized RBC's
↓ MCV, ↓ MCH

② Apathy - Sluggish dull child
decreased School Performance
Fatigue, weakness

Lab diagnosis :- ↓ Hb Normal range 14-16 gm/dl in Males
13-15 gm/dl in Females

Increased Total Iron binding capacity [TIBC]

Sr. Iron ↓, Sr. ferritin ↓

↓ MCV, ↓ MCH

Treatment

- Oral Iron therapy 100-200 mg/day
- Increase consumption of dietary Iron
- Treatment of underlying cause.
- IV Iron in severe cases

Hemosiderosis

- Excess iron due to ↑ absorption.
- Iron overload disorder due to accumulation of hemosiderin
- The pigment does not damage the parenchymal cells
- Found in Bantu tribe of South Africa, also called Bantu siderosis.

Hemochromatosis

· Excess Iron deposited in skin, Bone marrow, liver, spleen pancreas

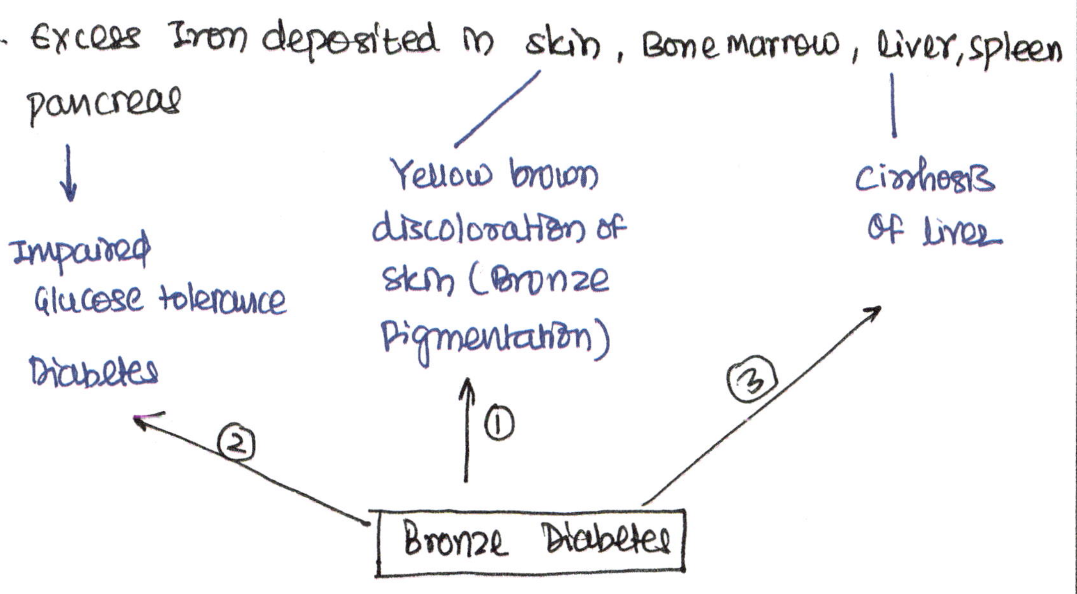

ZINC

SAQ :- Write about RDA, sources, functions and deficiency manifestations of zinc.

Trace element

Total body content - 2 gram < 60% skeletal muscle, 30% Bone

Rich areas - Prostate
Hippocampus area of Brain

Sources - Meat, Milk, Eggs, shell fish, grains, nuts, beans etc.

RDA - Adult - 10 mg
Pregnancy & lactation 10 - 20 mg

Biochemical functions

1. Important constituent of > 300 enzymes
   - Carbonic anhydrase
   - Superoxide dismutase
   - RNA polymerase
   - Carboxy peptidase
   - Alcohol DH, ALA dehydratase
   - ALP etc.
2. Antioxidant - Superoxide dismutase - Scavenges free radicals
3. Gustin - zinc metalloprotein - taste sensation
4. Wound healing
5. Storage and release of Insulin
6. Important in maintenance of vit A level in serum
7. Important role in reproduction.

Deficiency Manifestations

1. Acrodermatitis enteropathica - Inflammation and dermatitis involving acral parts of body - finger, nose, mouth, toes etc.
2. Poor wound healing
3. Alopecia
4. Impaired spermatogenesis
5. Loss of Taste acuity
6. Neurological symptoms.

COPPER

Trace element
Total body content - 100 mg
Widely distributed in various organs

Sources - Organ meat
Green leafy vegetables
Fruits, Nuts
Shellfish etc

RDA - Adult → 2 - 3 mg/day
Infant/children → 0.5 – 2 mg/day

Biochemical functions

1. Cofactor for enzymes – Cytochrome oxidase
   Catalase
   Ceruloplasmin
   ALA synthase
   Superoxide dismutase
   Tyrosinase
   Lysyl oxidase
2. Electron transport chain – Cytochrome oxidase
3. Incorporation of Iron in transferrin
   $Fe^{2+} \longrightarrow Fe^{3+} \longrightarrow$ Transferrin
4. Heme synthesis (ALA synthase)
5. Antioxidant (superoxide dismutase)
6. Melanin synthesis (Tyrosinase)
7. formation of vessel walls (lysyl oxidase) MCQ

Diseases related with copper

1. Wilson's disease
2. Anemia – Normochromic microcytic
3. CVS abnormalities – Aneurysm or Rupture of Aorta
4. Menke's disease – X-linked, Affect males, Kinky hair
5. Hypopigmentation – defect in Tyrosinase

Wilson's disease

Hepatolenticular degeneration

Mutation - Copper binding ATPase (ATP 7B)

Liver

↓ ATP 7B (blocked)

Bile

The copper binding ATPase is required for excretion of copper from liver to bile. Block in excretion causes copper accumulation in liver and Brain.

C/F :- Accumulation of copper in liver

Damages lenticular nucleus of brain

MCQ Kayser Fleischer ring - circular golden ring in the Descemets membrane of cornea

Blood ceruloplasmin level - decreases (Normal 18-35 mg/dl)

Symptoms - Jaundice, fatigue, muscle stiffness, tremors and behavioural changes.

Gold standard test :- Liver biopsy & quantitative copper estimation

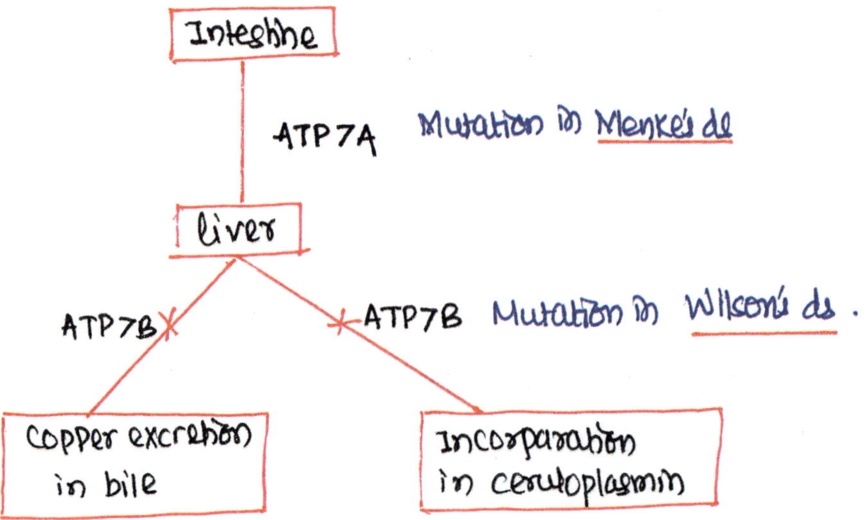

Treatment :- Penicillamine causes chelation & excretion of Copper.

Iodine

Que: Write about RDA, sources, functions and deficiency manifestations of Iodine.

RDA :- 100-150 µg/day in Adult, 200 µg/day in pregnancy

Sources :- Drinking water, seafood, Iodized salt vegetables & fruits grown on seaboard
High Altitudes are deficient in Iodine

Functions :-

Iodine is required for the synthesis of thyroid Hormones T_3 & T_4

Functions of T_3, T_4 :- ↑ Metabolic rate
↑ Protein synthesis
↑ Blood glucose level
↑ Lipid utilization
Maintenance of water & electrolyte balance.

Deficiency manifestation :- Simple endemic goitre OR Iodine deficiency goitre

Clinical case

A 25-year-old female patient came to OPD with chief complaints of swelling in neck, weight gain, Hair loss and feeling colder than usual. She also gave H/o irregular menstrual cycle and feeling tired, sluggish and weak.

What is your probable diagnosis? Iodine deficiency goitre

Swelling in Neck

Cells of thyroid gland grows and multiply to trap Iodine

Feeling colder

Iodine helps to generate body heat. Deficiency feels colder than usual.

Weight gain

low iodine and Thyroid Ⓗ
- slows metabolism & encourage food to be stored as fat rather than be burned as energy

Hair loss

Iodine helps in regeneration of skin & Hair follicles. Also help to sweat & hydrates skin cells

CRETINISM

Severe deficiency in pregnancy causes congenital iodine deficiency syndrome - Cretinism

Child is having — Mental retardation
Short stature
Thick skin
Early puberty

* Used for collection of blood sample for Glucose estimation
→ NaF : K.oxalate vacutainer

FLUORIDE

SAQ
Clinical case

Que:- Write about RDA, sources, functions and disorder of fluoride.

RDA : < 2 ppm

Source :- Drinking water

Functions :-

Fluoride forms protective layer of acid resistant fluoroapatite with hydroxyapatite of the enamel and prevent the tooth decay by bacterial acids.

Fluoride inhibits bacterial enzyme and reduces production of acids.

It is necessary for development of bones

Disorders

Dental caries :- prolonged deficiency of fluoride < 0.5 ppm during pregnancy and early childhood causes dental caries.

Toxicity :- ↑ fluoride

> 2 ppm :- Mild gastrointestinal symptoms

> 5 ppm :- Dental Fluorosis
Mottling and discoloration of teeths.

> 20 ppm :- Skeletal Fluorosis
Affect Bones
Genu valgum deformity of legs
Misalignment of knees in inward direction

Clinical case-1

A 43-year-old female was brought to the emergency department with chief complaints of carpopedal spasm in the left hand. She had H/o convulsion and stridor in the recent past. Her serum calcium level was 5 mg/dl and phosphorus level was 3 mg/dl. What is your probable diagnosis?

Δ Tetany (Hypocalcemia)

Carpopedal spasm - flexion of wrist joint and metacarpophalangeal joint
extension of Interphalangeal joint
when serum calcium <7 mg/dl

Calcium plays important role in stabilizing the RMP of neurons & preventing their spontaneous activation.

↓ calcium → ↑ Neuromuscular excitability, As a result neurons become unstable and fire spontaneous action potential

T/t:- Inj calcium gluconate 10%, 10 ml IV followed by maintenance drip

Clinical case-2

A 26-year-old pregnant female patient came to OPD for a routine check-up. She reports 6 months of progressive fatigue. Her investigations are as follows. Hb- 7 gm/dl

Mean corpuscular volume (MCV) <80 FL (Normal 80-100 FL)
RBC distribution width (RDW) - ↑ (indicates variability in RBCs size and low Iron status)
Total Iron binding capacity (TIBC) → 480 [250-450 μg/dl]
Serum ferritin levels → Low

What is your probable diagnosis?

Δ Iron deficiency Anemia

Clinical case-3

An infant was brought to the pediatric clinic with chief complaints of rash in perineal region, around mouth, in the toes, fingers. He also suffers from loose motion. Which mineral is deficient in this child?

Δ Acrodermatitis enteropathica - due to Zinc deficiency
(Dermatitis involving acral parts - finger, nose, toes, perineal area)

Clinical case-4

A 42-year-old man suffers from poor wound healing, rash and hair loss (alopecia). Rash typically involves the mouth, nose and fingers. He also complaints of loss of taste sensation and decrease sperm count. What is your probable diagnosis?

Δ Zinc deficiency manifestations

Zn↓

Diarrhea
Alopecia
Dermatitis

Clinical case-5

An 11-year-old boy was brought to the hospital with chief complaints of jaundice, fatigue, tremors and behavioral changes. Examination revealed hepatosplenomegaly, Kayser-Fleischer ring was noted around the cornea. What is your probable diagnosis?

Δ Wilsons disease - due to mutation in ATP 7B
block in the excretion of copper from liver to bile acid and ↓ incorporation of copper in ceruloplasmin.

Clinical case-6

An 8-year-old mentally retarded child was brought to the clinic with kinky, brittle and silvery hairs. What is your probable diagnosis?

Δ Menkes disease - due to mutation in ATP 7A
copper absorption from intestine decreases
X-linked recessive - Affect mostly Males
Hairs - sparse, kinky, brittle, colorless or silvery wire type
Neurological symptoms.

Clinical case-7

A child was brought to the clinic with dilated cardiomyopathy. What could be the probable nutrient deficiency that causes cardiomyopathy.

Δ Selenium deficiency - Keshan disease

Selenium dependent enzymes:
1. Glutathione peroxidase
2. Thioredoxin Reductase
3. Deiodinase

* MCQ

NUTRITION

MCI COMPETENCY BI 8.1-8.5

SAQs

1. Protein energy malnutrition [PEM]
2. BMR and factors affecting BMR
3. SDA - Specific dynamic action
4. Caloric value of food
5. Respiratory Quotient
6. Nitrogen balance
7. Balanced diet
8. Diet management in pregnancy
   Diabetes Mellitus
   Coronary artery diseases
9. Glycemic index
10. Dietary fibres
11. Obesity
12. Role of leptin (*) Imp

Clinical cases

1. Kwashiorkor :- Edema
   Moon face
   Distended abdomen
   Poor appetite
   Dermatitis
   Sparse hair, Flag sign

2. Marasmus :- No edema
   Old man face
   Dry atrophic skin
   Severe muscle wasting
   Shrunken abdomen

PROTEIN ENERGY MALNUTRITION [PEM]

LAQ:- Define and classify PEM, differentiate between Kwashiorkor and Marasmus.

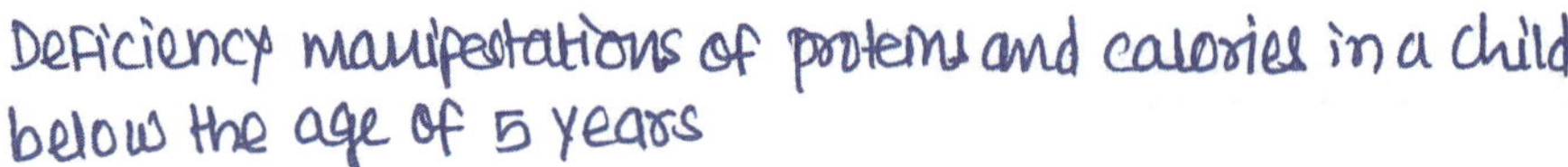

Deficiency manifestations of proteins and calories in a child below the age of 5 years

Malnutrition — undernutrition - underdeveloped / developing countries
overnutrition - Developed contries.

PEM : Definition :- The range of pathological conditions arising from lack of calories and proteins in a child below 5 years of age.

Classified as
1. Kwashiorkor :- Protein deficiency c̄ adequate calories
2. Marasmus :- Deficiency of calories in all forms
3. Marasmic - Kwashiorkor - features of both

Criterias for Assessment of PEM —
① weight by Height ratio
② Mid upper arm circumference
③ weight for Age

	Kwashiorkor	Marasmus
Age	Preschool child 1yr to 5yr	Infant < 1 yr
cause	Protein deficiency	calories deficiency
Weight / Height	60 - 80%	< 60%
Edema	Present	No edema
Hairs	sparse hair, Flag sign	No change in hairs
face	Moon face	old man face
skin	Dermatitis	Dry, Atrophic skin
Abdomen	Distended - Hepatosplenomegaly Ascitis	shrunken abdomen
liver	Fatty liver changes, liver enlargement	No liver involved
Apetite	Poor appetite	voracious feeder
Muscle	Muscle wasting mild or absent	Severe muscle wasting
Sr. Albumin level	0.5 to 2 gm/dl	> 2 gm/dl

Causes of PEM :- Poor socioeconomic status
Improper weaning
Chronic infection
Malabsorption

Complications :- Repeated infections
Hypoglycemia
Hypothermia
Dehydration
Shock
Vitamin and mineral deficiencies

Clinical case [very Important]

1. A 3 year old girl child was brought to the pediatric OPD c̄ chief complaint of failure to thrive, diarrhea and respiratory tract infection. On examination B/L edema over ankle and knee joints, Moon face, Sparse hair and easily pluckable, Abdominal distention

Child belongs to poor socio-economic status & resides in slum area.

- Hepatosplenomegaly
- Ascitis
- fatty liver
  ↓ mobilization of TAG due to ↓ lipoproteins.

Diagnosis - Kwashiorkor

Cause of edema ⇒ ↓ Albumin <2 gm/dl
↓
↓ oncotic pressure
↓
Retention of fluid in interstitial space
↓
Edema

T/t :- Protein rich diet 3-4 gm/kg body weight
t/t of complications
vitamin & Mineral supplementation

2. An infant brought to the pediatric OPD c̄ chief complaints of failure to thrive & diarrhea. The child belongs to poor family.
On examination - shrunken abdomen c̄ prominent bones
Dry & Atrophic skin
Old Man face
Signs of vitamin A deficiency
No edema

Δ Marasmus

T/t : 150-200 cal/kg/day Calories should be provided
t/t complications
vitamins, minerals

BASAL METABOLIC RATE [BMR]

Important SAQ: BMR and factors affecting it.

③ Important factors for calculating calorie requirement of individual

① BMR - Basal metabolic rate

② SDA - specific dynamic action

③ Calorie requirement for physical activity

BMR

Definition:- Amount of energy required by an individual in awake state at complete physical and mental rest in post-absorptive phase.

Energy required to maintain vital activities - Respiration
circulation
Brain function
Transport across memb.
renal reabsorption etc.

Measurement

① Direct measurement :- By measuring the Heat evolved

② Indirect measurement :- By measuring oxygen consumption per unit time

⇓

By Benedict Roth Apparatus

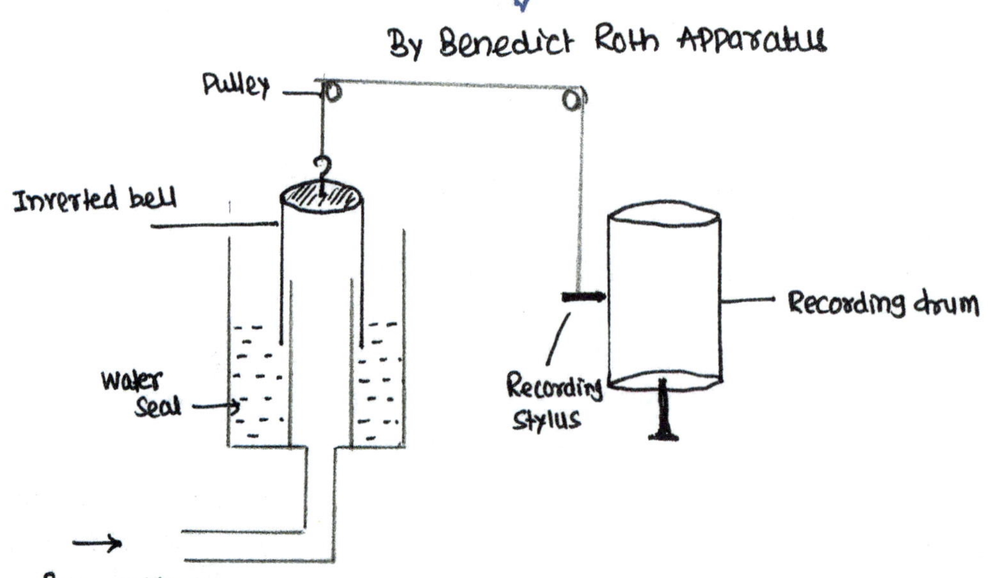

The apparatus is closed system circuit in which the subject breathes, in oxygen from a metal cylinder of about 6 lit capacity and CO_2 produced is absorbed by soda lime.

The subject breathes oxygen present in a cylinder for 6 min. The volume of O_2 used is recorded c̄ pen on a graph of recording drum.

If for 6 min → Y litre of O_2 consumed

Heat generated = 4.8 Y Kcal

↑ Calorific value of one lit O_2

Normal value of BMR

Males → 33-38 Kcal/m^2/Hr

Females → 32-35 Kcal/m^2/Hr

FACTORS AFFECTING BMR

1. **Surface area**

BMR is directly proportional to the surface area of the subject. greater heat loss and higher heat production if surface area is more

childhood, BMR is more

2. **Age**

As age advances, BMR decreases by 2% per decade of life because metabolic rate slows down

3. **Muscle mass**

Males → More muscle mass → More BMR

Females → More adipose tissue + estrogen → Less BMR.

4. **Climate**

BMR increases in cold climate

↑ Heat production by the body.

5. **Sleep** - BMR is less during sleep.

6. **Starvation** – BMR decreases in starvation

7. **Thyroid hormones** - Thyroxine increases BMR

Hyperthyroidism / Thyrotoxicosis → ↑↑ BMR

Hypothyroidism / Myxedema → ↓↓ BMR

8. **Fever**

1°C ↑ in temp. will ↑ BMR by 10-12%.

9 **Exercise**

cardiac output increases, BMR increases.
exercise increases muscle mass and power up your metabolism burning calories at faster rate, even when at rest.

10 **Genetics**

Some families have higher BMR than others
Due to some genetic disorder having influence on BMR

Clinical Significance

1. Important in calculation of energy requirement of individual.
2. Hypothyroidism BMR ↓↓
   Hyperthyroidism BMR ↑↑

Specific dynamic action [SDA]

Definition:- Amount of energy expediture above the BMR due to cost of processing food for use and storage.

SDA for proteins 30%
SDA for lipid 0-5%.
SDA for carbohydrates 5-10%.

Hence for mixed diet SDA = 10% of total calories

Also known as Thermic effect of food / calorigenic action
Additional 10% calories should be added to the total energy need towards SDA
SDA of food is due to energy required for digestion, absorption, transport, storage of food and metabolism.
Higher SDA for protein indicates that it is Not good source of energy
SDA of protein is mainly to meet the energy requirements for deamination, synthesis of urea, biosynthesis of proteins etc.

SDA of carbohydrate is attributed to energy expenditure for conversion of glucose to glycogen.

SDA of lipid is due to its storage, mobilization and oxidation

Caloric value of food

Calorie :- The amount of heat required to raise the temp of 1 kg of water by 1°C

Measurement : by Bomb calorimeter

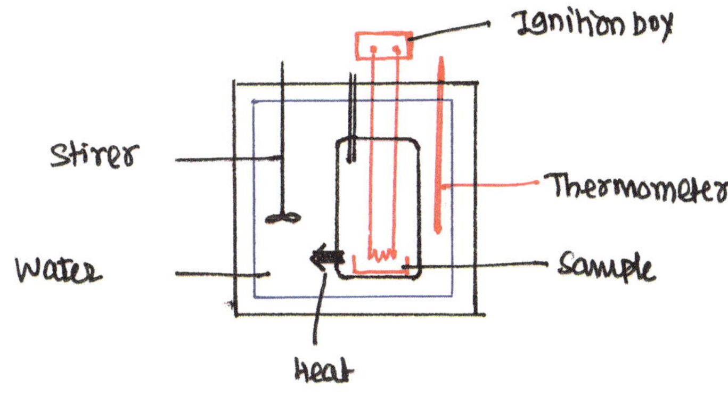

Caloric value 1 gm carbohydrate 4 Cal

1 gm fat 9 Cal

1 gm protein 4 Cal

Respiratory Quotient

Definition :- The ratio of volume of CO_2 produced to the volume of O_2 used during oxidation of foodstuff in given time

RQ for different food - carbohydrate :- 1

fat :- 0.7

protein :- 0.8

for mixed diet :- 0.85

Significance :- It is significantly lowered in DM & starvation because ↑ rate of utilization of lipids.

It is index of metabolic state of an individual.

Nitrogen balance

The difference between nitrogen intake and nitrogen output

Positive nitrogen balance

When protein intake is more that its output

Example :- Pregnant women
Growing children
Steroid therapy (Anabolic)

Negative nitrogen balance

When output exceeds input of nitrogen.

Example :- Fever, infections, Malnutrition, Hyperthyroidism, Nephrotic syndrome, Protein losing enteropathy.

BALANCED DIET

Definition :- It is a diet that contains different kind of foods in quantities and proportions so that the requirement for calories, proteins, vitamins, minerals and other nutrients is adequate with small provision is made for extra nutrients to withstand short duration of leanness.

ICMR Food Pyramid

No Alcohol

No tobacco

Group	Particulars	Cal/day	Protein g/day	Fat g/d	Calcium mg/d	Iron mg/d
Man	Sedentary work	2320	60 gm/day	25	600	17
	Moderate work	2730		30		
	Heavy work	3490		40		
Woman	Sedentary work	1900	55	20	600	21
	Moderate work	2230		25		
	Heavy work	2850		30		
	Pregnant woman	+350	+23	30	1200	35
	Lactation (up to 6 months)	+600	+19	30	1200	21

RDA for Indians by ICMR

Dietary Advice for optimal health in pregnancy

MCI competency: BI 8.3

Calorie requirement ⇒ 350 additional calories/day

Protein ⇒ 55+23 gm/day

Fat ⇒ 30 gm/day

Calcium ⇒ 1200 mg/day

Iron ⇒ 35 mg/day

Vitamin C rich fruits – Amla, Guava, orange should be added to improve the absorption of Iron

Add green leafy vegetables, legumes and nuts to daily diet

In case of nausea – take small and frequent meals

Expose to direct sunlight atleast for 15 min in morning.

Use up to 30 gram oil (20 gm vegetable + 10 gm ghee) per day

Use Iodised salt in food preparation

No Alcohol, No tobacco

Iron and folic acid supplementation daily for 100 days in pregnancy → 100 mg elemental iron, 500 mcg of folic acid → for prevention of Anemia and Neural Tube defect

Adequate vitamins and mineral supplementation as per RDA

Advice for optimal health in Diabetic Mellitus : Diet Management

MCI BI 8.3

1. Have fresh fruits and green leafy vegetables → for antioxidants, fibres, vitamins & minerals.

   Avoid fruit juice drinks & canned vegetables.

2. Have skimmed milk & low fat curd for calcium, vit D & protein → Avoid whole milk, Ice cream ↓ High fat, High sugar

3. Fish, Egg white, beans, seeds are good source of proteins → Avoid Red meat

4. Nuts, seeds, oats, sprouts High in vitamin, fibres, antioxidants → cookies, cakes, fast-food, fried food processed food should be Avoided

- Regular check up for plasma glucose level
- Glycated Hb will tell about average blood sugar for preceding 2-3 month. [HBA1C]
- Check for lipid profile - TG, cholesterol, HDL, LDL, VLDL
- Increase high fibre diet : Fruits, vegetables, sprouts etc.
- Increase consumption of ω3 fatty acids - α-linolenic acid
- Maintain healthy weight
- Regular physical exercise
- Regular check up for eyes - to Avoid complication
- Take care of your feet - diabetic neuropathy.
  - wash daily, Apply moisturizer, dry your feet between toes
- Say No to tobacco & Alcohol
- Manage your stress, Learn relaxation techniques.
- Regular dental checkup - take care of your teeth

use sugar free OR strictly stop consuming sweets

Coronary Artery Diseases - Dietary Advice

Fresh fruits and vegetables → vitamin, nutrients, antioxidants
Reduces weight
lowers cholesterol level
Maintain blood glucose level

Whole grains → Nutrient dense c̄ an excellent source of fibres. Fibres are important to reduce the absorption of cholesterol from intestine.

Healthy fat → Polyunsaturated fatty acid, α- linolenic acid is important for reverse transport of cholesterol by HDL

PUFA & MUFA found in → Olive oil
Sesame oil
Mustard oil
Safflower oil
Sunflower oil
Fish oil

↓ saturated fat
Avoid - coconut oil
Butter
Palm oil

Avoid - Trans fatty acids - present in partially hydrogenated oil, ↑ shelf life of food
used in processing & packaging of food
↑ LDL, ↓ HDL
↑ Risk of Atherosclerosis

- Avoid smoking
- Avoid Tobacco in any form
- Avoid Alcohol consumption

Healthy proteins → Egg white, soyabean, peas & lentils, fish

Avoid High sodium food → potato chips, pickle, processed meat canned soups

Check for lipid profile → Cholesterol
TG
HDL, LDL, VLDL

Maintain healthy weight → by regular exercise (brisk walk)

Regular Yoga & meditation practices & Relaxation techniques.

Obesity and overweight

$$\text{Body mass index (BMI)} = \frac{\text{weight in kg}}{\text{height in m}^2}$$

overweight ⇒ BMI ≥ 25 [25.0 to 29.9]

Obesity ⇒ BMI ≥ 30 [30.0 and higher]

obesity :- chronic complex disease define by excessive fat deposits that can impair health.

Causes :- Imbalance of energy intake and energy expenditure

Mutifactorial origin – Many causes

- Obesogenic environment → Eating habit and lack of physical activity.
- psycho-social factors
- Genetic variente

Also can be due to Medications
Diseases
Immobilization
Genetic syndromes
Hormonal imbalance etc

Common health consequences OR Health Risks

- cardiovascular diseases - Atherosclerosis, Hypertension
- Diabetes - type 2 DM, Resistance to insulin
- cancers
- Neurological disorders - Depression, social isolation
- chronic Resp. diseases
- Digestive disorders - Heart burn, Gallbladder diseases, liver diseases
- strokes - Atherosclerosis and ↓ Blood supply to Brain
- Sleep apnea - start and stop breathing in sleep
- Osteoarthritis - due to stress on weight bearing joints

Role of Leptin [leptos means 'thin' or light or small]

- It is a protein hormone made by adipocytes
- It regulates long term energy balance between food intake and energy expenditure.
- Leptin helps to inhibit hunger response when it doesn't need energy (calories).
- Leptin levels are high in obesity and it results in Leptin-resistance.

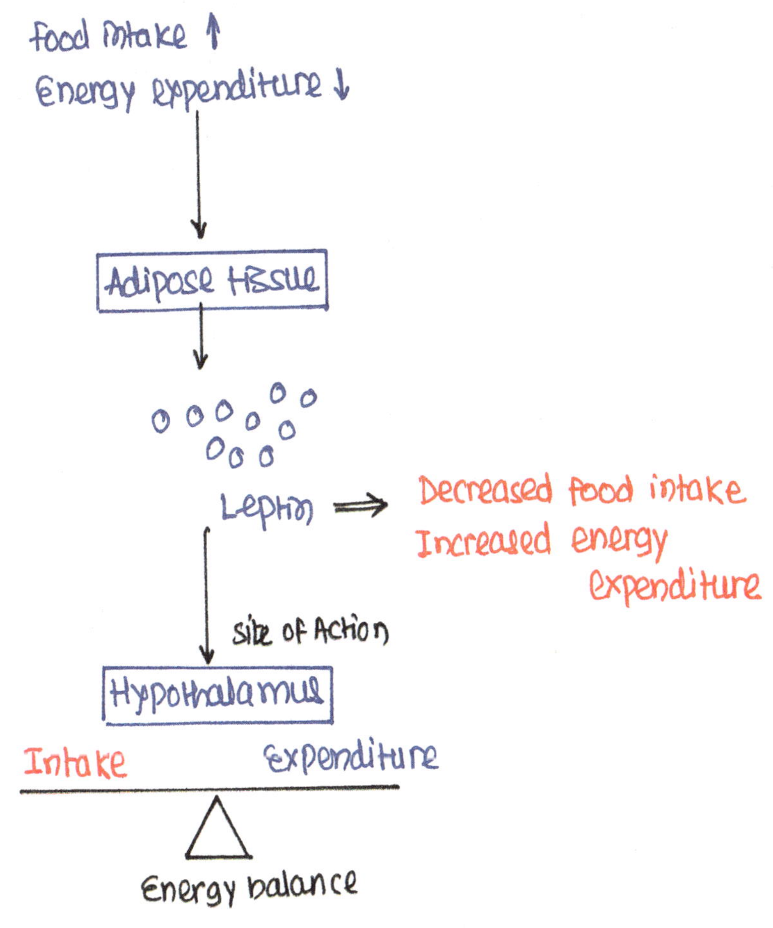

Role of Ghrelin:- produced by stomach →
- ↑ apetite
- ↓ energy expenditure
- ↑ gastric secretions
- ↑ gastric motility

Management of obesity

1. Support behavior of children – Healthy eating
   Physical activity
   Sedentary behavior and sleep
   Limit screen time
2. limit consumption of sugar sweetened beverages and energy dense foods.
3. Avoid tobacco & Alcohol
4. Engage in regular physical activities, walking, outdoor games.
5. Increase consumption of dietary fibres, legumes, pulses, oats
6. Decrease saturated fat, Trans fatty acids etc
7. Avoid High sodium food
8. Healthy proteins :- egg white, soyabean, peas & Lentils, fish
9. Regular checkup for – lipid profile
   HbA1C
   Blood pressure
10. Regular Yoga, Meditation practices and Relaxation technique.

ORGAN FUNCTION TESTS

MCI COMPETENCY BI 6.13-6.15

LAQ :- Liver function test
Kidney function test
Thyroid function test

SAQ :- ① Clearance tests
② Enzyme markers for liver disease

Clinical cases

1. Acute renal failure → ↑ urea, ↑ creatinine
2. Jaundice – Prehepatic
   Hepatic
   Posthepatic
3. Hypothyroidism → ↓ T3, ↓ T4, ↑ TSH
4. Hyperthyroidism → ↑ T3, ↑ T4, ↓ TSH

Laboratory Interpretation

Interpret the data and give the type of Jaundice. What are the causes of condition ?

Parameter	observed value
Total bilirubin	18.2 mg/dl
Direct bilirubin	10.4 mg/dl
Indirect bilirubin	7.8 mg/dl
SGOT	1856 IU/L
SGPT	2022 IU/L
ALP	250 IU/L
fouchet test	positive
Hays sulfur test	positive

Bilirubin

Direct Indirect Bilirubin
Both ↑

Van den Bergh – Biphasic

↑↑↑↑ SGOT
↑↑↑↑ SGPT
Normal → ALP
Urine – positive for bile Pigment & bile salt.

A Hepatocellular Jaundice
cause- Viral Hepatitis

KIDNEY FUNCTION TESTS [KFT]

KFT

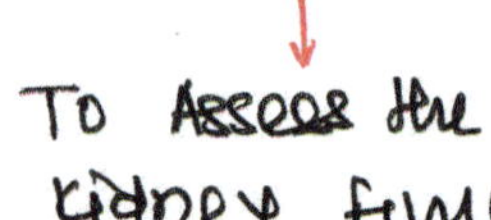

To Assess the kidney disease

1. **Serum parameters**
   - Urea
   - Creatinine
   - Na^+
   - K^+
2. **Urine Analysis**
   - Physical properties
   - Chemical tests

To Assess the kidney functions

1. **Glomerular functions**
   - Clearance tests
2. **Glomerular permeability**
   - Urinary protein estimation
3. **Tubular functions**
   - Concentration test
   - Dilution test

Functions of kidney

① Maintenance of acid-base balance
② Blood pressure regulation
③ Maintenance of water and electrolyte balance
④ Excretion of waste products - Urea, creatinine, uric acid etc.
⑤ Synthesis of vit-D
⑥ Synthesis of erythropoietin
⑦ Renin - Release of Aldosterone

Urine formation

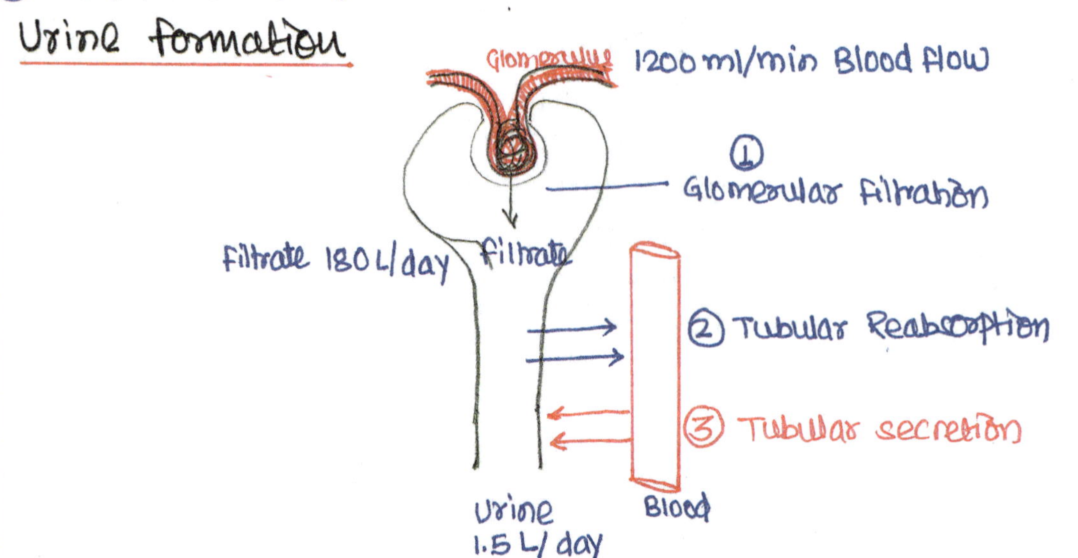

Tests for Glomerular Filtration

Clearance tests

Clearance is defined as amount of plasma/blood that would be completely cleared off a particular substance in unit time.

Substances/parameters used for clearance test

(4) Cr-EDTA

1. INULIN (Best)

- Freely filtered through glomerulus
- Neither absorbed
- Nor secreted by renal tubules
- Has to be given IV
- Not affected by diet
- Values nearer to GFR 125 ml/min

2. CREATININE

- Synthesized in our body
- Freely filtered through glomerulus
- Not reabsorbed
- Secreted through renal tubules in some extent
- Values: slightly higher than (N)
  - Male 135-145 ml/min
  - Female 125-135 ml/min
- Not affected by diet

3. UREA

- Freely filtered
- 40% reabsorbed
- Affected by protein diet.
- Less values of GFR because of tubular absorption.

$$\text{Clearance [ml/min]} = \frac{UV}{P}$$

U_c = Urinary creatinine in mg/dl

V = Volume of urine ml/min

P_c = Plasma/serum creatinine mg/dl

$$\text{Creatinine clearance} = \frac{U_cV}{P_c}$$

Clearance

- >90% → Normal
- 60-90% → Mild renal impairment
- 30-60% → Moderate renal impairment
- <30% → Severe renal impairment
- <15% → End stage renal failure

Tests for glomerular permeability - Urinary microalbumin

24 Hr Urinary albumin excretion	<30 mg/24 Hr	Normal
Microalbuminuria	30-300 mg/24 Hr	
Macroalbuminuria	>300 mg/24 Hr	

Also done by dipstick method.

Tests for Tubular function - Measure the specific gravity

Ⓝ specific gravity 1.010 to 1.025

Concentration test

overnight fluid deprivation
↓
specific gravity of urine in Morning

2 samples in gap of 1 Hr
↓
concentrating ability of kidney is measured

Normal :- Atleast one sample should have sp gravity >1.025

Dilution test

Fluid load is given
1200 ml water in 30 min
↓
4 samples in 1 Hr interval
specific gravity is measured

Normal - Atleast one sample should have sp. gravity <1.010

Serum parameter

1. Serum Urea :- End product of Protein metabolism

Normal range → 10-40 mg/dl

Urea estimation by → Urea - GLDH Method
Berthelot's method

↑ Urea level - causes

a. Prerenal
- Dehydration
- Vomiting
- Diarrhea
- Burns
- Severe infection

b. Renal
- Glomerulonephritis
- Nephrotic syndrome
- Renal failure

c. Post-renal
- Tumor
- Stricture
- Stone
- Tuberculosis

} Ureter, bladder, Urethra

2. Serum creatinine

More sensitive indicator than Urea

Normal range Male 0.7 to 1.4 mg/dl
female 0.6 to 1.3 mg/dl

↑ creatinine → Glomerulonephritis
Nephrotic syndrome
Pyelonephritis
Obstruction to Urinary tract

3. Serum Na^+ → 135 to 150 mmol/L
4. Serum K^+ → 3.5 to 5.0 mmol/L
5. Serum cystatin C – Glomerular filtration marker
Ⓝ level 0.8 to 1.2 mg/L

Urine Analysis

Physical properties – Volume
Color
Odour
Appearance
PH
Specific gravity
Total solids.

Chemical tests

Glucose → Benedict's test
proteins → Heat coagulation test
Bile salts → Hay's sulfur test
Bile Pigments → Fouchet's test
Blood → Benzidine test
Ketone bodies → Rothera's test

Liver function Test [LFT]

LAQ: Describe the various tests for assessment of liver functions.

Mnemonic - ABCDEFGHI

A → Albumin
A/G ratio: Albumin : Globulin ratio
α₁ antitrypsin

B → Bilirubin → Direct / Indirect
Bile pigment → Bilirubin, Biliverdin in urine
Bile salt → Na & K salts of Glycocholate, Taurocholate

C → Cholesterol
coagulation profile
ceruloplasmin

D → Dye → Bromosulphthalein BSP dye to Assess excretory function *MCQ

E → Enzyme markers – Hepatic markers – SGOT, SGPT, LDH4, LDH5
Obstructive markers – ALP, GGT, 5'NT

F → Ferritin
Fetoprotein (AFP)

G → Globulins
Galactose tolerance test

H → Hippuric acid excretion test
Haptoglobulin
HbsAg - detection of infection by Hepatitis B virus

I → Immunological tests - IgM & IgG for detection of infection

Functions of the liver

1. Synthetic	2. Metabolism	3. Detoxification & excretion	4. Storage
Albumin	Blood glucose regulation	Ammonia	Vit A, D, E, K
cholesterol	Fat metabolism	Unconjugated bilirubin	Vit B12
Prothrombin	Protein Catabolism	Cholesterol	Iron
TAG		Drugs	Glycogen
Lipoproteins			

1. Albumin, serum A/G ratio

Normal range → 3.5 to 5 gm/dl

Synthesized by the liver

Half life of albumin is 20 days

Albumin level decreases in chronic liver damage

A/G Ratio. Albumin : Globulin ratio 1.5 to 2

Clinical significance → Reversed in ↓ synthesis – cirrhosis of liver

↑ excretion – Nephrotic syndrome

→ Important to differentiate between True or False increase or decrease in sr protein level.

For example in hemoconcentration or hemodilution A/G ratio is unaffected.

Globulins : α, β, γ globulins (N) 2-3 gm/dl

↑ in Chronic liver ds

2. Ammonia → Increase in Hepatic encephalopathy

Cirrhosis of liver

Urea cycle disorders

3. α_1-antitrypsin → Deficiency mainly affects lungs, may affect liver

4. Bilirubin → Total Bilirubin → up to 1.2 mg/dl

Direct bilirubin → 0.2 to 0.4 mg/dl

Indirect bilirubin → 0.2 to 0.7 mg/dl

Van den bergh Reaction (Diazo Reaction)

Bilirubin + Diazotised sulfanilic acid ⟶ Azobilirubin (pink to purple in color)

Conjugated bilirubin → Immediate purple color → Direct positive [Obstructive Jaundice]

Unconjugated bilirubin → Gives color after 30 min → Indirect positive

Need solubilizer – caffeine, methanol [Hemolytic Jaundice]

Hepatic Jaundice → Biphasic van den Bergh Reaction

Direct +ve & Indirect +ve

5. Bile pigment :- Bilirubin and biliverdin in urine

Detected by fouchet test :- Hepatic OR obstructive Jaundice

3 ml urine + 1 ml $MgSO_4$ + 2 ml $BaCl_2$, filter it and Add 1-2 drops of fouchet reagent on ppt. on filtre paper	Pista green color of ppt on filtre paper	Indicates presence of bile pigment in urine.

6. Bile salts :- Na and K salts of Glycocholate and Taurocholate

3 ml urine and sprinkle sulfur powder on surface of urine	Powder sink to bottom	Bile salt present
3 ml Dist. water and sprinkle pinch of sulfur powder	Powder floats	Bile salt Absent

7. Urobilinogen :- presence indicates hemolytic Jaundice

detected by Ehrlich's aldehyde test

8. serum cholesterol :- Desirable level is below 200 mg/dl

↓↓ in cirrhosis of liver

↑↑ in obstructive liver diseases.

Indicates synthetic as well as metabolic function of liver

9. Coagulation profile

Prothrombin Time (PT) → Increases in Acute liver injury (Because clotting factors are synthesized in liver)

↓

short term marker of liver damage

half life 6 Hours

10. Ceruloplasmin :- ↓ in Wilson's disease

11. Bromosulphthalein excretion test :- Indicates excretory function of liver

IV injection ⟶ excretion of BSP is measured in serum

→ < 5% After 45 min (N)

→ Retention ↑ in liver damage

12. Enzyme Markers

Hemolytic Jaundice	Hepatic Jaundice	Obstructive Jaundice
Normal SGOT	SGOT ↑↑↑↑	SGOT ↑
SGPT	SGPT ↑↑↑↑	SGPT ↑
ALP	ALP ↑	ALP ↑↑↑
GGT	GGT ↑	GGT ↑↑↑
5'NT	5NT ↑	5NT ↑↑↑

SGPT/ALT → Serum Glutamate Pyruvate Transaminase OR Alanine Transaminase

Ⓝ Range 5-35 IU/L

SGOT/AST → Serum Glutamate Oxaloacetate transaminase OR Aspartate transaminase

Ⓝ Range 5-40 IU/L

ALP → Alkaline phosphatase

Ectoenzyme, present on cell surface of biliary canaliculi

↓

Obstruction to flow of bile causes irritation of bile canaliculi

↓

Release of ALP in blood stream

↑↑ 10-12 times of Normal

* MCQ

GGT → γ-Glutamyl Transferase – specific for Alcoholic liver disease

13. Ferritin :- Indicates storage of Iron in liver

↓↓ in Anemia and ↑↑ in Hemosiderosis

14. α-fetoprotein :- slight ↑ in cirrhosis

Marked ↑ in AFP → Hepatic cancer

Maternal ↑ in AFT → Neural tube defect.

15. Galactose Tolerance Test → To Assess metabolic function of liver

Galactose is given

↓

Rate of Galactose excretion is measured

Generally utilized in half to one hour

Retention of Galactose indicates damage to metabolic function of liver

Classification of LFT

Biochemical parameters

1. **Excretory function**
   - Bilirubin
   - Bile pigment
   - Bile salt
   - Urobilinogen
2. **Enzyme markers**
   - SGPT
   - SGOT
   - ALP, GGT, 5'NT
   - LDH4, LDH5
3. **Metabolic function**
   - Galactose tolerance test
4. **Synthetic function**
   - Total proteins
   - Albumin
   - Globulins
   - A/G ratio
5. **Special tests**
   - Alpha fetoproteins
   - Ferritin
   - Ceruloplasmin
   - Immunoglobulins

Clinical Aspects

1. **For liver dysfunction**
   - Sr. Total protein
   - Sr. Albumin
   - Sr. Bilirubin — Total, Direct
   - SGOT
   - SGPT
   - ALP
2. **Enzymatic markers.**
   - For Hepatocellular injury
     - SGOT
     - SGPT
   - For obstruction
     - ALP
     - GGT
     - 5'NT

Indications for LFTs →
1. Jaundice
2. Cirrhosis of liver
3. Alcoholic liver de
4. Metastasis to liver
5. Toxins

Clinical cases - LFT/ KFT

1. An old man was brought to the hospital with chief complaints of abdominal pain, pruritus and yellow colored urine. Laboratory investigations revealed,

SGOT - 89 IU/L
SGPT - 93 IU/L
ALP - 1527 IU/L
Direct bilirubin - 18 mg/dl
Indirect bilirubin - 0.7 mg/dl

What is your probable diagnosis?

Mention the causes of above condition.

Δ Obstructive Jaundice

causes - Common bile duct stones
Gall stone
Cholangiocarcinoma
Cancer of head of pancreas
Stricture of CBD

2. An 18-year-old boy was brought to the clinic with chief complaints of nausea, vomiting, flu like symptoms and yellow skin & sclera. Lab investigation shows,

SGOT - 1829 IU/L
SGPT - 1601 IU/L
ALP - 190 IU/L
Direct Bilirubin - 8.2 mg/dl
Indirect bilirubin - 9.1 mg/dl

What is your probable diagnosis?

Δ Hepatocellular Jaundice or Infective Jaundice or Viral Hepatitis

3. A 40-year-old diabetic patient brought to the hospital with decreased urine output. His plasma glucose was 279 mg/dl, serum urea - 410 mg/dl and creatinine → 15.4 mg/dl. His creatinine clearance was less than 30%. What is your probable diagnosis?

Δ Acute Renal Failure due to diabetic Nephropathy

THYROID FUNCTION TESTS

LAQ:- Describe various tests done for the assessment of thyroid gland.

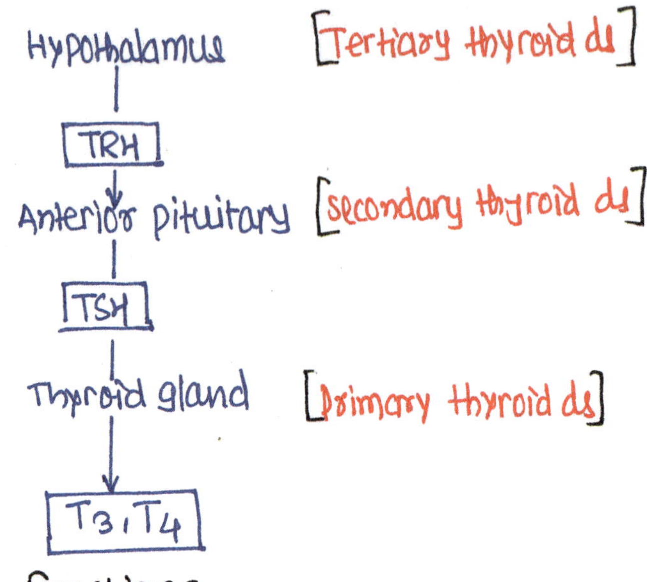

Endocrine gland in front region of Neck

↑ BMR	↑ Protein synthesis	↑ Blood Glucose	↑ Lipid utilization	Water and electrolyte balance

Normal levels and parameters done for assessment of thyroid gland

- Thyroid stimulating hormone (TSH) → 0.5 to 5.0 MIU/L
- free T4 ⟶ 0.7 to 1.8 ng/dl
- free T3 ⟶ 260 to 480 pg/dl
- Total T4 ⟶ 4 to 12 μg/dl
- Total T3 ⟶ 80 to 220 ng/dl

Thyroid binding globulins

Radioiodine uptake - I-131 is used for Radioiodine uptake studies * MCQ

Antibodies to - TPO, TRAb, TGAb

USG - Ultrasonography

CT-Scan

FNAC

* MCQ Serum calcitonin - is a marker for Medullary carcinoma of thyroid gland. It is a hormone secreted by parafollicular cells of thyroid gland.

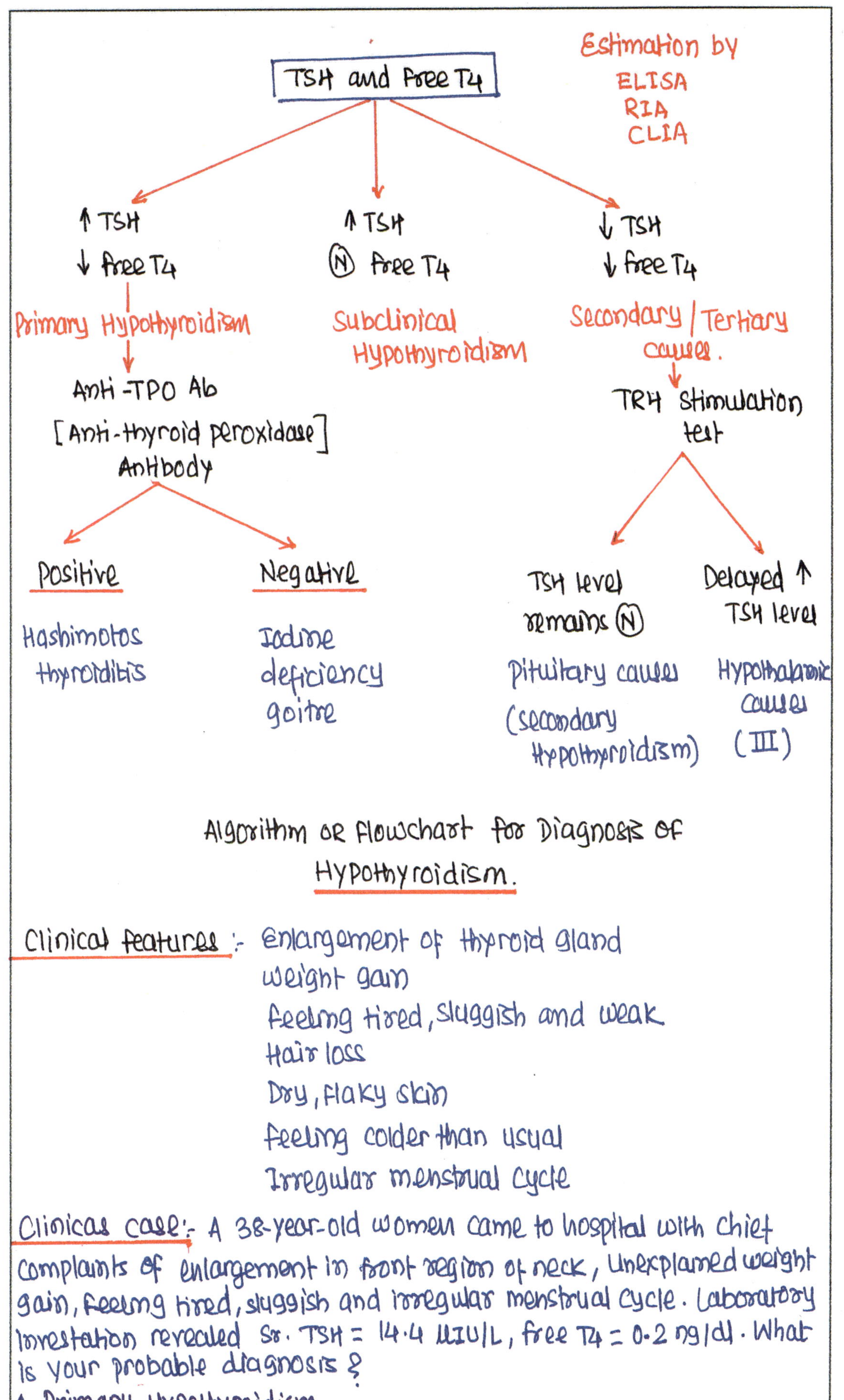

Algorithm or Flowchart for Diagnosis of Hypothyroidism.

Clinical features :- Enlargement of thyroid gland
weight gain
feeling tired, sluggish and weak
Hair loss
Dry, Flaky skin
feeling colder than usual
Irregular menstrual cycle

Clinical case :- A 38-year-old women came to hospital with chief complaints of enlargement in front region of neck, Unexplained weight gain, feeling tired, sluggish and irregular menstrual cycle. Laboratory investation revealed Sr. TSH = 14.4 μIU/L, free T4 = 0.2 ng/dl. What is your probable diagnosis ?

Δ Primary Hypothyroidism

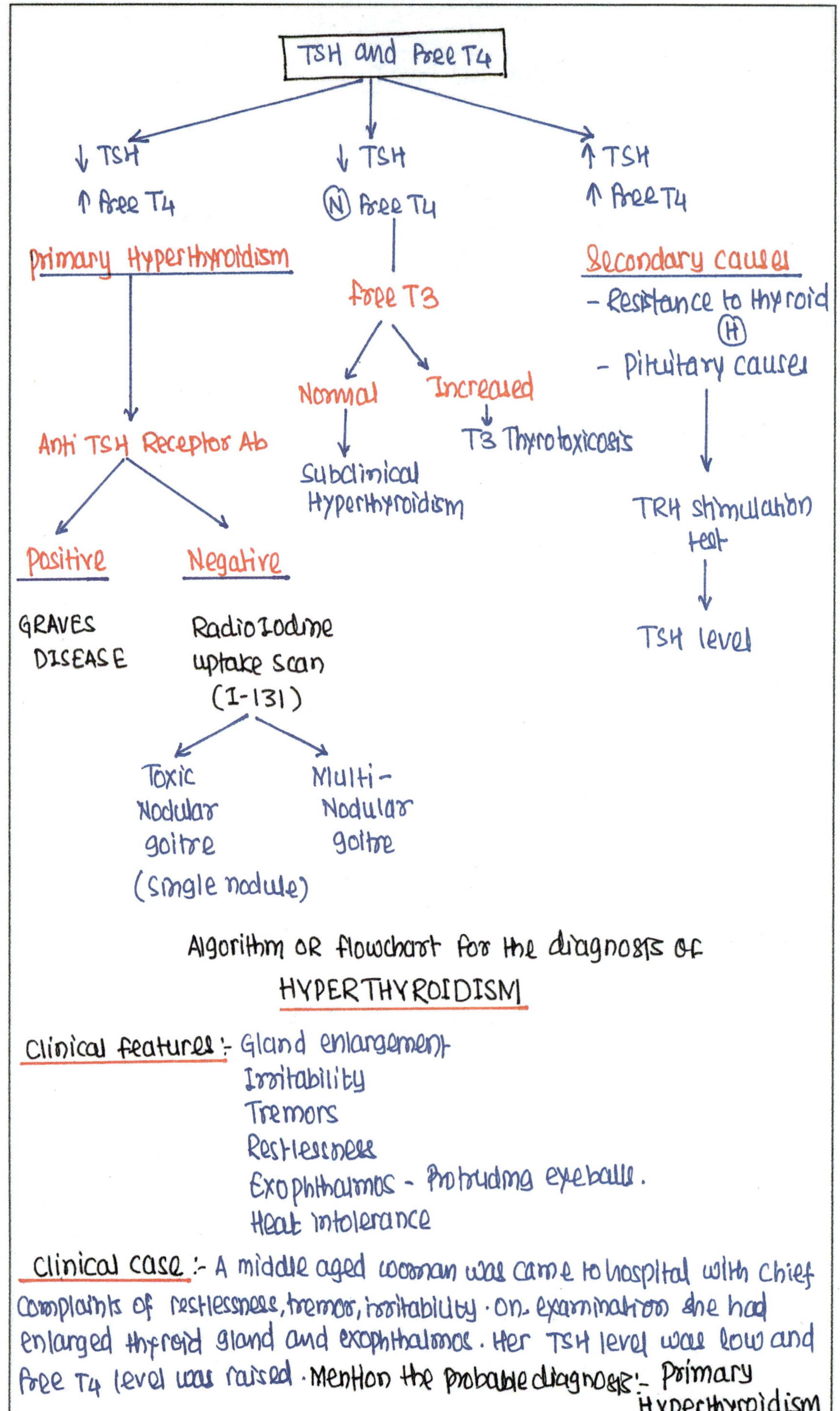
TSH and Free T4
↓ TSH
↑ Free T4
Primary Hyperthyroidism
Anti TSH Receptor Ab
Positive
Negative
GRAVES DISEASE
Radio Iodine uptake scan (I-131)
Toxic Nodular goitre (single nodule)
Multi-Nodular goitre
↓ TSH
N Free T4
Free T3
Normal
Increased
Subclinical Hyperthyroidism
T3 Thyrotoxicosis
↑ TSH
↑ Free T4
Secondary causes
- Resistance to thyroid H
- Pituitary causes
TRH stimulation test
TSH level
Algorithm OR flowchart for the diagnosis of HYPERTHYROIDISM
Clinical features :- Gland enlargement
Irritability
Tremors
Restlessness
Exophthalmos - Protruding eyeballs.
Heat intolerance
Clinical case :- A middle aged woman was came to hospital with chief complaints of restlessness, tremor, irritability. On examination she had enlarged thyroid gland and exophthalmos. Her TSH level was low and free T4 level was raised. Mention the probable diagnosis :- Primary Hyperthyroidism

Clinical Case

A 50-Year-Old woman with enlarged thyroid gland was sent by the endocrinologist for the thyroid function tests. She was Tested for Serum FT3, free T4 and TSH levels in the Biochemistry laboratory.

		Ⓝ Range
serum free T3 →	100 pg/dl	[260 – 480 pg/dl]
free T4 →	0.5 ng/dl	[0.7 – 1.8 ng/dl]
TSH →	48 µIU/L	[0.5 – 5.0 µIU/L]

Name the condition :- Primary hypothyroidism [Myxoedema]

[↓ free T3, ↓ free T4, ↑ TSH]

Mention the other TFTs :- Radioiodine uptake

Antibodies – Anti TR Ab

Anti TPO Ab

USG, CT-Scan, FNAC

Which mineral and vitamin is required for T3 & T4 synthesis?

↓ Iodine

↓ Tyrosine

Clinical case-

An 8-Year-Old child was brought to the pediatric clinic with chief complaints of mental retardation, short stature, thick skin and early puberty. His mother was suffered from severe Iodine deficiency during his pregnancy. What is most probable diagnosis in the child?

Δ **Cretinism** :- Congenital iodine deficiency syndrome due to deficiency of Iodine in pregnancy.

ACID BASE BALANCE

MCI COMPETENCY BI 6.7-6.8

Blood Buffer - Bicarbonate
Phosphate
protein

Respiratory system:- Hypo/Hyperventilation
Hemoglobin

Renal mechanism
1. Excretion of H^+ ions
2. Reabsorption of HCO_3^-
3. Excretion of titrable acid
4. Excretion of ammonium ions

Normal Range:- $pH = 7.35$ to 7.45
$PCO_2 = 35$ to 45 mmHg, $HCO_3^- = 22$ to 28 mmol/L

Disorders

Acidosis ($pH < 7.35$)
- Metabolic: $\downarrow HCO_3^-$
- Respiratory: $\uparrow pCO_2$

Alkalosis ($pH > 7.45$)
- Metabolic: $\uparrow HCO_3^-$
- Respiratory: $\downarrow pCO_2$

Que:- Interpret the data and give the type of acid-base disturbance. What are the causes of condition?

Blood pH	7.20
PCO_2	60 mmHg
HCO_3^-	26 mmol/L

A - - - - - - - - - - - -

Blood pH	7.59
PCO_2	22 mmHg
HCO_3^-	24 mmol/L

A - - - - - - - - - - - -

Blood pH	7.29
PCO_2	42 mmHg
HCO_3^-	12 mmol/L

A - - - - - - - - - - - -

ACID BASE BALANCE

LAQ :- Write about different Mechanisms related with blood pH regulation.

pH :- It is defined as negative logarithm of hydrogen ion concentration

$$pH = -\log[H^+]$$

Normal range 7.35 to 7.45

Acid : donate protons

strong acid HCl → donate protons & ionizes completely

$$HCl \longrightarrow H^+ + Cl^-$$

weak acid H_2CO_3 → does not ionizes completely

$$H_2CO_3 \rightleftharpoons H^+ + HCO_3^-$$

Base :- Accept protons

$$HCO_3^- + H^+ \longrightarrow H_2CO_3$$

$$NH_3 + H^+ \longrightarrow NH_4^+$$

Buffer : solution which resist the change in pH

The gram of acid or alkali which is required to change a pH of one unit of 1 litre buffer solution is known as buffering capacity of that solution.

Alkali Reserve :- HCO_3^- concentration 22-28 mmol/L

- It is bicarbonate concentration [HCO_3^-] to overcome the acids synthesized in our body is called as alkali Reserve.
- It is 20 times higher than acids.

PK :- pH at which acid is half ionized

Salt : Acid → 1:1

If PK value is nearer to body pH (7.4), buffer is more effective

Henderson-Hasselbalch Equation

$$pH = pKa + \log\frac{[Base]}{[Acid]}$$

$$pH = pKa + \log \frac{[HCO_3^-]}{[H_2CO_3]}$$

$[HCO_3^-]$ → Metabolic Regulation

$[H_2CO_3]$ → Respiratory regulation

For Bicarbonate buffer

$$pH = 6.1 + \log \frac{24}{1.2}$$
$$= 6.1 + \log 20$$
$$= 6.1 + 1.3$$
$$= 7.4$$

Regulation of blood PH

I. First line of defence

➡ Blood Buffers

1. Bicarbonate
2. Phosphate
3. Protein

II. Second line of defence

➡ Respiratory regulation

1. Hypoventilation
2. Hyperventilation
3. Hb

III. Third line of defence

➡ Renal mechanism

1. Excretion of H^+
2. conservation HCO_3^-
3. Excretion of Acid
4. Excretion of NH_4^+

Blood Buffers

Bicarbonate	Phosphate	Protein
$HCO_3^- : H_2CO_3$ 20 : 1	$Na_2HPO_4 : NaH_2PO_4$ 4 : 1	Albumin, Hb
PK → 6.1	PK → 6.8	PK → 6.1
Most important buffer in ECF Because 1. Quantity (Alkali Reserve) 2. Under physiological control CO_2 by Lungs HCO_3^- by kidneys	Most Potent buffer intracellularly Active at wide range of PH	Mainly because of imidazole group of Histidine.

Respiratory Mechanism

① LUNGS

When pH > 7.45 (Alkalosis)

↓ PCO_2

↓

Lungs – Hypoventilation

↓

CO_2 Retention

↓

↑↑ pCO_2

↓

↑ H_2CO_3

↓

↓ pH

When pH < 7.35 (Acidosis)

↑ PCO_2

↓

Lungs – Hyperventilation

↓

Exhalation of CO_2

↓

↓↓ CO_2

↓

↓ H_2CO_3

↓

↑ pH

② Hemoglobin

Tissue (↓ O_2)

$CO_2 + H_2O \xrightarrow{CA} H_2CO_3$

↓

$H^+ + HCO_3^-$

Hb → HHb

Lungs (O_2 ↑↑)

$HHb + O_2 \rightarrow HbO_2 + H^+$

HCO_3^- → H_2CO_3

↓ CA

$CO_2 + H_2O$

↓

Exhaled out

Renal Mechanism

1. Excretion of H^+
2. Recovery of HCO_3^-
3. Excretion of titrable acids
4. Excretion of NH_4

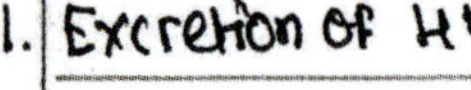

1. Excretion of H^+

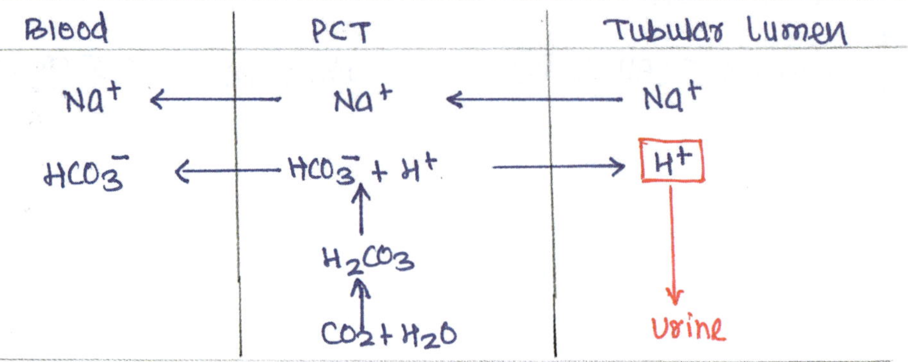

2. Recovery of HCO_3^- / Reabsorption

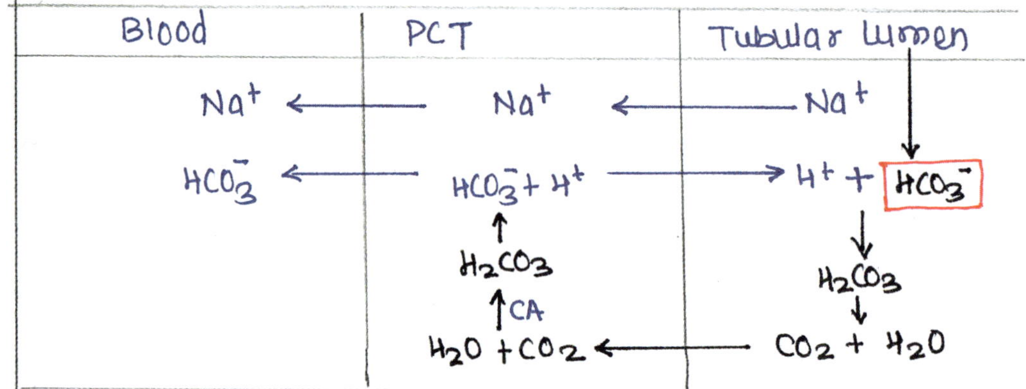

3. Excretion of Titrable acid

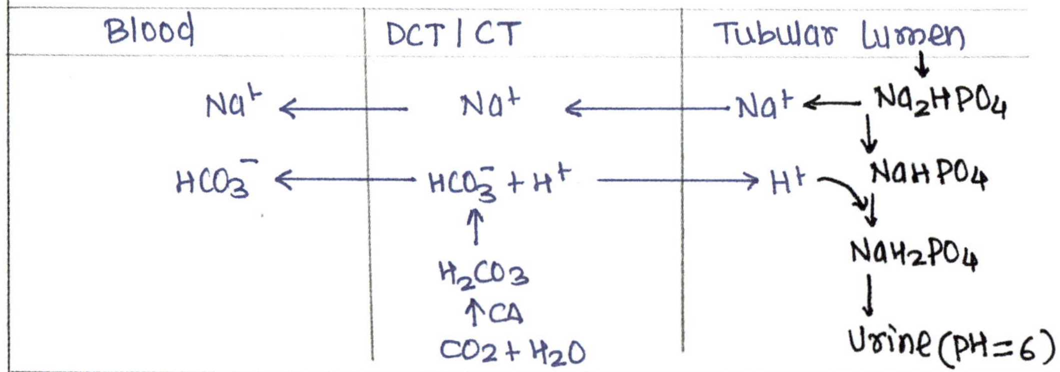

4. Excretion of Ammonium ions NH_4^+

Blood | DCT / CT | Tubular Lumen

Glutamine

↓ Glutaminase

Glutamate + NH_3 ⟶ NH_3

Na^+ ⟵ Na^+ ⟵ Na^+

HCO_3^- ⟵ $HCO_3^- + H^+$ ⟶ H^+

↑

H_2CO_3

↑ CA

$H_2O + CO_2$

NH_4^+

↓

Urine

Acid - Base disorders

Mnemonic :- parameter which goes with PH is the primary defect.

1 2 3 4 5 6 7 8 9 10 11 12 13 14

PH
Ⓝ 7.35 to 7.45

< 7.35 Acidosis   Alkalosis > 7.45

HCO_3^- Normal 22 to 28 mmol/L

HCO_3^- < 22 indicates Metabolic acidosis

HCO_3^- > 28 indicates Metabolic Alkalosis

pCO_2 Normal 35 to 45 mmHg

pCO_2 > 45 indicates Respiratory Acidosis

pCO_2 < 35 indicates Respiratory Alkalosis

Disorder	Primary defect	causes
Metabolic acidosis	HCO_3^- < 22 mmol/L	Diabetic ketoacidosis Renal tubular acidosis Renal failure, Diarrhea Aminoaciduria
Metabolic Alkalosis	HCO_3^- > 28 mmol/L	vomiting upper GI tract obstruction cushing syndrome Adrenal tumor
Respiratory Acidosis	pCO_2 > 45 mmHg	pneumonia Asthma COPD - obstructive diseases Narcotic & sedatives paralysis of Resp Muscles
Respiratory Alkalosis	pCO_2 < 35 mmHg	Hypoxia High Altitude Hysteria septicemia, meningitis

$$pH = pKa + Log \frac{[HCO_3^-]}{[H_2CO_3]}$$

$[HCO_3^-]$ → Metabolic regulation

$[H_2CO_3]$ → Respiratory regulation pCO_2

If HCO_3^- decreases, H_2CO_3 also decreases to maintain ratio

If H_2CO_3 decreases, HCO_3^- will also decrease to maintain ratio

Metabolic Acidosis

↓ HCO_3^- —— Resp. compensation ——→ compensated by Hyperventilation
blow out CO_2
↓ pCO_2
↓ H_2CO_3

Metabolic Alkalosis

↑ HCO_3^- —— Resp. compensation ——→ Hypoventilation
Retention of CO_2
↑ pCO_2
↑ H_2CO_3

Respiratory Acidosis

↑ pCO_2 —— Metabolic compensation ——→ Renal conservation of HCO_3^-
↑ HCO_3

Respiratory Alkalosis

↓ pCO_2 —— Metabolic compensation ——→ Renal excretion of HCO_3^-
No Reabsorption of HCO_3^-
↓ HCO_3

Assessment of Acid-base disorders

Sample collected in Heparinized syringe from Radial artery / Femoral artery

↓

Arterial Blood Gas Analyzer

pH
HCO_3^-
pCO_2
Anion Gap

1. Metabolic Acidosis

pH < 7.35, $HCO_3^- < 22$ mmol/L

Anion gap = $[Na^+ + K^+] - [HCO_3^- + Cl^-]$

measured cations 95%   Measured Anions 86%

↓

Unmeasured Anions → Proteins, sulfate, Phosphate, Organic acids

Ⓝ 8 to 18 mmol/L

Causes

Metabolic Acidosis (MA)

High Anion gap MA	Normal Anion gap MA (* MCQ)
Diabetic Ketoacidosis	Diarrhea
Lactic acidosis	Renal tubular acidosis I, II, IV
Renal failure	Intestinal fistula
Organic aciduria	
Aminoaciduria	
Aspirin	

Compensated by → Respiratory mechanism → Hyperventilation
- blow out CO_2
- ↓ pCO_2
- ↓ H_2CO_3

Treatment:- Sodium Bicarbonate
Monitor K^+ level

2. Metabolic Alkalosis

pH > 7.45, $HCO_3^- > 28$ mmol/L

Causes:

Chloride Responsive
- vomiting
- upper GI tract obstruction

Chloride Resistant
- Cushing syndrome
- Adrenal tumor

Compensated by Respiratory mechanism - Hypoventilation
Retention of CO_2
↑ pCO_2
↑ H_2CO_3

Treatment :- Saline infusion (NaCl)
Chloride infusion
Electrolyte Monitoring
t/t of cause.

3. Respiratory Acidosis : ↑ PCO_2
$pH < 7.35$, $pCO_2 > 45$ mmHg

Causes :- pneumonia
Asthma
Chronic obstructive pulmonary diseases
Narcotics and sedatives - depression of Resp centres
paralysis of Resp. muscles.

Compensated by renal reabsorption of HCO_3^- (Metabolic)

Treatment → Bronchodilator and steroids to reverse airway obstruction
ventilatory support
Oxygen support
t/t of cause

4. Respiratory Alkalosis :- ↓ PCO_2
$pH > 7.45$, $pCO_2 < 35$ mmHg

Causes - Hypoxia
High Altitude
Hysteria
Hyperventilation
Septicemia
Meningitis
febrile conditions

Compensated by renal excretion of HCO_3^- (Metabolic)

Treatment → Breath into paper bag & breathing practice to slow down Resp. rate
t/t of underlying cause of Hyperventilation

Clinical case-1

An infant was brought to the hospital withe severe vomiting. On ultrasonography diagnosis of upper GI tract obstruction (Pyloric stenosis was given. On ABG Analysis, PH of blood was 7.60, PCO_2- 42 mmHg and HCO_3^- was 32 mmol/L.

Mention the acid-base disorder?

Δ Metabolic Alkalosis

Clinical case-2

A child was suffered from meningitis (fever c̄ neck rigidity) His ABG Analysis showed, Blood PH 7.62, PCO_2- 24 mmHg and HCO_3^-- 24 mmol/L.

Mention the acid-base disorder?

Δ Respiratory Alkalosis

PH: 7.62, PCO_2: 24 mmHg

Clinical case-3

An 18-Year-old boy was brought to the emergency dept in semiconscious state with kussmaul breathing (deep and rapid breathing). On laboratory investigations, his blood glucose was 512 mg/dl, dipstick method shows positive test for glucose and ketone bodies. ABG Analysis shows. PH: 7.22, HCO_3^-: 12 mmol/L and PCO_2: 30 mmHg.

Mention the probable diagnosis?

Δ Diabetic Ketoacidosis with Metabolic acidosis with respiratory compensation

Kussmaul breathing: compensatory mechanism- Hyperventilation to blow out CO_2 to maintain HCO_3^-/H_2CO_3 ratio.

- HCO_3^- is utilized to buffer the acide (ketone bodies), hence decreases causing metabolic acidosis.

MOLECULAR BIOLOGY

MCI COMPETENCY BI 7.1-7.3

LAQ :- Describe the various types of RNAs

Replication of DNA in eukaryotes c̄ reference to prokaryotes

Type of DNA damage and repair mechanisms.

SAQ :- ① tRNA structure

② Watson and crick model of DNA

③ Telomers and Telomerase

④ Transcription in prokaryotes

⑤ Transcription in Eukaryotes

⑥ post transcriptional modifications

⑦ RNA splicing

⑧ post translational modifications

⑨ Inhibitors of protein synthesis

⑩ Regulation of Gene expression c̄ Lac operon concept

⑪ vectors of rDNA technology

⑫ DNA vaccine

⑬ recombinant Insulin

⑭ Transgenic animals.

⑮ Genetic code

Two very very Important LAQs

1. Describe the process of protein biosynthesis in eukaryotes c̄ post translational modifications and inhibitors.

2. Recombinant DNA technology c̄ its applications in Medicine.

New LAQ

Rt PCR technique for identification of SARS cov-2

BAQs - Gene transfer methods

Inhibitors of transcription

DNA polymerase

RNA polymerase

RNA editing

RNA Types

LAQ :- Describe the various types of RNAs

SAQ :- mRNA, rRNA, tRNA

Important diagram - tRNA structure

- Single stranded
- Polymer of Ribonucleotides
- Connected by 3'- 5' phosphodiester linkage
- Composed of

N-Base	Nucleoside	Nucleotide
Adenine	Adenosine	AMP- Adenosine monophosphate
Guanine	Guanosine	GMP- Guanosine monophosphate
Cytosine	Cytidine	CMP- Cytidine monophosphate
Uracil	Uridine	UMP- Uridine monophosphate

- Synthesized by RNA Polymerase I, II, III

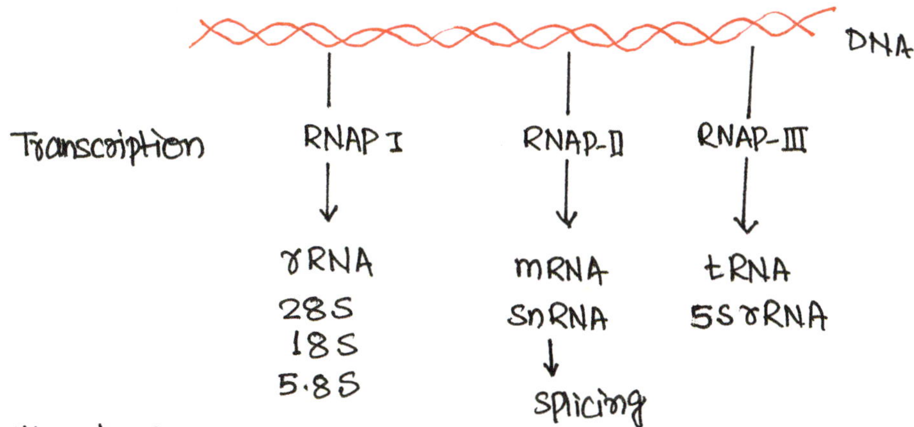

Other types

Small Nucleolar RNA - facilitates modification of RNAs.

Small cytoplasmic RNA - Protein synthesis and targeting

Small interfering RNA [siRNA] → Affect gene expression, Used by the Scientist to Silence gene expression

Micro RNA → Regulates gene expression, causes degradation of mRNA

m RNA - Messenger RNA

coding RNAs
5-10% of total RNAs

function :- carries information from DNA to ribosomes for protein synthesis

produced from hnRNA - heterogenous nuclear RNA by post-transcriptional modifications

Enzyme for synthesis – RNA polymerase II

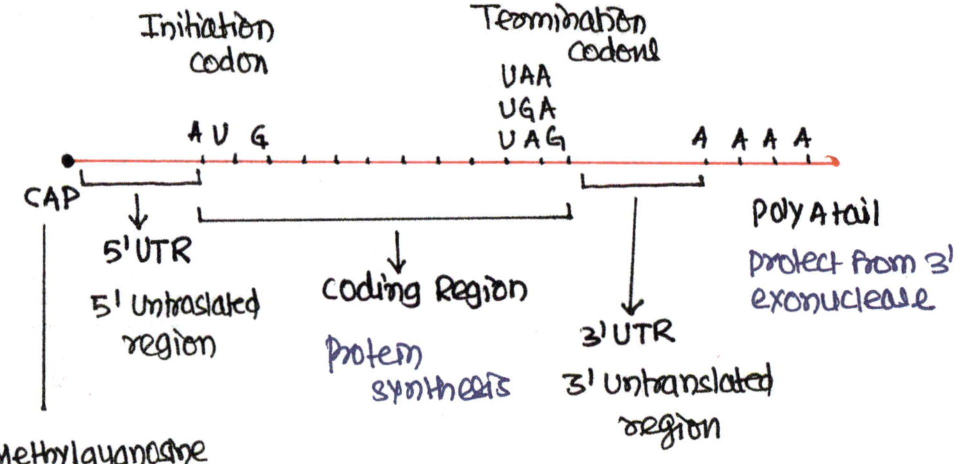

7 Methylguanosine triphosphate cap

- for transport of mRNA from Nucleus to cytoplasm
- promote binding of mRNA to Ribosome
- protect 5' end from exonuclease degradation

tRNA   Transfer RNA
Soluble RNA, Adapter Molecule

15 to 30% of total RNAs

II - Secondary structure - cloverleaf Model

III - Tertiary structure - L-shaped

Composed of 75 to 90 Nucleotides

From 5' to 3' end → D-arm, Anticodon arm, TΨC arm, Acceptor arm

D A T A

4 5 5 7 Nucleotides

Classified into 20 types depending on Aminoacid attached

Contain unusual bases - Thymine
dihydrouracil
Hypoxanthine

Unusual Nucleoside - Pseudouridine [C1-C5 glycosidic bond betn ribose & N-Base]

Function:- decode the message of mRNA and transfer of Amino acid to Ribosome

3'
A
C
C
5'

Acceptor arm: For attachment of AA to form Aminoacyl tRNA

D-Arm
Dihydrouracil Arm
|
help in recognition of specific aminoacyl t-RNA synthetase

TΨC Arm:-
Thymidine-pseudouridine cytosine
|
important for attachment of tRNA to Ribosome

Extra/variable arm

Anticodon arm
contain sequence of 3 nucleotides Complementary to mRNA

rRNA Ribosomal RNA

60-80% of total RNAs

catalytic RNA

formation of Ribosomes

Function :- site for protein synthesis

Involved in formation of peptide bond during chain elongation

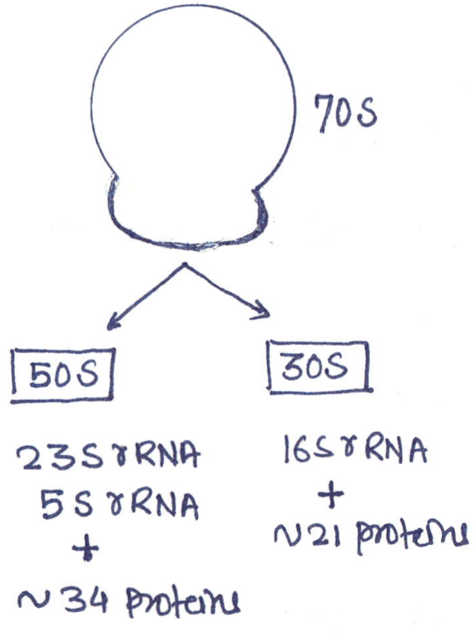

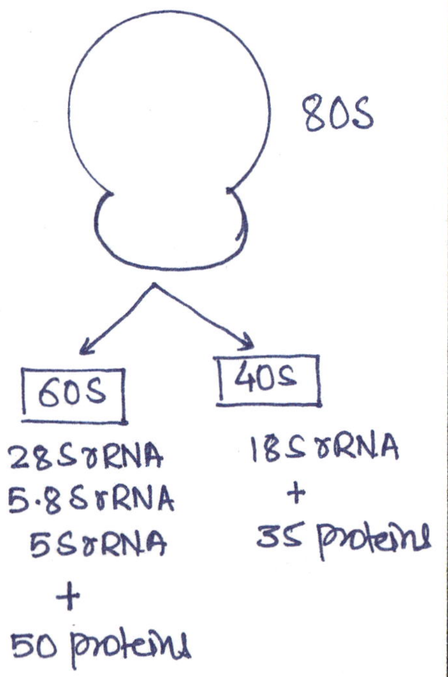

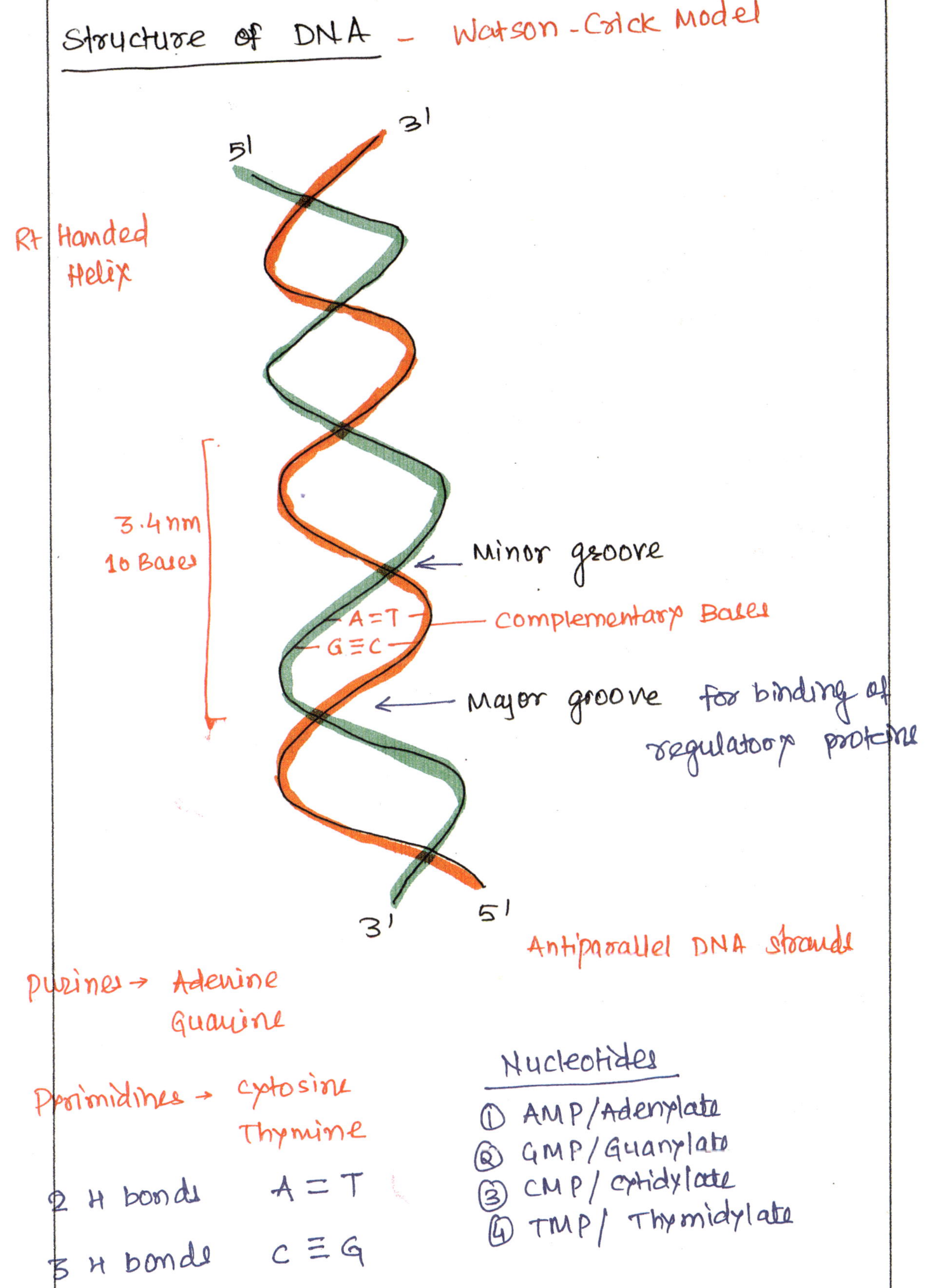
Structure of DNA - Watson-Crick Model
5'
3'
Rt Handed Helix
3.4nm
10 Bases
Minor groove
A=T
G≡C
Complementary Bases
Major groove for binding of regulatory proteins
3'
5'
Antiparallel DNA strands
Purines → Adenine
Guanine
Pyrimidines → Cytosine
Thymine
2 H bonds A = T
3 H bonds C ≡ G
Nucleotides
① AMP/Adenylate
② GMP/Guanylate
③ CMP/Cytidylate
④ TMP/Thymidylate

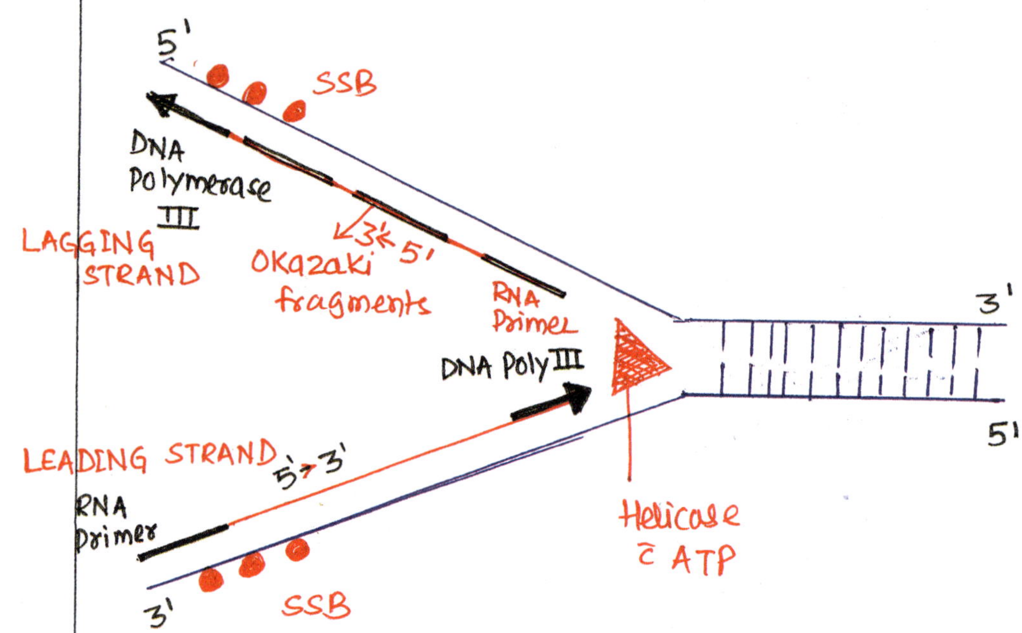

DNA Replication in Prokaryotes

1. semiconservative
2. Begins at origin and proceeds bidirectionally
3. synthesis occurs in 5'→ 3' direction
4. semidiscontinuous
5. Degraded by nucleases
6. Extraordinary degree of fidelity
7. Require DNA Polymerase
8. Requires enzymes and proteins
9. Replication proceeds in stages

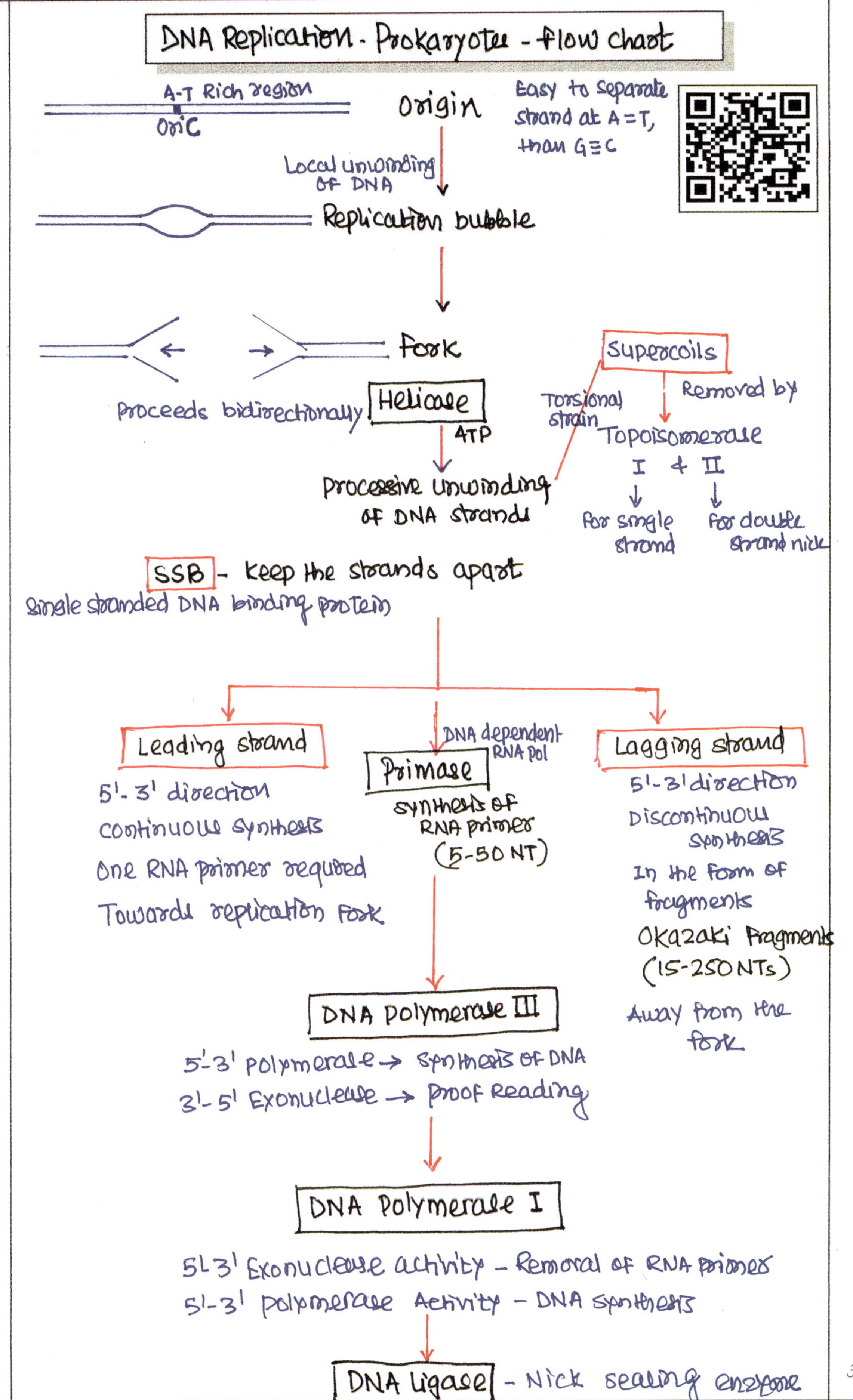
DNA Replication - Prokaryotes - flow chart
A-T Rich region
OriC
Origin
Easy to separate strand at A=T, than G≡C
Local unwinding of DNA
Replication bubble
Fork
Proceeds bidirectionally
Helicase
ATP
Processive unwinding of DNA strands
Supercoils
Torsional strain
Removed by
Topoisomerase
I & II
for single strand
for double strand nick
SSB - keep the strands apart
single stranded DNA binding protein
Leading strand
5'-3' direction
continuous synthesis
One RNA primer required
Towards replication fork
DNA dependent RNA pol
Primase
synthesis of RNA primer (5-50 NT)
Lagging strand
5'-3' direction
Discontinuous synthesis
In the form of fragments
Okazaki fragments (15-250 NTs)
Away from the fork
DNA Polymerase III
5'-3' polymerase → synthesis of DNA
3'-5' Exonuclease → proof Reading
DNA Polymerase I
5'-3' Exonuclease activity - Removal of RNA primer
5'-3' polymerase Activity - DNA synthesis
DNA ligase - Nick sealing enzyme

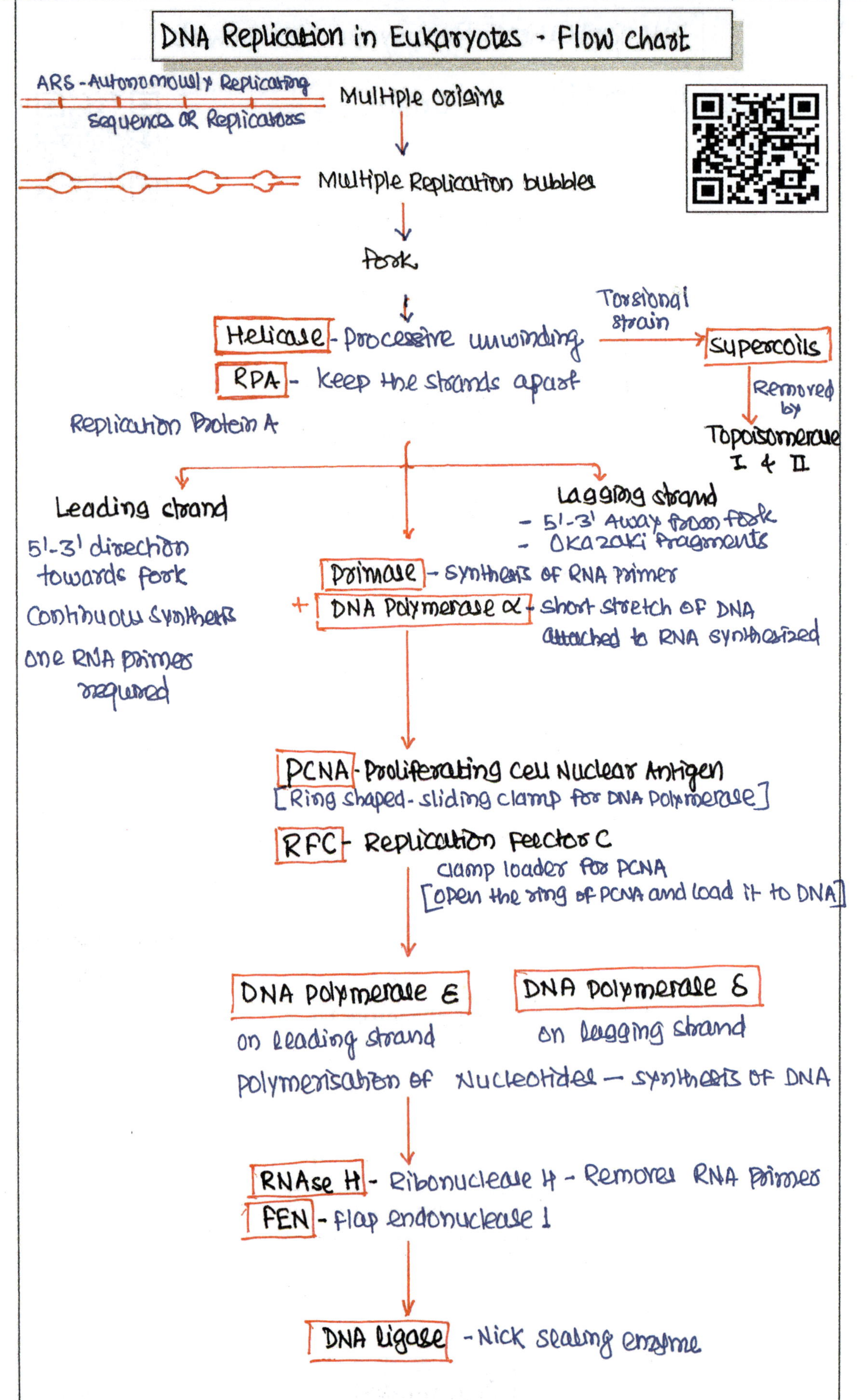
DNA Replication in Eukaryotes - Flow chart
ARS - Autonomously Replicating Sequence or Replicators
Multiple origins
Multiple Replication bubbles
Fork
Helicase - Processive unwinding
RPA - Keep the strands apart
Replication Protein A
Torsional strain
Supercoils
Removed by
Topoisomerase I & II
Leading strand
5'-3' direction towards fork
Continuous synthesis
one RNA primer required
Lagging strand
- 5'-3' Away from fork
- Okazaki fragments
Primase - Synthesis of RNA primer
+ DNA Polymerase α - short stretch of DNA attached to RNA synthesized
PCNA - Proliferating cell Nuclear Antigen
[Ring shaped - sliding clamp for DNA Polymerase]
RFC - Replication Factor C
clamp loader for PCNA
[open the ring of PCNA and load it to DNA]
DNA polymerase ε
on leading strand
DNA polymerase δ
on lagging strand
polymerisation of Nucleotides — synthesis of DNA
RNAse H - Ribonuclease H - Removes RNA primer
FEN - Flap endonuclease 1
DNA ligase - Nick sealing enzyme

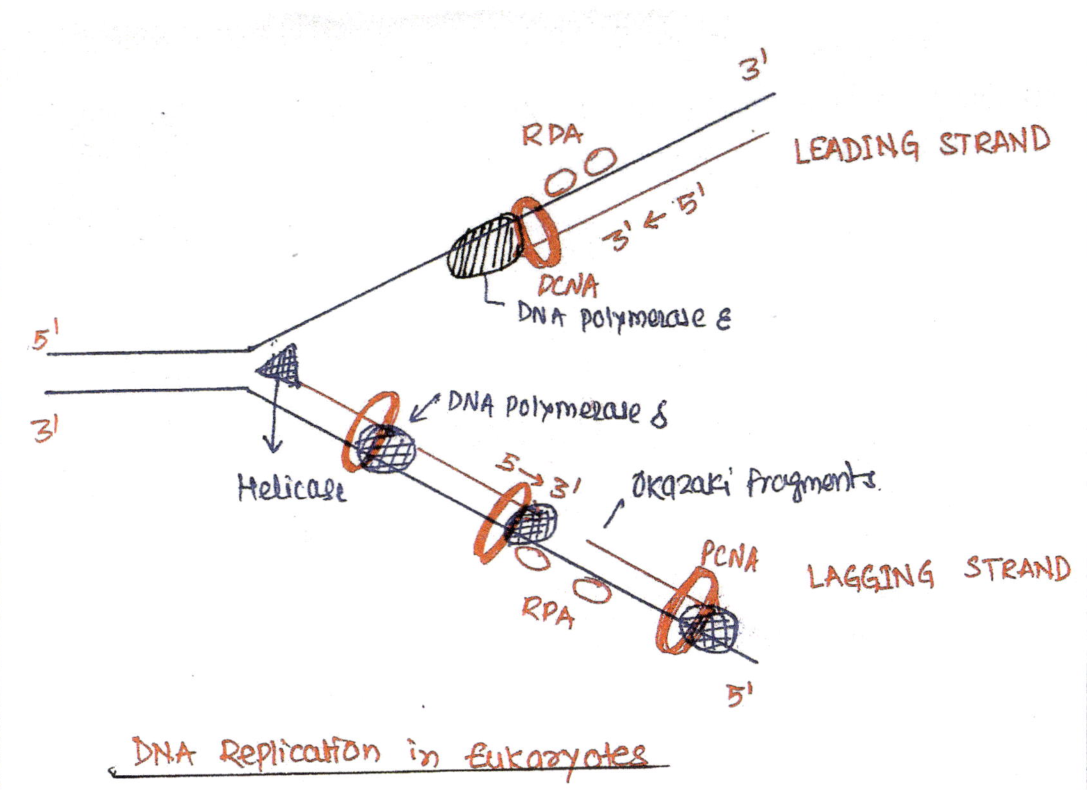

DNA Replication in Eukaryotes

DNA Polymerase

Prokaryotes	Eukaryotes	Function
I		Gap filling following replication Repair and recombination
II		DNA Proof reading & Repair
III	β	DNA Repair
	γ	mt-DNA Synthesis
	ε	Processive leading strand synthesis
	α	synthesis of short DNA, attached to RNA
	δ	Processive lagging strand synthesis

TELOMERES

SAQ :- Write about telomeres. Add a note on clinical significance of Telomerase.

- It is a 6 Nucleotide [TTAGGG] repeats at the end of chromosome
- 15000 bp
- Maintained by enzyme **Telomerase** - Protein + RNA
  complementary to region of Telomeres repeat.

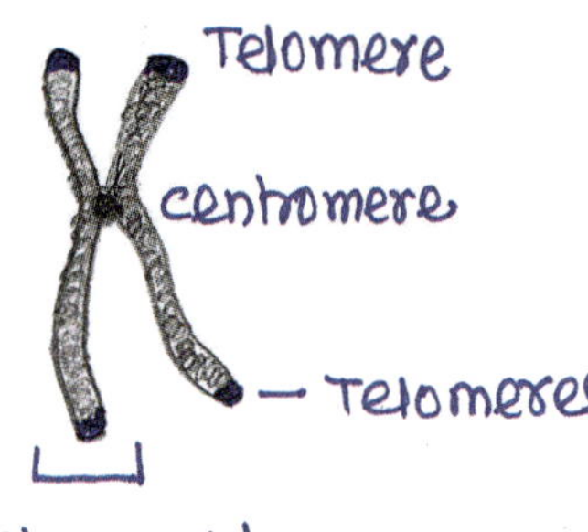

Chromosome

function

1. Prevents loss of DNA molecule (genetic information) at the end of chromosome
2. Prevents fusion of DNA
3. Protect the DNA from Nuclease, gives stability.

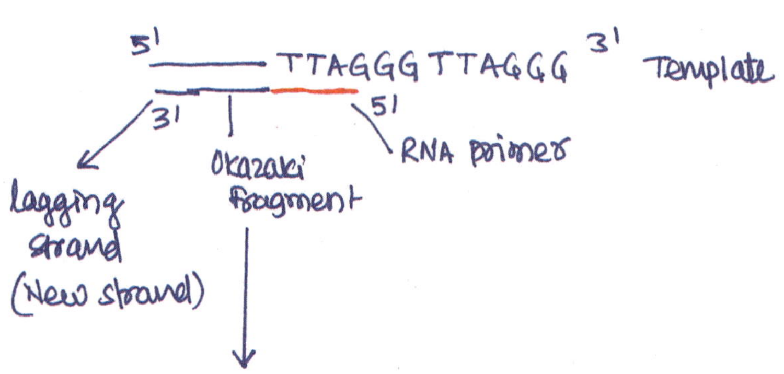

Removal of RNA primer leaves a small gap, that becomes significant after several cell division - shortening of chromosome

Telomeres

5' ———————— TTAGGGTTAGGG 3' Template

3' ———— 5'

New DNA

CAAUCCCAA
Telomerase — Telomere terminal transferase

↓

Lengthening of Template strand (Noncoding region) by translocation and synthesis

↓ Telomerase act as a template for extention of Telomeres

Primase and DNA polymerase causes synthesis of New DNA.

extended Telomeres

5' Template ———— TTAGGGTTAGGGTTAGGGTTA 3'

3' ————- - - - - - - - - - - ———— 5'

New DNA

↑ synthesis of New DNA by DNA polymerase

↑ primer

Telomerase

- Telomere terminal transferase [Protein + RNA]
- Absent in somatic cell, present in germ cell, stem cell.
- Increased activity in cancer cell
- Telomerase activity decreases with aging.
- complementary to Telomere repeat c̄ Reverse transcriptase activity
- The telomeres TTAGGG repeats at the end of DNA is maintained by Telomerase.
- It can be a potential target for cancer chemotherapy
- It corrects the end replication error.

DNA Repair

DNA damage - causes

1. free Radicals
2. Physical agent
   - UV Rays
   - X-Ray
3. chemical agents
   - Aflatoxin
   - Alkylating agent
   - Hydrazine
   - Ozone

Types of DNA damage

1. **Single base alterations** :- spontaneous deamination
   - Cytosine ⟶ Uracil
   - Adenine ⟶ Hypoxanthine
   - Depurination
   - Deletion or insertion of nucleotide
   - Base analog incorparation
   - Alkylating agents
2. **Two base alterations** :- T-T dimer formation (by UV rays)
   - Alkylating agents
3. **Chain breaks** :- Ionizing radiation
   - Radioactive disintegration
   - Oxidative damage
4. **Cross linkage** :- between bases of same or opposite strand
   - between DNA & protein

DNA Repair Mechanism

1. **Mismatch Repair** :- Repair of mismatched base
2. **Base Excision Repair** :- For deamination, depurination or damage by alkylating agents.
3. **Nucleotide Excision Repair** :- UV light induced T-T dimer repair OR T-C cross linkage
4. **Double strand break Repair** :- Break in DNA strand
5. **Direct Repair** :- For modified bases, damage by alkylating agents
   T-T pyrimidine dimer formation

1. Mismatch Repair – Metyl directed mismatch repair

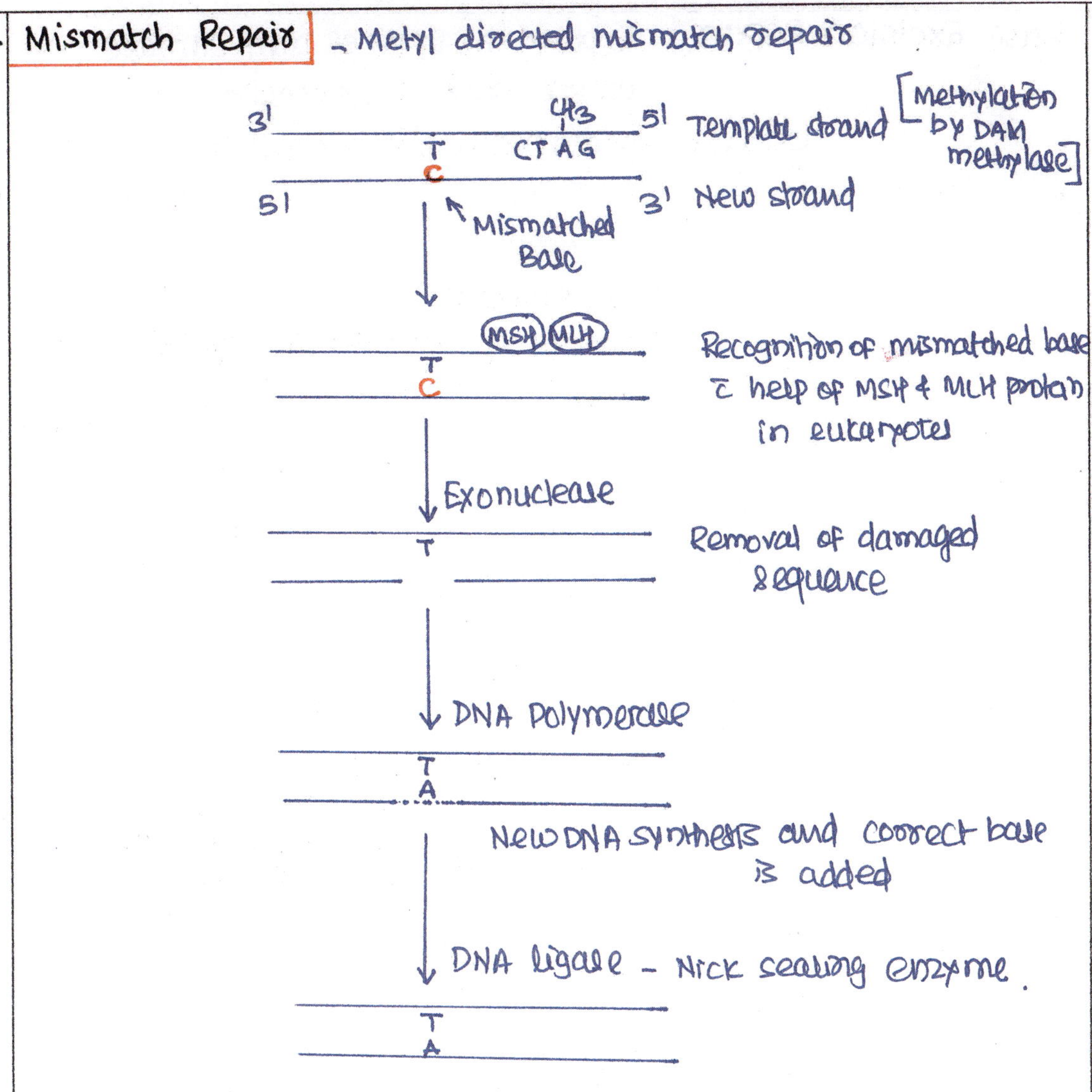

Clinical Significance

Faulty mismatch repair causes – Hereditary Nonpolyposis Colon cancer [HNPCC] (Lynch syndrome)

Autosomal dominant

due to mutation in MSH & MLH proteins

colon cancer
endometrial cancer
ovarian cancer etc.

4 Basic steps in DNA Repair

1. Recognition
2. Removal
3. Synthesis
4. Ligase

2. Base Excision Repair

- Repairs the type of DNA damage where base is damaged.

Causes - spontaneous deamination

Cytosine ⟶ Uracil

Adenine ⟶ Hypoxanthine

Depurination

Damage by alkylating agents.

G — Template

C — New DNA

↓ spontaneous deamination C ⟶ U

G

U — damaged DNA

1. Recognition ↓ Uracil DNA Glycosidase — cleave Uracil

G

Only Ribose-Ⓟ backbone remains — Abasic site

2. Removal ↓ AP endonuclease

3. Synthesis ↓ DNA Polymerase

4. Ligase ↓ DNA ligase

G

C

3. Nucleotide Excision Repair

- Repairs UV light induced
  T-T dimer (Pyrimidine dimer)
  T-C cross linkage

T-T — UV-Rays induced dimer formation

1. Recognition ↓ Exciunclease binds to lesion

(T-T)

2. Removal of segment ↓ Removal of affected segment.

3. Synthesis ↓ DNA Polymerase
4. Ligation ↓ DNA Ligase

Clinical Significance :- Faulty Nucleotide excision Repair

↓

Xeroderma Pigmentosum

- UV-rays induced T-T Pyrimidine dimer formation (covalent bonding)
- person is sensitive to sunlight
  - Multiple freckles on sun exposed part of body
- Precancerous condition

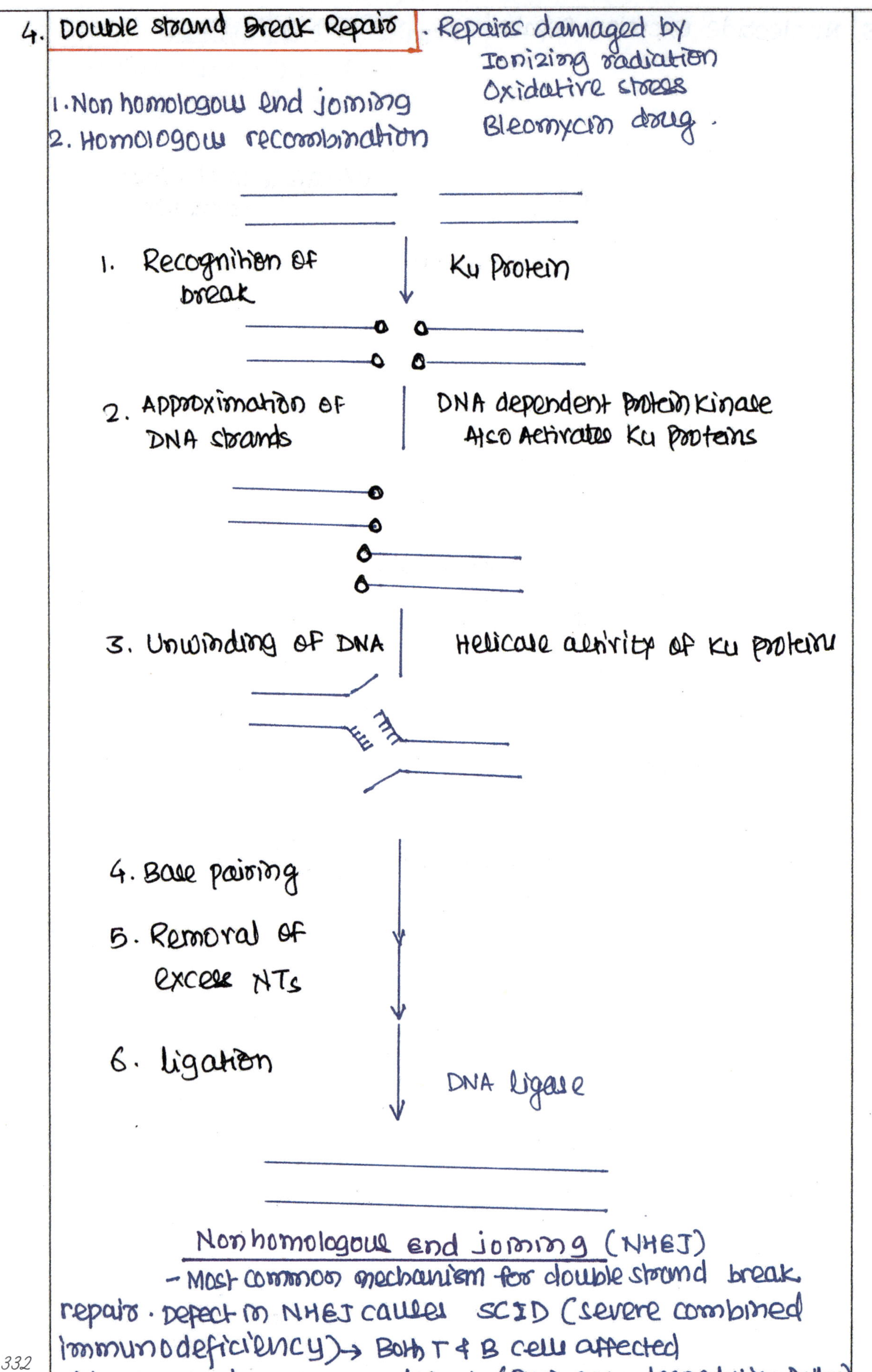
4. Double strand Break Repair
. Repairs damaged by
Ionizing radiation
Oxidative stress
Bleomycin drug.
1. Non homologous end joining
2. Homologous recombination
1. Recognition of break
Ku Protein
2. Approximation of DNA strands
DNA dependent Protein Kinase
Also Activates Ku Proteins
3. Unwinding of DNA
Helicase activity of Ku proteins
4. Base pairing
5. Removal of excess NTs
6. ligation
DNA ligase
Nonhomologous end joining (NHEJ)
- Most common mechanism for double strand break repair. Defect in NHEJ causes SCID (severe combined immunodeficiency) → Both T & B cells affected
Adenosine deaminase defect (Purine NT degradation Pathway)

Homologous Recombination

- Repair mechanism for double strand break repair
- lost DNA is synthesized by using homologous part of gene.
- homologous part of sister chromatid act as a template for synthesis of DNA
- occurs in S and G_2 phase of cell cycle

Break in doublestrand of DNA

Homologous DNA

Homologous recombination

Nonhomologous end joining	Homologous recombination
1. Break ends are directly ligated without need for homologous template	Required homologous DNA as a template
2. Fast process	Slow process
3. In G_0/G_1 phase of cell cycle	occurs in S and G_2 phase
4. Deletion of NTs occurs, hence Mutagenic Disease - defect causes SCID	occurs with high fidelity less mutations. Disease:- Defect causes ATLD, Bloom syndrome etc.

Faulty double strand break repair (Homologous Recombination)

1. <u>Ataxia telangiectasia</u> - Affect Nervous system and immune system
Ataxia → Progressive difficulty in coordinating movements.

Telangiectasia → small clusters of enlarged blood vessels.

2. <u>Fanconi's Anemia</u> - Affect the bone marrow
↓ production of all types of cells
Aplastic anemia, leukemia

3. <u>Bloom's syndrome</u> → short stature, sun sensitive skin rash
↑ susceptibility to cancer

4. <u>Breast cancer susceptibility</u> - BRCA-1, BRCA-2

5. Direct Repair - Repairs pyrimidine dimer formation → Mainly by NER
damage by Alkylating agent
Modified bases

Guanine

↓ Damage — Alkylating agent

Methyl Guanine

↓ Methyl transferase

Guanine.

TRANSCRIPTION IN PROKARYOTES

LAQ: Describe the process of transcription in prokaryotes. Enumerate inhibitors of transcription.

Definition:- synthesis of RNA from DNA

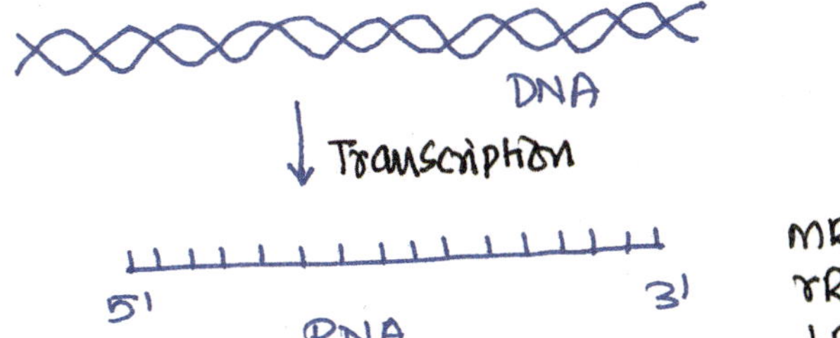

central dogma of molecular biology — DNA → (Transcription) RNA → (Translation) Proteins

RNA Polymerase

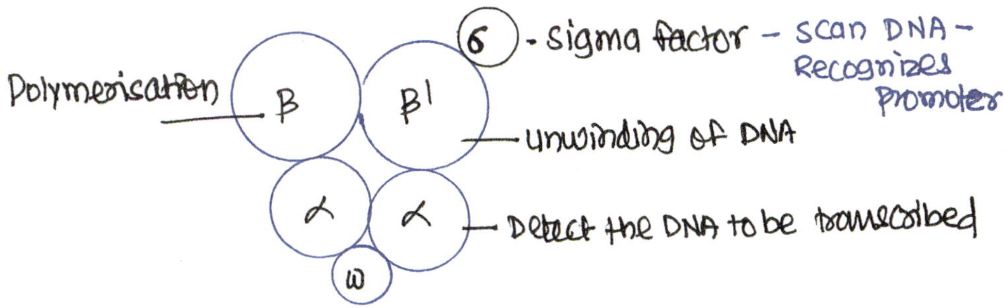

Holoenzyme RNA Polymerase

Steps in transcription process -
1. Initiation
2. Elongation
3. Termination

Promoter :- Region on coding strand of DNA, where RNA polymerase can bind.

Template strand :- strand from which RNA is synthesized.

Coding strand :- opposite strand of template DNA, Nucleotide sequence is same as that of new RNA except Thymine in DNA

Upstream elements :- before the transcription start site

downstream elements :- sequences after the transcription start site.

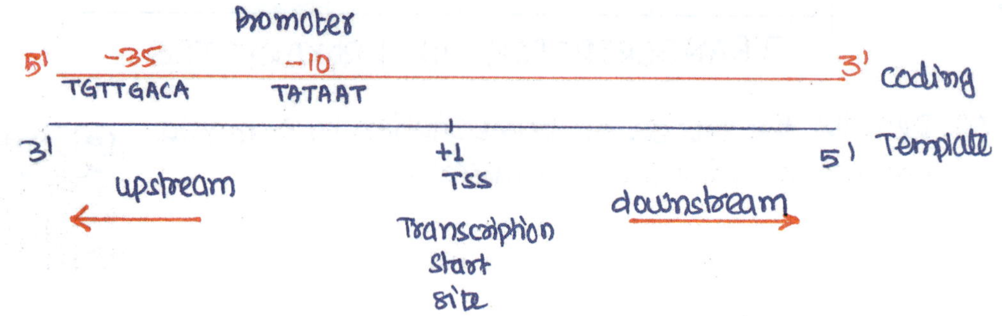

Consensus sequences in promoter: Helps RNAP to recognize & bind to DNA

−10 bp upstream – TATAAT (TATA box OR Pribnow box)

−35 bp upstream – TGTTGACA – Initial site for attachment of RNA polymerase

INITIATION

1. Formation of closed complex

Attachment of RNA polymerase to the Promoter region

Sigma factor (σ) – Scan the DNA
Recognizes −35 bp sequence
helps in attachment of RNA Polymerase to DNA

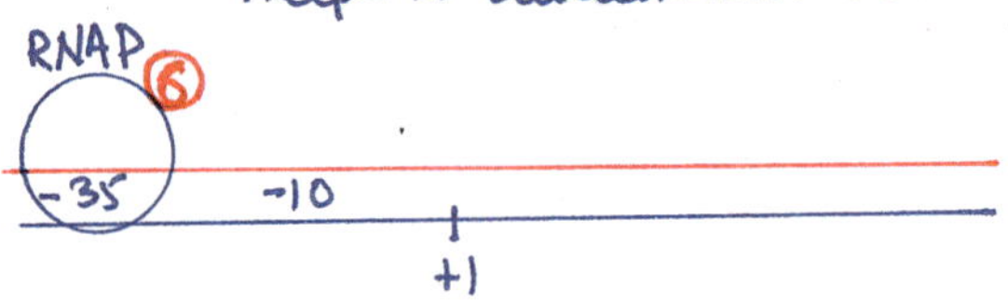

2. Formation of open complex

Local unwinding of DNA in the region of TATA box

3. Initiation of RNA synthesis: Synthesis of RNA from template strand, complementary to template strand by RNA polymerase
No Need of Primer, short stretch of RNA is synthesized – 5-10 bp of RNA

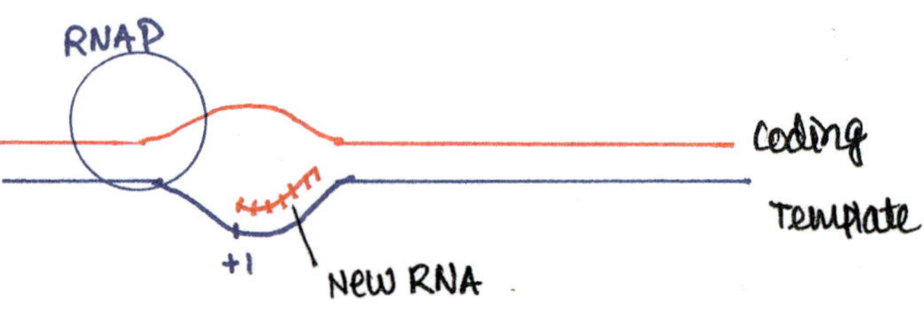

4 Promoter clearance

RNA polymerase is shifted to transcription site and moves away from the promoter region

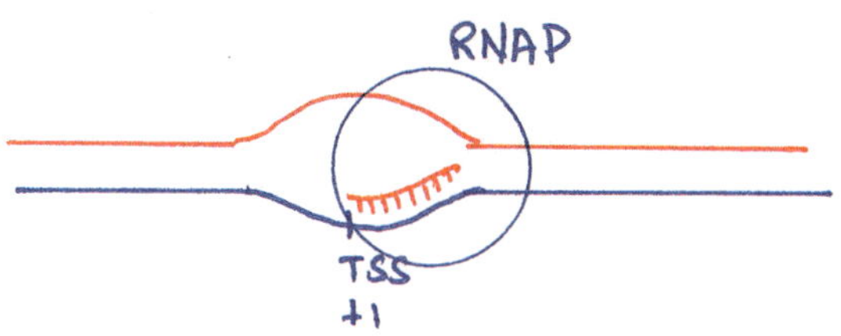

ELONGATION

Ribonucleotides are added and chain will grow continuously

formation of 3'- 5' phosphodiester linkage

No proofreading by RNA polymerase

Protein bridge - provides only one nucleotide as a template for RNA synthesis

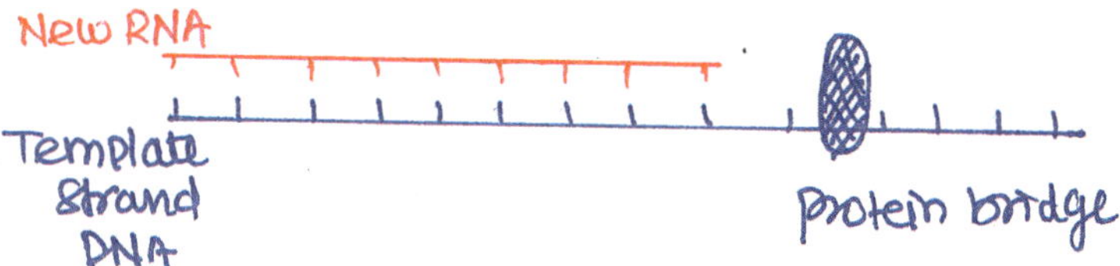

Nucleotides are added to 3' end of RNA molecule.

TERMINATION

1. Intrinsic termination :-

UUUUU Poly U in RNA

3' AAAAA 5'

Poly A in DNA Template

5' RNA

G C G C G C

C G C G C G

G-C rich sequences in RNA

Complementary base pairing

↓

Hairpin/ Loop like structure

↓

Termination of transcription

Intrinsic signals - Poly U in RNA and Hairpin due to G-C rich region signals the termination of transcription.

2. Rho protein dependent termination

RNA activated Helicase activity - requires ATP

Goes in rotating manner and cut between DNA & RNA.

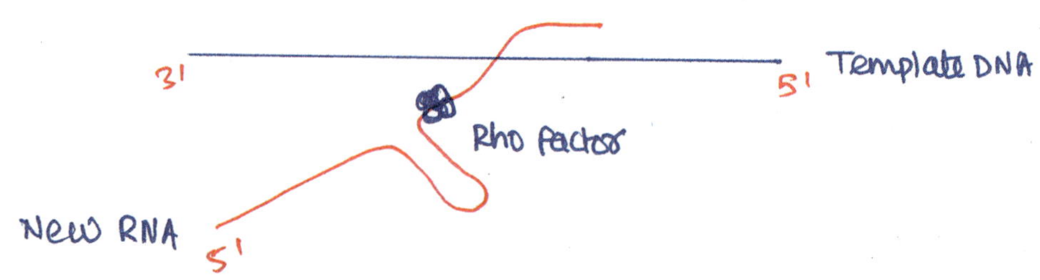

Inhibitors of prokaryotic transcription

Actinomycin D - Antibiotic, intercalates into double helical DNA and blocks the movement of RNA polymerase.

Rifampicin :- Antitubercular drug used for prevention and treatment of tuberculosis.

Binds to β-subunit of RNA polymerase and blocks the transcription.

TRANSCRIPTION IN EUKARYOTES

LAQ :- Describe the process of transcription in eukaryotes. Add a note on post transcriptional modifications.

Definition :- It is process of synthesis of RNA from DNA.

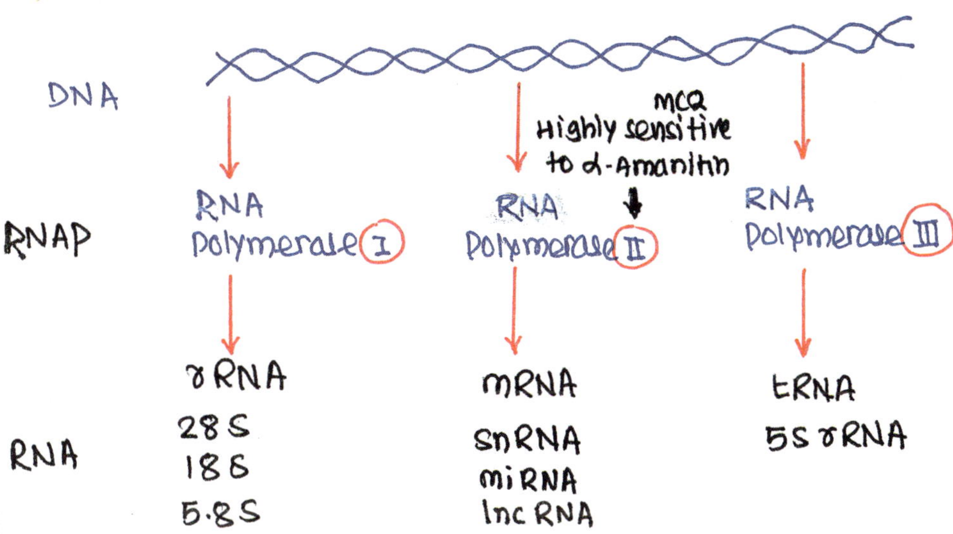

Requirement for transcription

1. Activators
2. Basal transcription factors - TBP, TF - II A, II B, II F, II E, II H
3. Coregulators ⇒ TF II D [TBP + 14 TAF]
   - TBP → TATA binding Protein
   - TAF → Transcription Associated factors.

INITIATION

Regulatory elements — Promoter Proximal elements — Promoter

5' Enhancer OR Repressor — -70 CAAT — -25 TATAAA — 3' coding strand

3' — +1 TSS — transcribed region — 5' Template

Upstream elements

1. **Promoter** :- Help RNA Polymerase to recognize sequence in DNA site for attachment of RNA Polymerase

① -25 bp upstream - TATAAA — TATA BOX
Initiator element (Inr) OR
Downstream promoter elements
} directs RNAP to the correct site

② Proximal promoter element '-70 sequence'
CAAT / GC region (-90)
} specify the frequency of initiation

③ Enhancer / Repressor elements → Increases or decreases the transcription process.

2. **Formation of Preinitiation Complex (PIC)**

TBP → Recognizes promoter region and Helps in binding of TF II D

↓

TF II D → Bend DNA to ~100°, Allow all factors to work simultaneously on bent DNA

↓

TF II A → stabilizes association between TBP and TF II B

↓

TF II B

↓

TF II F c̄ RNA Polymerase II

↓

TF II E

↓

TF II H Helicase activity – unwinding of DNA
Activates RNA polymerase by Phosphorylation of carboxyl terminal domain
Enzyme involved in Nucleotide excision repair.

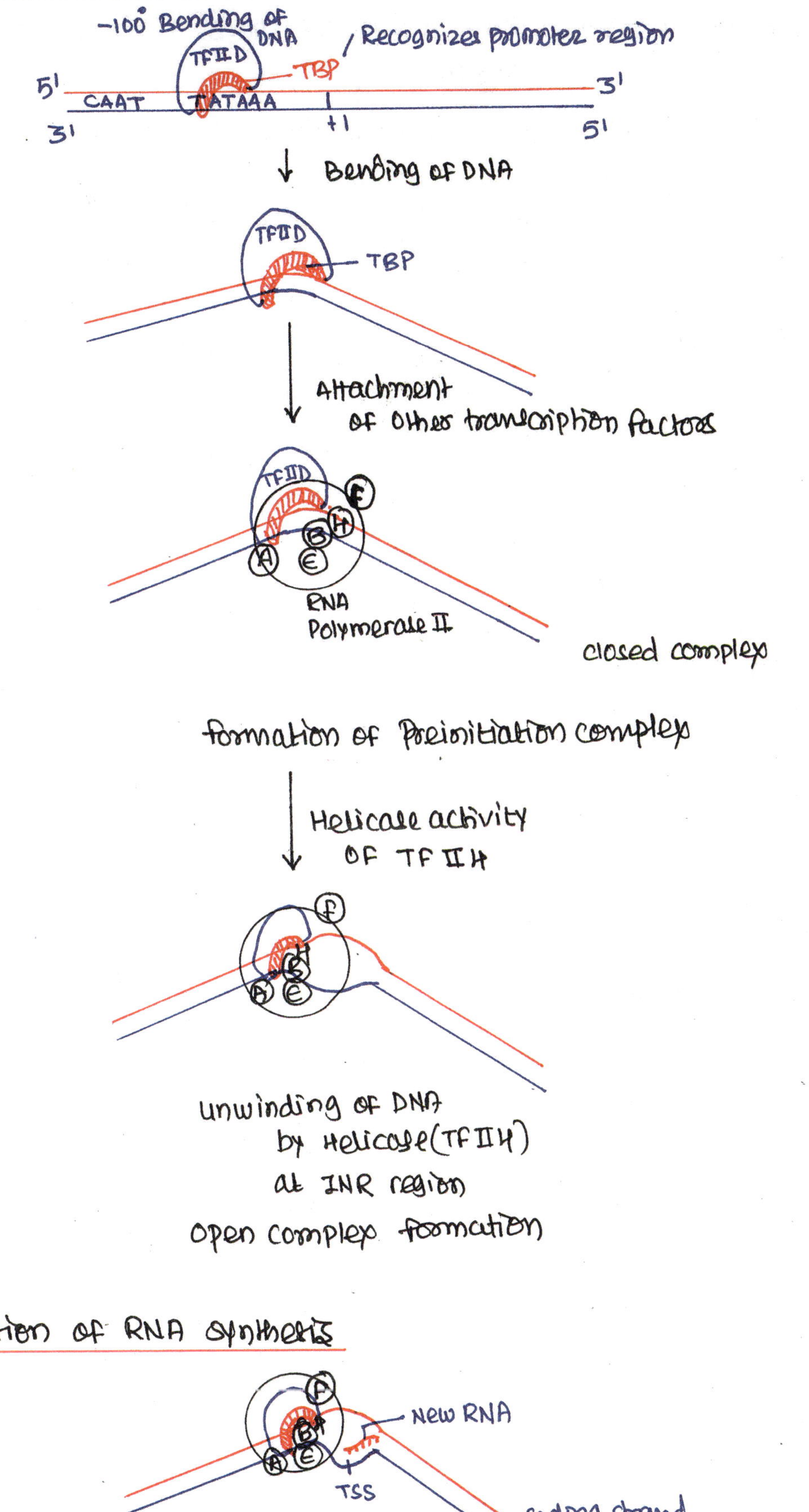

3. Initiation of RNA synthesis

Short stretch of RNA complementary to template strand, in 5'- 3' direction is synthesized

4. Promoter clearance : RNA polymerase moves away from promoter region
↓
All other transcription factors released
↓
only TFIIF in association c̄ RNA polymerase remains.

ELONGATION

New nucleotides are added to 3' end of growing RNA
formation of 3'- 5' phosphodiester linkage
synthesis will occur in 5'-3' direction
Base pairing is complementary to template DNA strand.

TERMINATION

Intrinsic sequences in growing RNA signals the termination

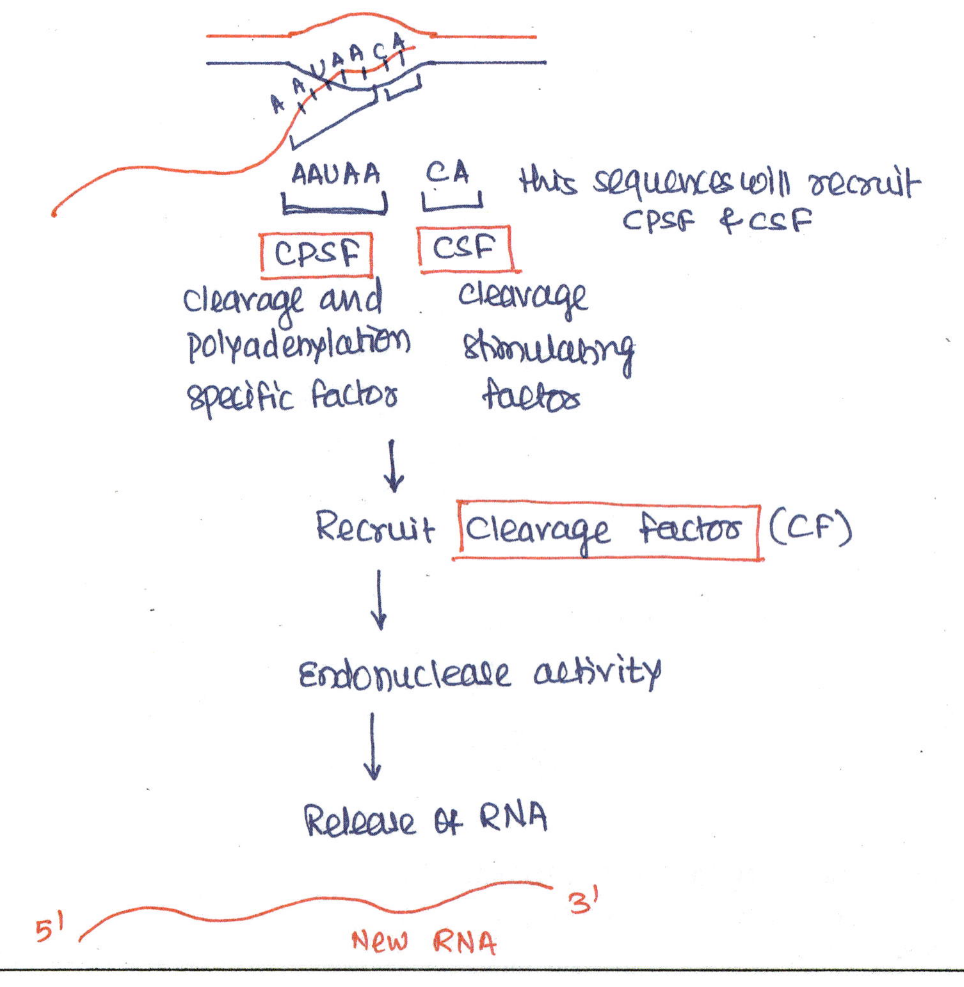

Posttranscriptional Modifications

SAQ:- Describe the Posttranscriptional modifications in RNA

Definition:- After synthesis of RNAs, it undergoes some modification or changes known as post-transcriptional modifications

hnRNA (Primary transcript) ⟶ mRNA

1. 5' capping :- Attachment of 7 methylguanosine triphosphate to 5' end of RNA c̄ triphosphate linkage.

Function :- Protect from 5' exonuclease activity and give stability to mRNA
Help in translocation of mRNA from Nucleus to cytoplasm
Help in binding of mRNA to Ribosomes.

7 Methylguanosine triphosphate
5' ... AAAAAAAA 3'

2. 3' Poly A tail :- Added by polyadenylate polymerase
Attachment of multiple adenine molecule (80-250 NT) to the end of mRNA

Function :- Protect the 3' end from 3' exonuclease activity
Help in recruitment of 40S ribosomal unit to mRNA

3 splicing :- Joining of two ends (Joining of two exons)
Removes the introns (Non-coding sequences) and join the exons (coding sequences)

Two types

① . Introns Mediated (Self Mediated)

No ATP Required
No Factor/enzyme required
Intrinsic signals for splicing
Group I and Group II introns

②. Spliceosome mediated splicing

- for Group III introns
- Required protein + RNA
- 5 RNAs (snRNA) + 60 proteins
SNURP - Small Nuclear Ribonucleoproteins

1. Self Mediated splicing

Steps - 1. Recognition of Sequence
2. 5' cut
3. Lariat OR Loop formation
4. 3' cut
5. Joining of Exons

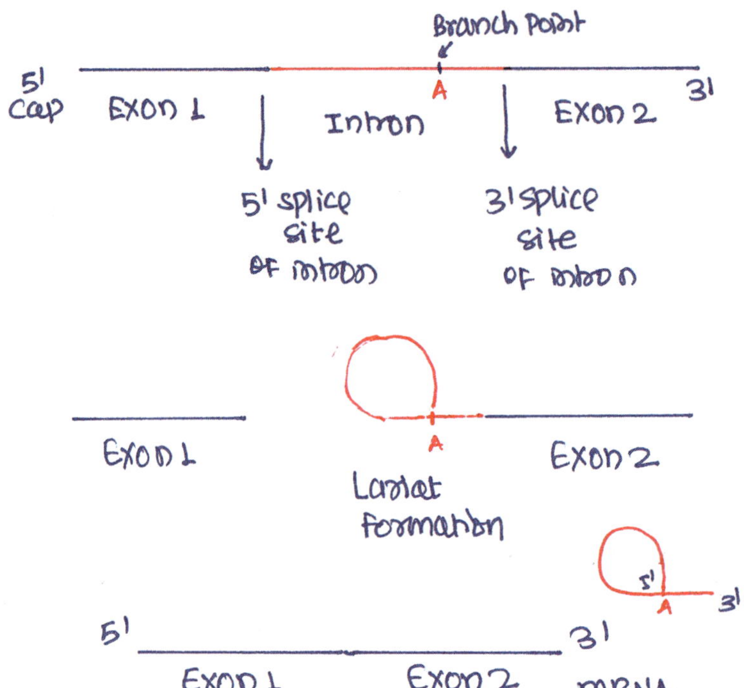

2. Spliceosome mediated splicing

5 snRNA + 60 proteins ⟶ SNURP

U_1
U_2
U_4
U_5
U_6

Steps

1. Attachment of U_1 snRNA to 5' exon-intron junction
2. U_2 bind to branch point
3. Inactive complex of snRNA U_4, U_5, U_6 bind
4. This complex is activated by removal of U_4
5. U_1 is replaced by U_6
6. Lariat or loop formation occurs around active complex
7. Removal of introns
8. Exons are joined together.

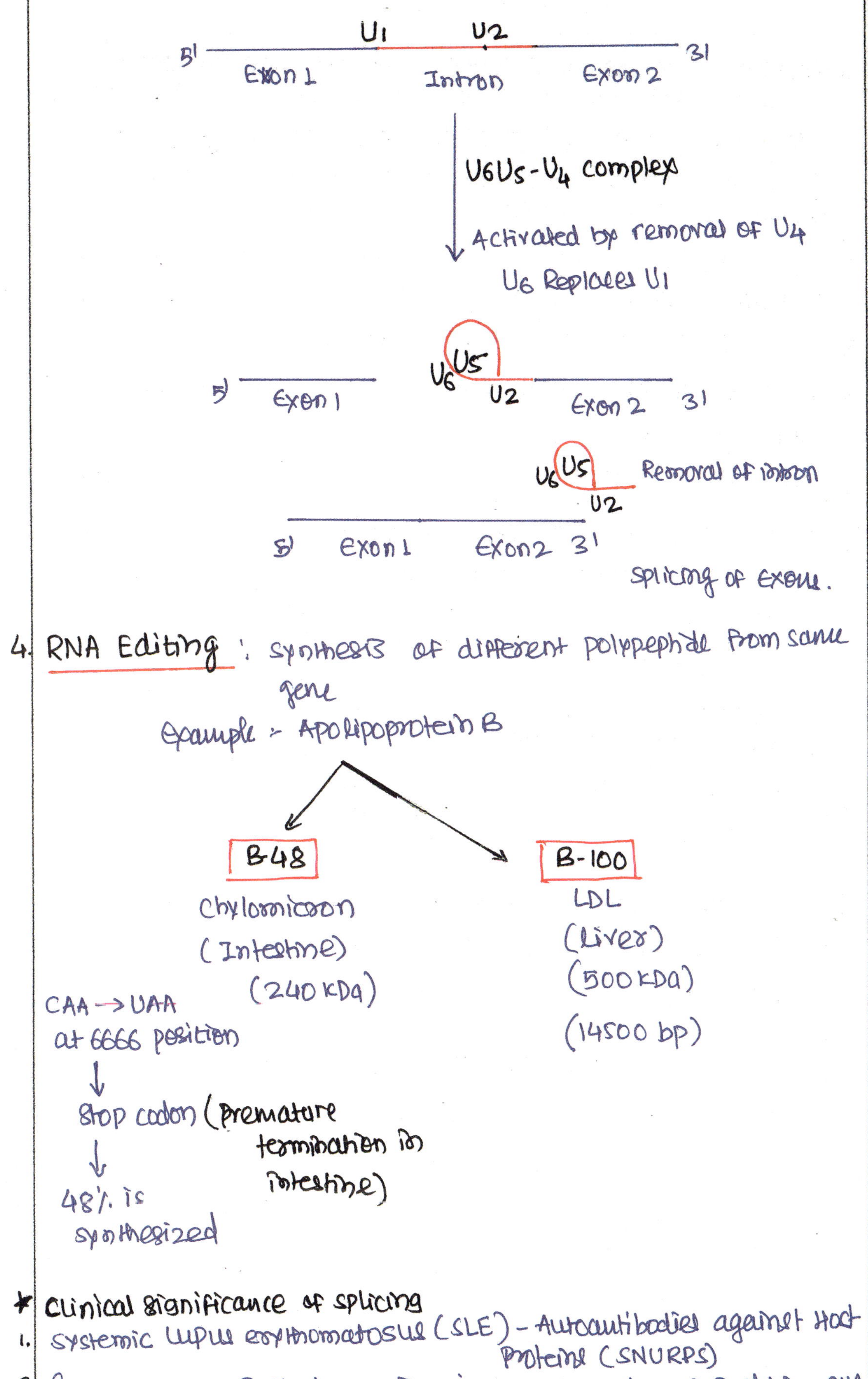
U1
U2
5'
3'
Exon 1
Intron
Exon 2
U6U5-U4 complex
Activated by removal of U4
U6 Replaces U1
U6
U5
U2
5'
Exon 1
Exon 2
3'
U6
U5
U2
Removal of intron
5'
Exon 1
Exon 2
3'
Splicing of exons.
4. RNA Editing : Synthesis of different polypeptide from same gene
Example :- Apolipoprotein B
B-48
Chylomicron
(Intestine)
(240 kDa)
B-100
LDL
(Liver)
(500 kDa)
(14500 bp)
CAA → UAA
at 6666 position
Stop codon (premature termination in intestine)
48% is synthesized
* Clinical significance of splicing
1. Systemic lupus erythromatosus (SLE) - Autoantibodies against Host Proteins (SNURPS)
2. Some cases of β-thalassemia - incorrect splicing of β-globin mRNA

GENETIC CODE

Definition:- A set of three-letter combination of Nucleotides called codon, each of which correspond to a specific AA or stop signal

AUG — mRNA

1. The code is a triplet codon :- each codon consisting of 3 successive nitrogen bases
2. The code is non-overlapping - The codons do not overlap but are read sequentially
3. The code is commaless - No punctuation
4. Non-ambiguous - particular codon will always code for same AA
5. The code has polarity - Always read in 5-3' direction
6. Degenerate - More than one codon may specify same AA known as degeneracy of genetic code

   UCU
   UCC
   UCA
   UCG } Code for serine

7. In most organism AUG is the start OR initiation codon
8. Some are stop codons UAG - amber
   UAA - ochre
   UGA - opal
9. The code is universal - It is same in all organism
   some exception - mitochondria gene

   Each codon is specific for only one AA

Francis Crick - proposed Wobble hypothesis for degeneracy of genetic code - Freedom is allowed in the base pairing at the 3rd position

TRANSLATION IN PROKARYOTES

LAQ: Describe the process of protein synthesis in prokaryotes and write about inhibitors.

Definition:- Information present on mRNA in the form of genetic code is utilized for the synthesis of proteins/peptides is known as translation.

mRNA

5' — 1 2 3 4 5 6 7 8 9 10 — 3'

codons - Triplet base pair sequences present on mRNA

↓ Translation

AA-1 – AA-2 – AA3 – AA4 – AA5 – AA6 – AA7 – AA8 – AA9 – AA10

protein/ peptide

Requirements

RNAs - mRNA - codons
tRNA - carrier of aminoacids
rRNA - part of Ribosomes
catalyst

Amino acids - 20 amino acids

Factors → IF 1, 2, 3, 5, 6 - Initiation
Elongation - EF Tu, EF-G
Termination - RF 1, 2, 3

Ribosome - site for protein synthesis

70S → 50S + 30S

50S: 23S rRNA, 5S rRNA + ~34 proteins

30S: 16S rRNA + ~21 proteins

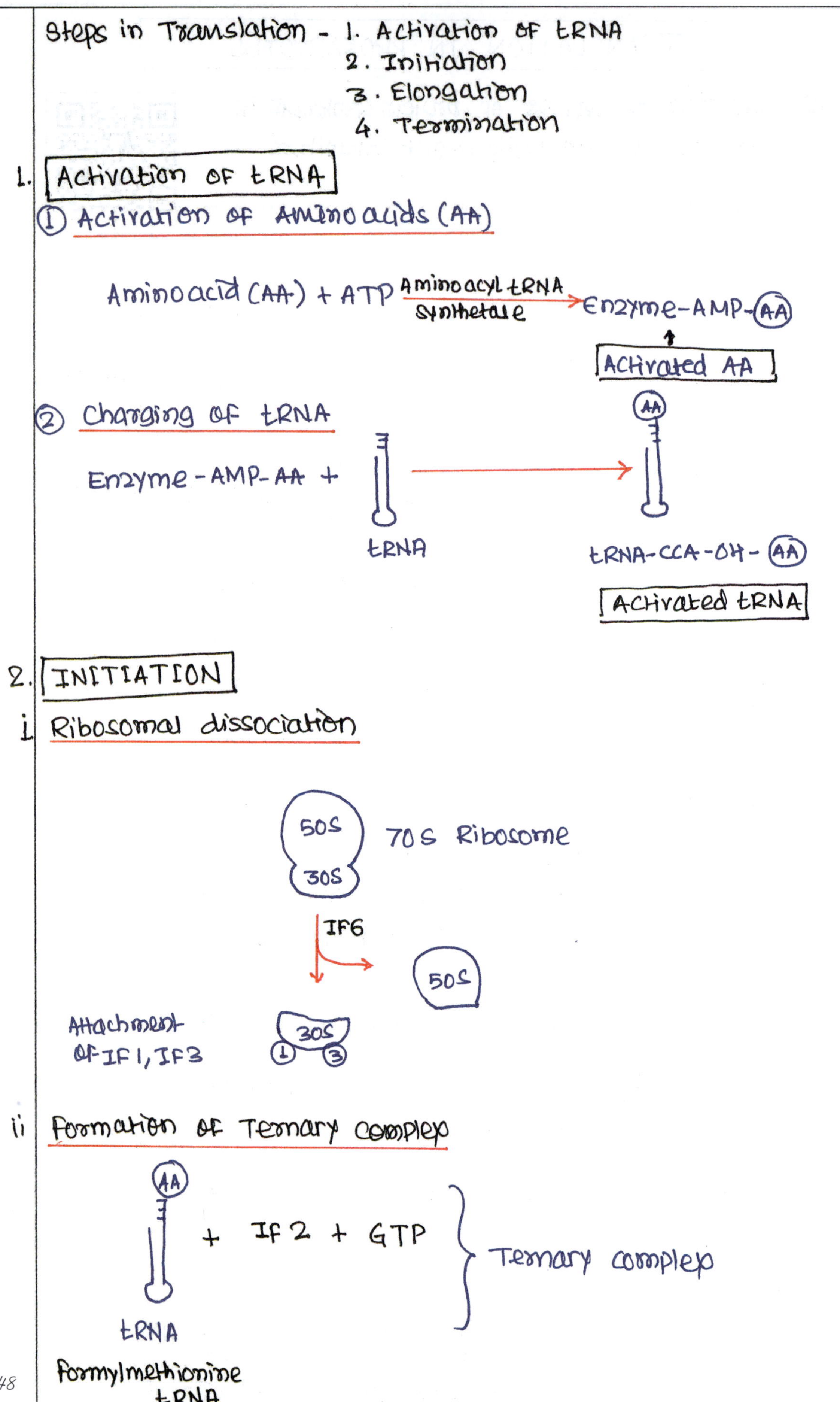
Steps in Translation - 1. Activation of tRNA
2. Initiation
3. Elongation
4. Termination
1. Activation of tRNA
① Activation of Amino acids (AA)
Aminoacid (AA) + ATP
Aminoacyl tRNA synthetase
Enzyme-AMP-AA
Activated AA
② Charging of tRNA
Enzyme-AMP-AA +
tRNA
AA
tRNA-CCA-OH-AA
Activated tRNA
2. INITIATION
i. Ribosomal dissociation
50S
30S
70S Ribosome
IF6
50S
Attachment of IF1, IF3
30S
1
3
ii Formation of Ternary complex
AA
tRNA
+ If 2 + GTP
Ternary complex
Formylmethionine tRNA

iii. formation of 30S Initiation complex

30S Initiation complex

iv. formation of 70S Initiation complex

70S Initiation complex

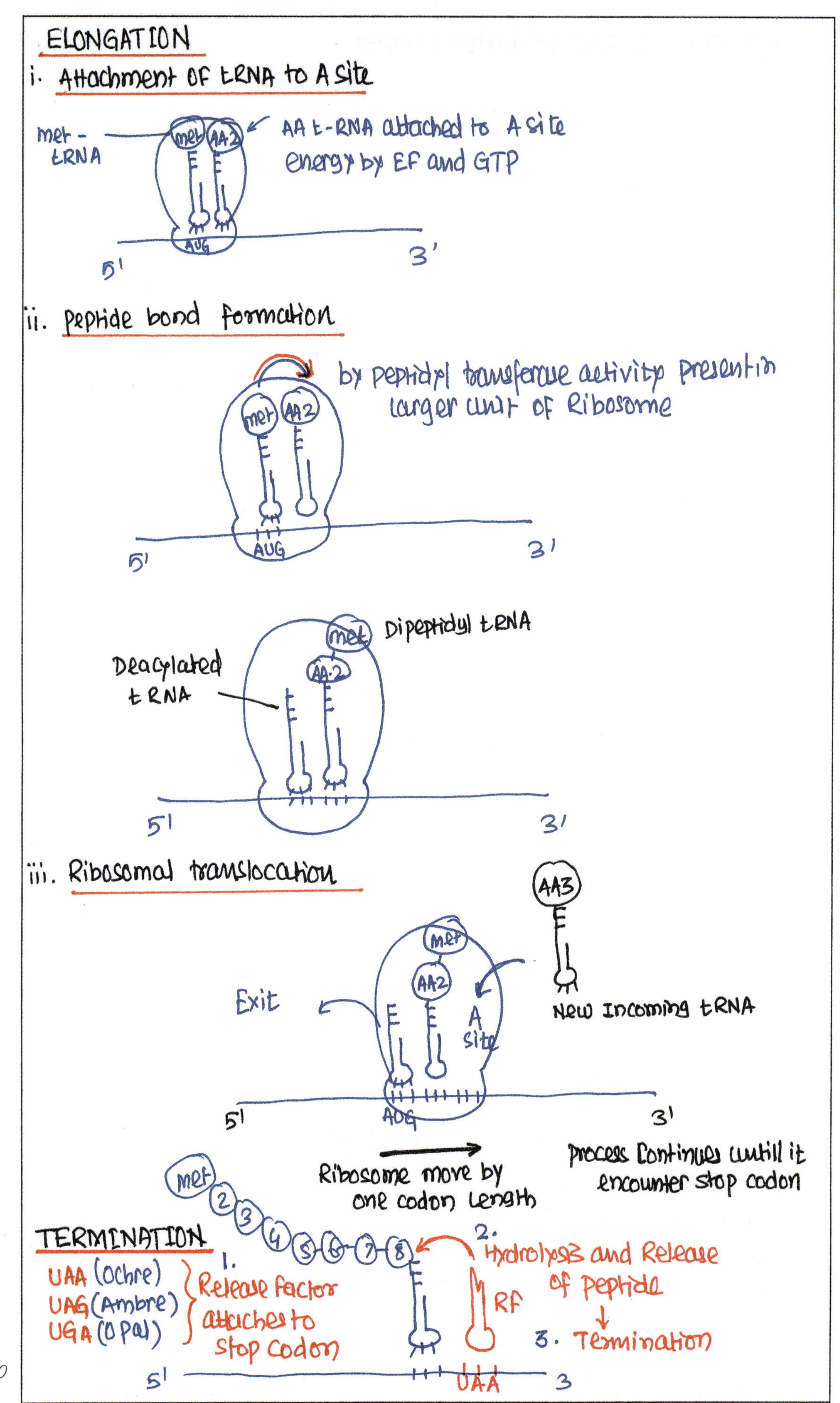
ELONGATION
i. Attachment of tRNA to A site
met-tRNA
met
AA2
AA t-RNA attached to A site
energy by EF and GTP
AUG
5'
3'
ii. Peptide bond formation
by peptidyl transferase activity present in larger unit of Ribosome
met
AA2
AUG
5'
3'
met
AA 2
Dipeptidyl tRNA
Deacylated tRNA
5'
3'
iii. Ribosomal translocation
AA3
met
AA2
Exit
A site
New Incoming tRNA
AUG
5'
3'
Ribosome move by one codon length
Process continues until it encounter stop codon
TERMINATION
UAA (ochre)
UAG (Ambre)
UGA (OPal)
1. Release factor attaches to stop codon
met
2
3
4
5
6
7
8
RF
2. Hydrolysis and Release of peptide
3. Termination
5'
UAA
3

PROTEIN BIOSYNTHESIS

TRANSLATION IN EUKARYOTES

Very Important LAQ :-

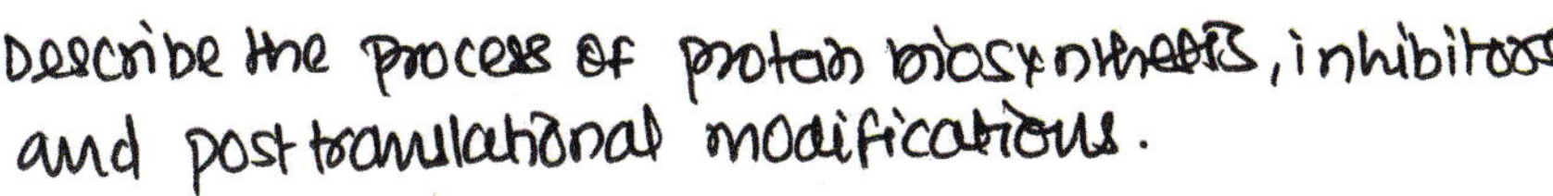

Describe the process of protein biosynthesis, inhibitors and post translational modifications.

Definition :- Information present on mRNA in the form of genetic code is utilized for the synthesis of proteins/peptide is known as translation.

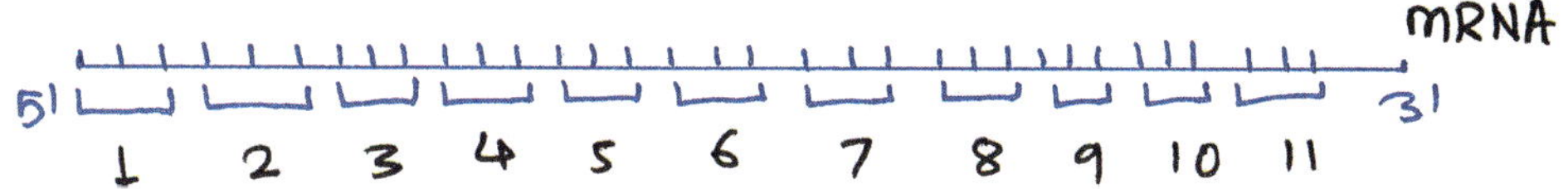

codons → Triplet base pair sequences present on mRNA

Peptide / protein

Requirements -
- mRNA
- tRNA
- rRNA c̄ Ribosomes
- factors - Eukaryotic Initiation factor
  - Eukaryotic elongation factor
  - Release factor 1, 2, 3
- Amino acids

Steps

1. **Activation of tRNA**

$$\text{Amino acid (AA)} + \text{ATP} \xrightarrow[\text{synthetase}]{\text{Aminoacyl tRNA}} \text{Enz-AMP-(AA)} \text{ [Activated AA]}$$

↓ tRNA

tRNA-CCA-OH-(AA)

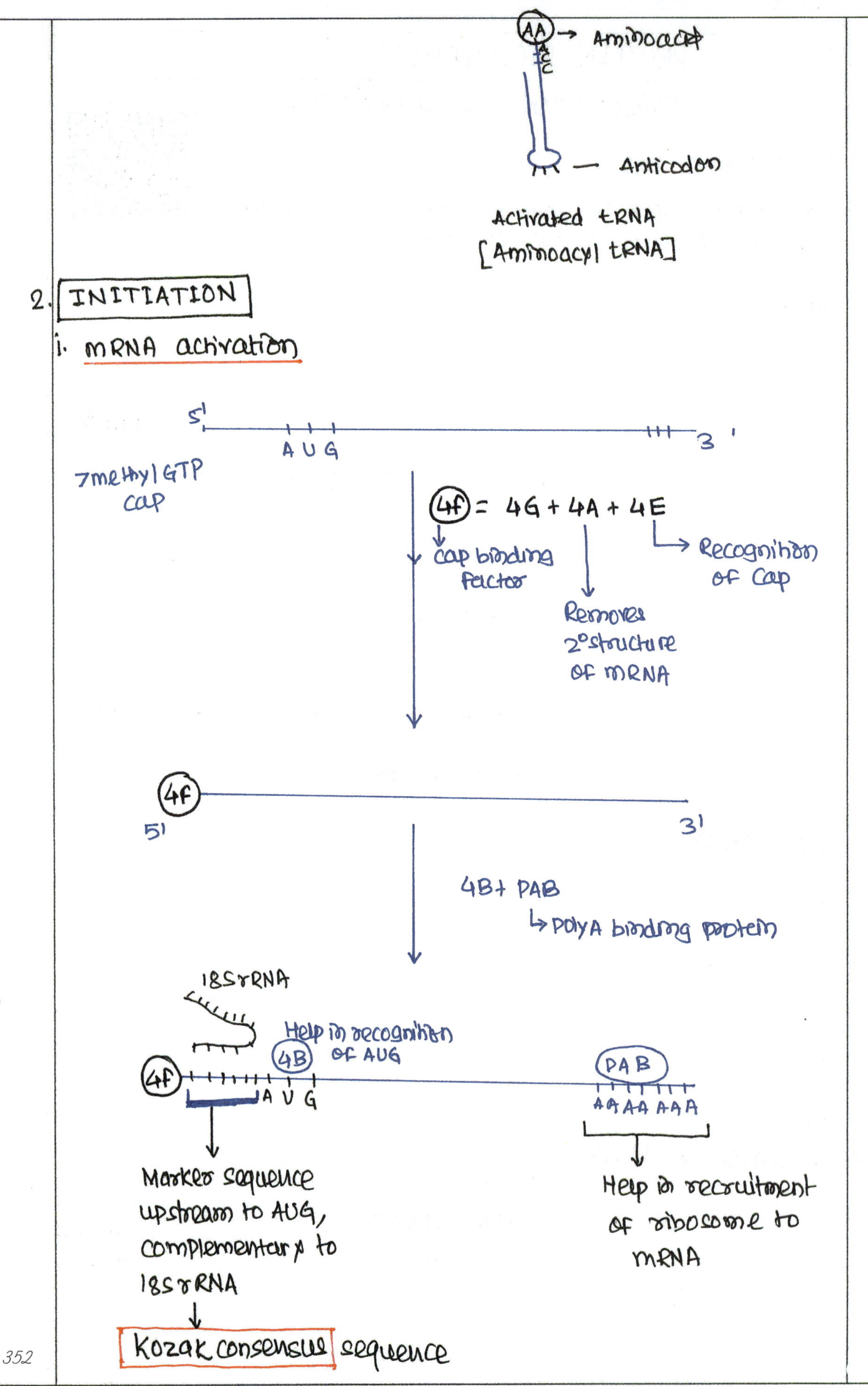
AA
Aminoacid
A
C
C
Anticodon
Activated tRNA
[Aminoacyl tRNA]
2. INITIATION
i. mRNA activation
5'
A U G
3'
7methyl GTP cap
4f = 4G + 4A + 4E
cap binding factor
Removes 2°structure of mRNA
Recognition of Cap
4f
5'
3'
4B+ PAB
polyA binding protein
18S rRNA
Help in recognition of AUG
4B
PAB
4f
A U G
AAAAAAA
Marker sequence upstream to AUG, complementary to 18S rRNA
Kozak consensus sequence
Help in recruitment of ribosome to mRNA

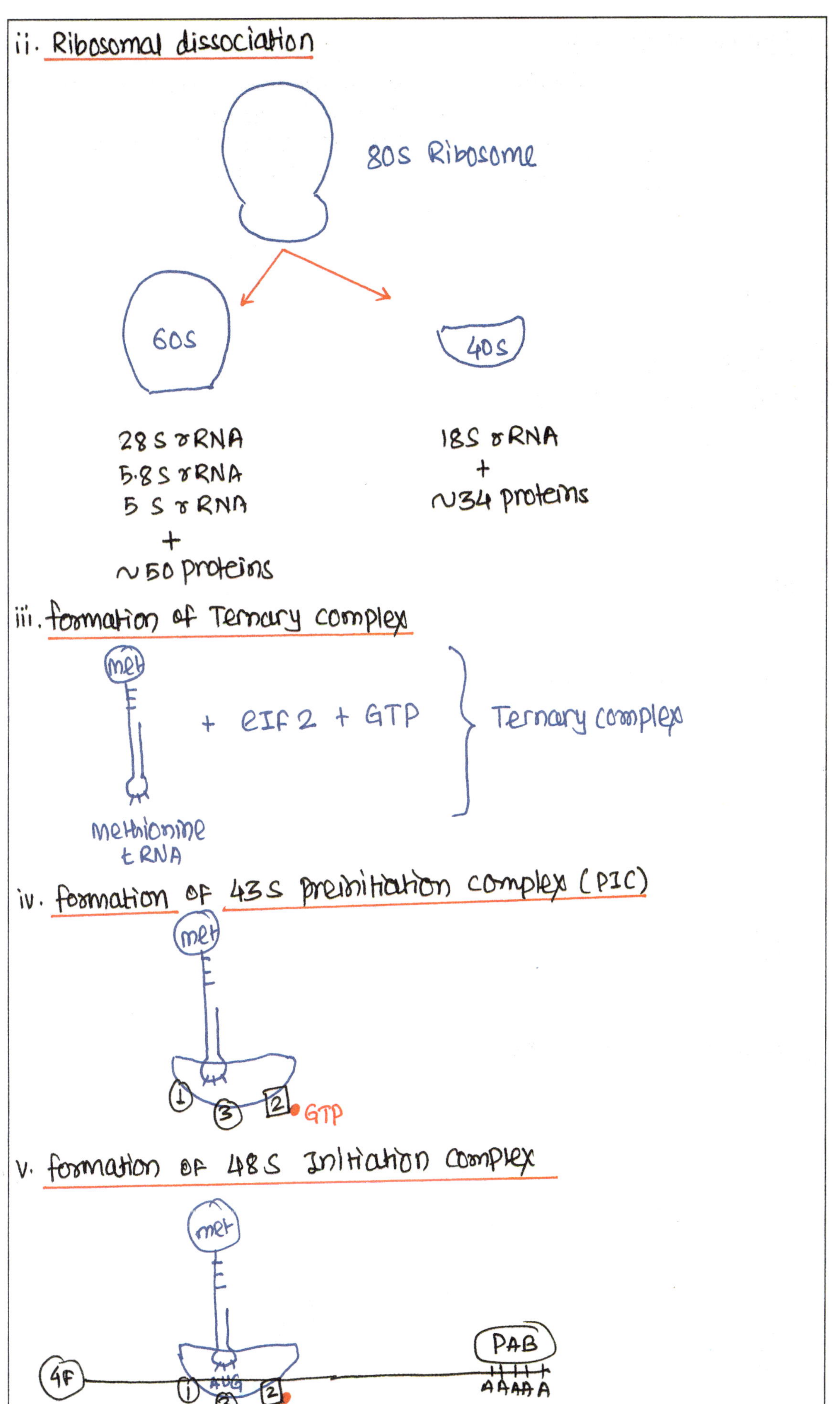
ii. Ribosomal dissociation
80S Ribosome
60S
40S
28S rRNA
5.8S rRNA
5 S rRNA
+
~50 proteins
18S rRNA
+
~34 proteins
iii. formation of Ternary complex
met
+ eIF2 + GTP
Ternary complex
methionine tRNA
iv. formation of 43S preinitiation complex (PIC)
met
1
3
2
GTP
v. formation of 48S Initiation complex
met
4F
AUG
1
3
2
PAB
AAAAA

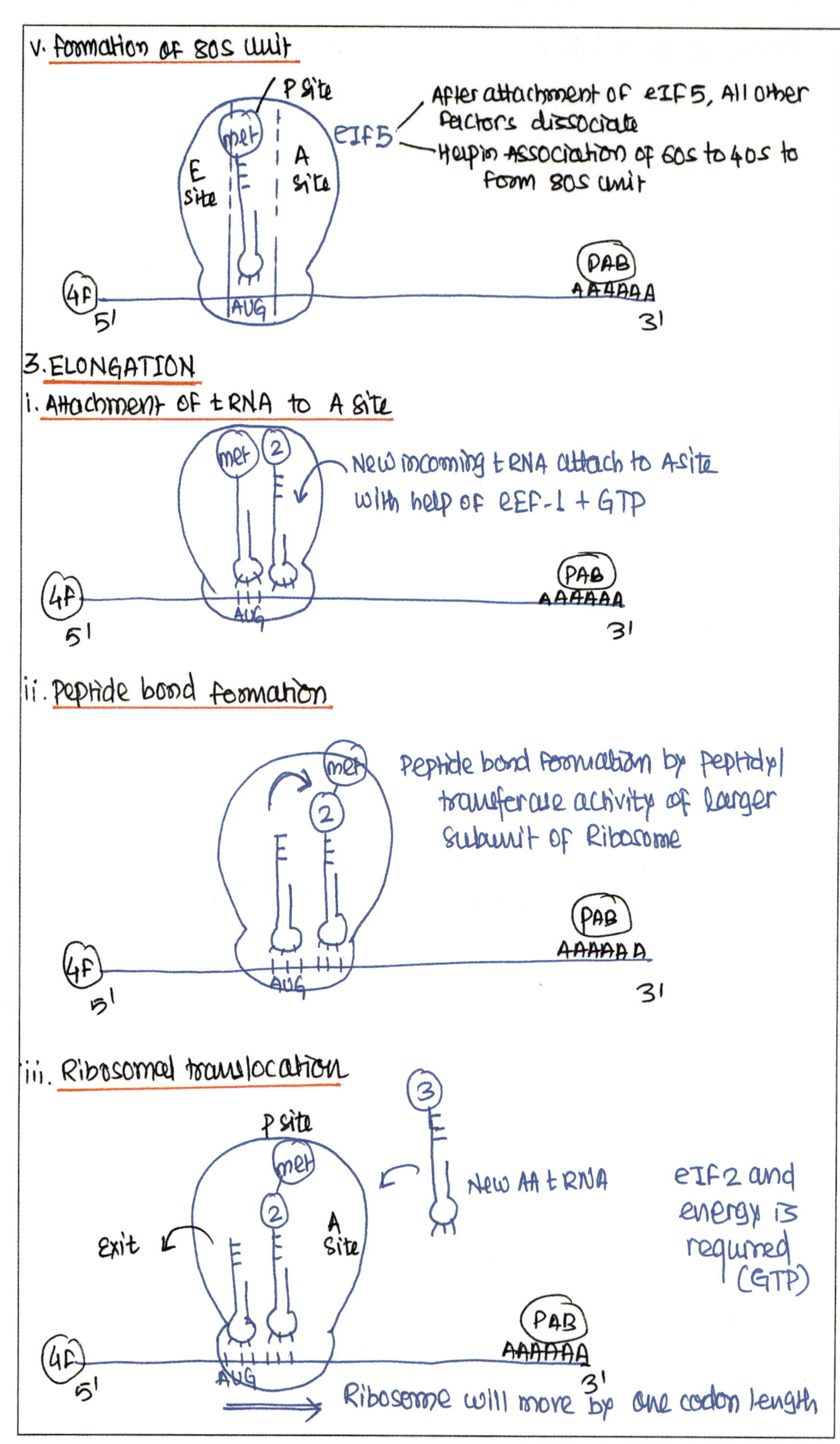
v. formation of 80S unit
P site
met
E site
A site
eIF5
After attachment of eIF5, All other factors dissociate
Help in association of 60s to 40s to form 80S unit
4F
5'
AUG
PAB
AAAAAA
3'
3. ELONGATION
i. Attachment of tRNA to A site
met
2
New incoming tRNA attach to A site with help of eEF-1 + GTP
4F
5'
AUG
PAB
AAAAAA
3'
ii. Peptide bond formation
met
2
Peptide bond formation by peptidyl transferase activity of larger subunit of Ribosome
4F
5'
AUG
PAB
AAAAAA
3'
iii. Ribosomal translocation
P site
met
2
A site
Exit
3
New AA tRNA
eIF2 and energy is required (GTP)
4F
5'
AUG
PAB
AAAAAA
3'
Ribosome will move by one codon length

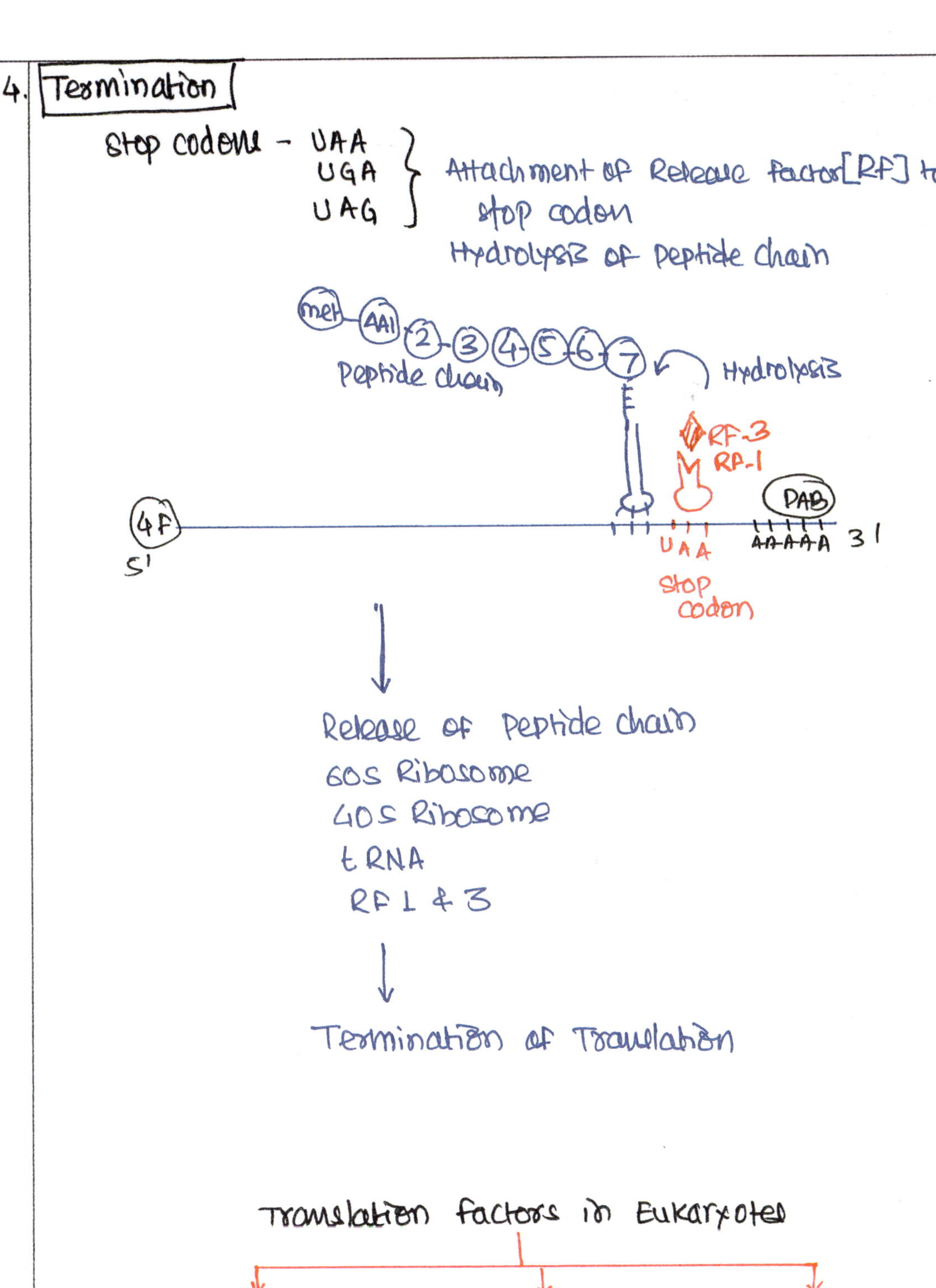

Translation factors in Eukaryotes

① Initiation factors	② Elongation factors	③ Termination factors
eIF 1,2,3	eEF 1α	eRF-1
eIF 4A, 4B, 4G, 4F, 4E	eEF 2	eRF-3
eIF 5		
eIF 6		

Translation factors	functions.
eIF2	facilitates binding of initiating Met-tRNA to 40S
eIF1A eIF3	first factor to bind 40S unit, prevent reassociation of ribosome, facilitates subsequent steps.
eIF4B	Scanning of mRNA to locate AUG
eIF4A	Removes secondary structures in mRNA to permit binding to 40S unit
eIF4E	Binds to the 5' cap of mRNA, Part of eIF4F complex
eIF4G	Binds to eIF4E and PAB, part of eIF 4f complex
eIF 5	Promotes dissociation of several IFs from 40S. Association of 60S subunit to form 80S initiation complex.

Difference between eukaryotic & prokaryotic translation

	Eukaryotic translation	Prokaryotic translation
Initiator AA	Methionine	formyl methionine
Initiation factors	Many (~13)	Few (3)
Ribosomes	80S→ 40S & 60S	70S→ 30S + 50S
mRNA bind to small subunit	Involves Cap at 5' end Initiation factors Kozak sequence	shine dalgarno sequence upstream of AUG
Elongation factors	eEF1A eEF2	EF-Tu EF-G
Termination	eRF 1 eRF 3	RF 1 RF 2 RF 3

Regulation of Translation

- Phosphorylation of eEF 2
- Regulation of eIF4E

Inhibitors of Protein synthesis

SAQ:- Enumerate Inhibitors of translation c̄ Mechanism.

- Prokaryotes
- Eukaryotes
- Both

Prokaryotes	
1. Tetracyclin, Doxycyclin	Bind to 30S unit and prevent attachment of aminoacyl tRNA at A site. Elongation ✗. Antibiotic – used for Atypical pneumonia, Rickettsia
2. Chloramphenicol	Bind to 50S subunit, Inhibit transferase. Elongation ✗ – used for Meningitis, typhoid
3. Erythromycin	Bind to 50S subunit, Inhibit translocation of Ribosome, Elongation ✗ – used for gram Negative infection
4. Streptomycin, Gentamycin, Neomycin	Bind to 30S subunit. causes misreading of mRNA, Initiation ✗. Inhibit translocation, Elongation ✗. used for – gram –ve organism, Tb, local infection
Eukaryotes	
1. Diphtheria Toxin	ADP ribosylation of eEF-2, Prevent translocation. Blocks elongation
2. Ricin	Bind to 60S subunit, Removes one adenine nucleotide from larger unit of ribosome
3. Cycloheximide	Bind to 60S subunit, Inhibit peptidyl transferase. Block elongation

MCQ

Puromycin – Inhibit **both** Prokaryotic & eukaryotic translation

It is structurally simillar to tyrosinyl tRNA. bind to A site and incorporate into growing peptide chain. Blocks elongation process.

Post-translational Modifications

SAQ:- Enumerate post-translational modifications c̄ examples.

Proteolytic cleavage

1. Removal of initiator amino acid - Methionine
Protein becomes functionally active.

2. Proteolysis of proproteins
- by removing small peptide
Example - pancreatic enzymes clotting factors.

3. Proteolysis of preproproteins
Example - removal of signal sequences.
Imp for protein targeting

Covalent Modification

1. Glycosylation :- Attachment of carbohydrates - gives active protein and heterogenicity
2. Formation of disulfide bond between two cysteine residue.
3. Addition of prosthetic group
- attachment of nonprotein group
4. Hydroxylation of proline & Lysine
- Important in collagen synthesis
5. Gamma carboxylation
Req. vit K, Activation of clotting factors II, VII, IX, X
6. Phosphorylation
Activation & deactivation of enzymes.
7. Prenylation - Addition of isoprenyl group - for anchoring of protein to plasma membrane
8. Acetylation - Addition of acetyl group to histone protein

LAC OPERON - Regulation of Gene Expression in Prokaryotes.

SAQ: Write a note on regulation of gene expression in Prokaryotes c̄ Lac operon concept.

LAC Operon

↓ Lactose

↳ Linear array of genes involved in metabolic pathway

Described by Jacob & Monod

Deals with Lactose metabolism in E. Coli

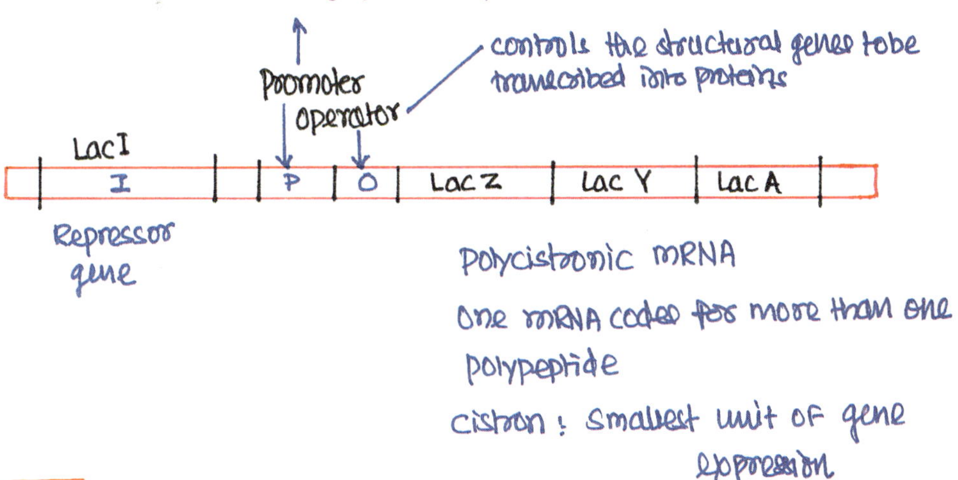

Polycistronic mRNA

One mRNA codes for more than one polypeptide

Cistron: Smallest unit of gene expression

3 Modes

1. When Glucose Available → Repressed (switched off)
2. Glucose & Lactose both available → Repressed [catabolite Repression]
3. Only lactose available → Induced (increase activity)

1. When only glucose is available

bacteria utilizes glucose as a source of energy

operon is switched off - Repressed

because Lac I (Repressor gene) is a constitutive gene constantly expressed in a cell at fairly constant rate

Repressor protein will bind to operator region and prevent the binding of RNA Polymerase to promoter region

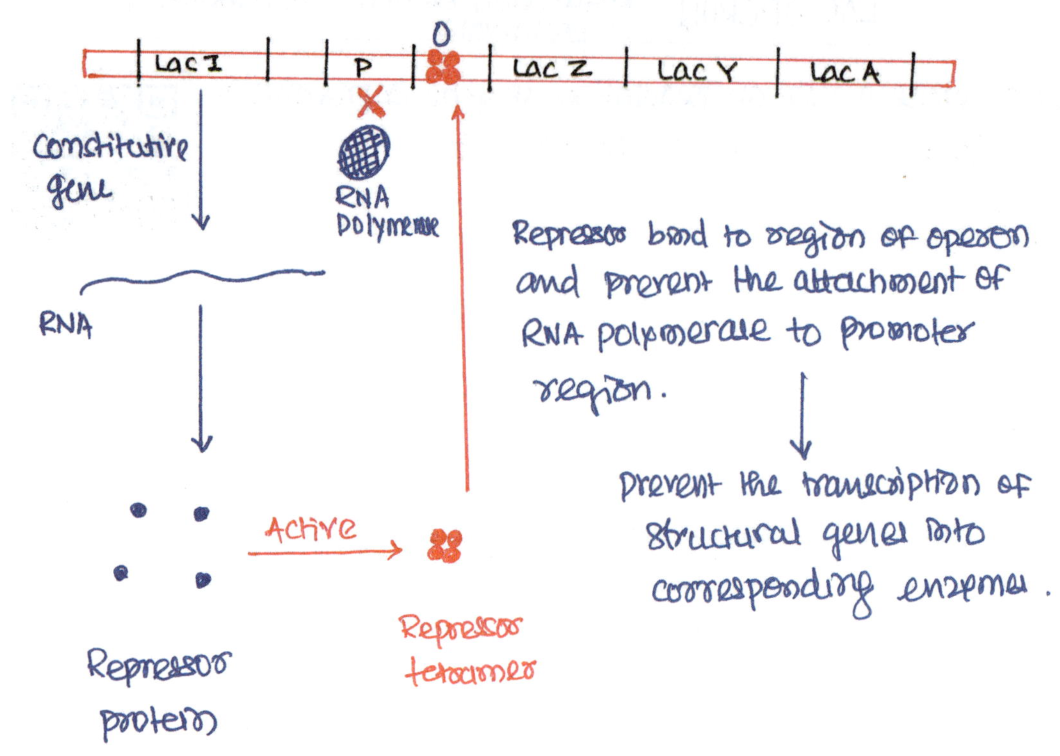

2. When glucose and lactose both available

bacteria prefers Glucose as a source of energy.

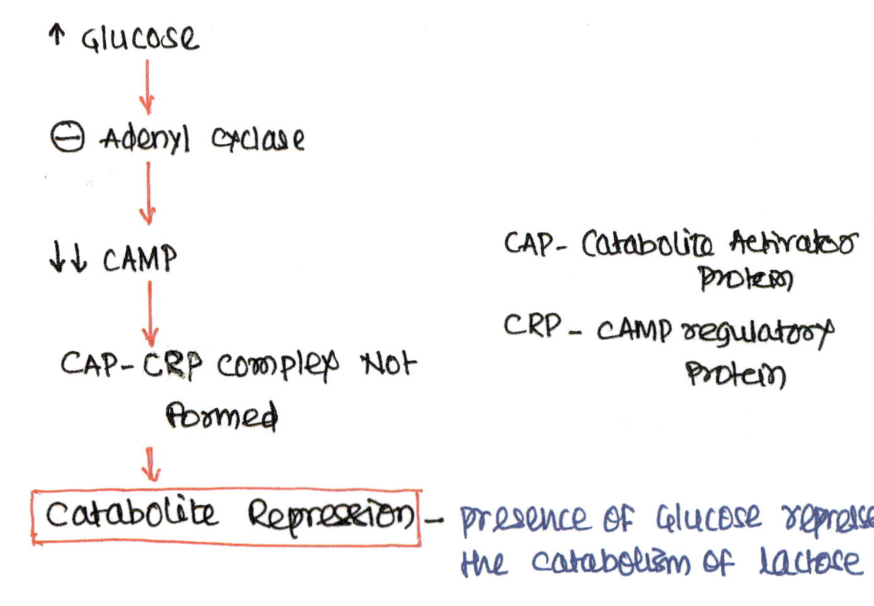

RNA polymerase required CAP-CAMP complex for binding to promoter region.

Repression of lac operon in presence of Glucose

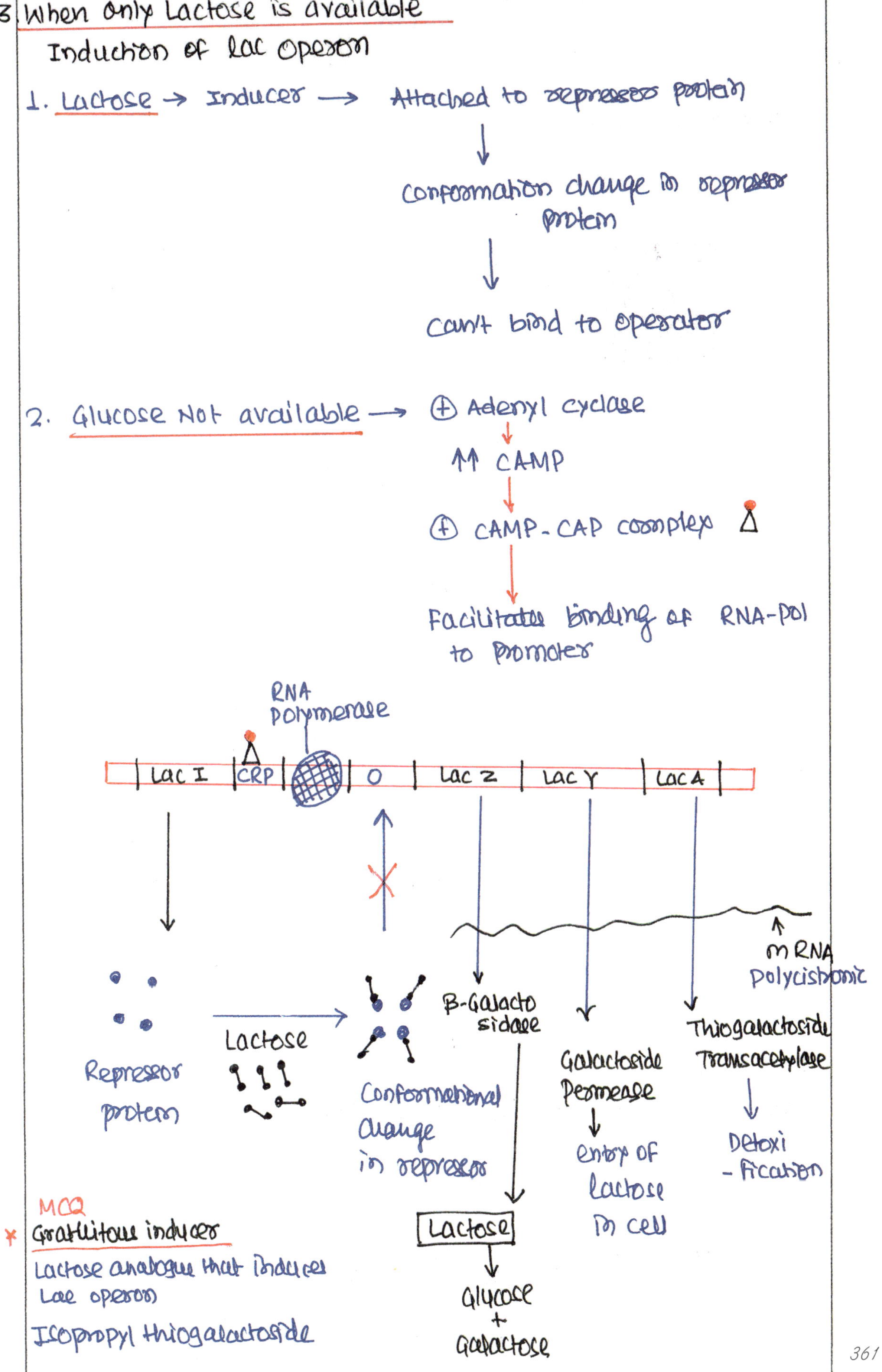
3. When only Lactose is available
Induction of lac operon
1. Lactose → Inducer → Attached to repressor protein
↓
Conformation change in repressor protein
↓
Can't bind to operator
2. Glucose Not available → ⊕ Adenyl cyclase
↑↑ cAMP
⊕ cAMP-CAP complex
Facilitates binding of RNA-Pol to promoter
RNA Polymerase
Lac I
CRP
O
Lac Z
Lac Y
Lac A
mRNA Polycistronic
Repressor protein
Lactose
Conformational change in repressor
B-Galactosidase
Lactose
Glucose + Galactose
Galactoside Permease
entry of Lactose in cell
Thiogalactoside Transacetylase
Detoxi-fication
MCQ
* Gratuitous inducer
Lactose analogue that induces Lac operon
Isopropyl thiogalactoside

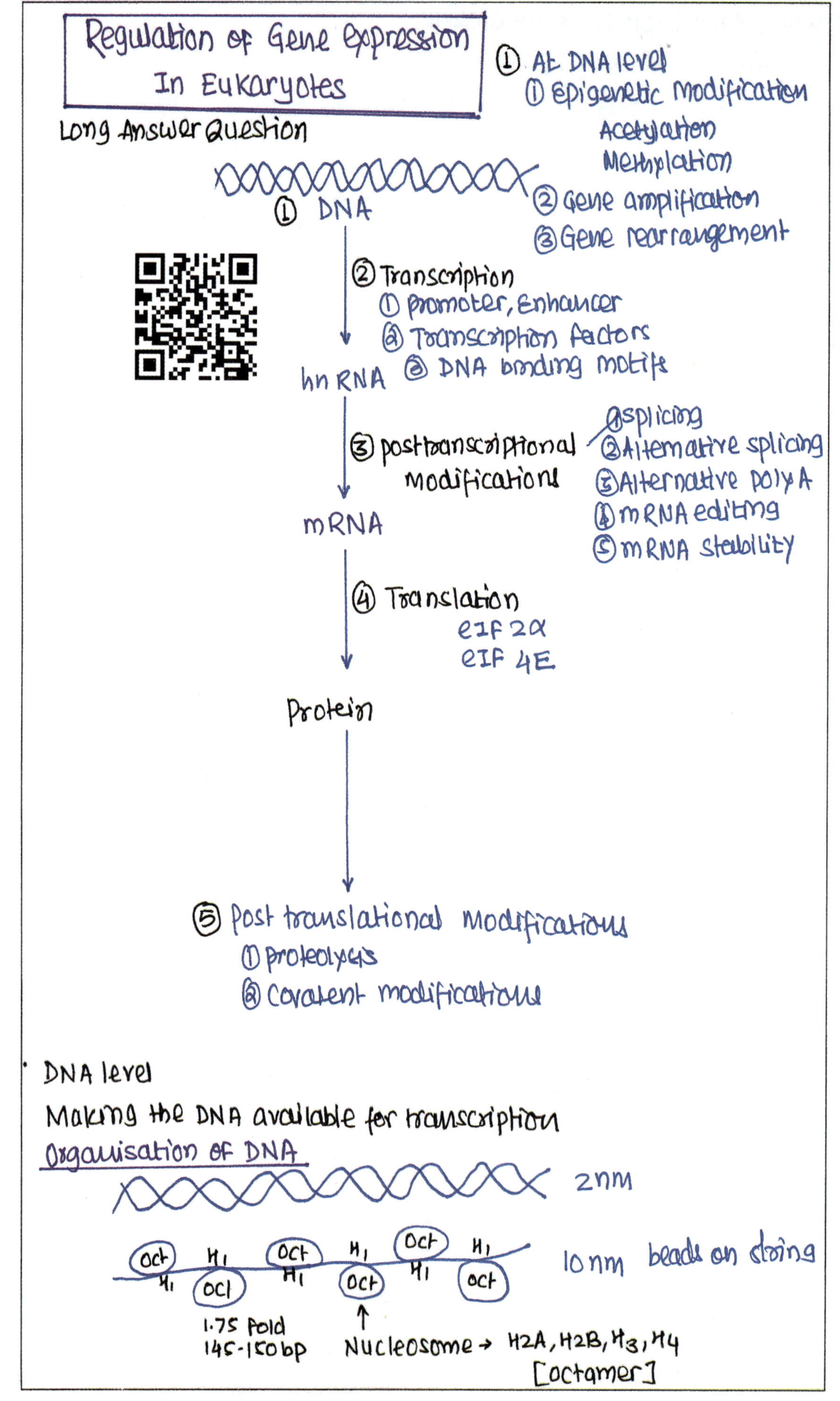
Regulation of Gene Expression In Eukaryotes
Long Answer Question
① At DNA level
① Epigenetic modification
Acetylation
Methylation
② Gene amplification
③ Gene rearrangement
① DNA
② Transcription
① promoter, Enhancer
② Transcription factors
③ DNA binding motifs
hn RNA
③ posttranscriptional modifications
① splicing
② Alternative splicing
③ Alternative poly A
④ mRNA editing
⑤ mRNA stability
mRNA
④ Translation
eIF 2α
eIF 4E
Protein
⑤ Post translational modifications
① proteolysis
② Covalent modifications
DNA level
Making the DNA available for transcription
Organisation of DNA
2nm
Oct
H1
10nm beads on string
1.75 fold
145-150bp
Nucleosome → H2A, H2B, H3, H4
[octamer]

30 nm chromatin fibril → 6 to 7 Nucleosomes

300 nm Noncondensed loop
↑
(Nuclear scaffold Associated) form
↓
700 nm condensed loop

1400 nm Metaphase chromosome

(Above gene)

Epigenetic modification :- Chemical modification in DNA without altering the NT sequence.

Acetylation of Lysine Ⓗ tails → ↓ positive charge on Histones

Interaction between Ⓗ & DNA decreases

Makes DNA available

↑ Gene expression

Methylation of → deoxycytosine in CG rich-promoter causes DNA to organise and DNA will not be available for transcription.

↓ Gene expression

Phosphorylation of Ⓗ → organisation & compactness of DNA

Ribosylation of Ⓗ → Important in Repair mechanism

Gene Amplification :- Making the copies of DNA

DHF Reductase gene is amplified on prolong therapy with Methotrexate. Hence we have to increase the dose after prolong therapy.

folic acid —DHF Reductase (⊖)→ DHF —DHF Reductase (⊖)→ THF

Methotrexate

Gene for
DHF Reductase

A B

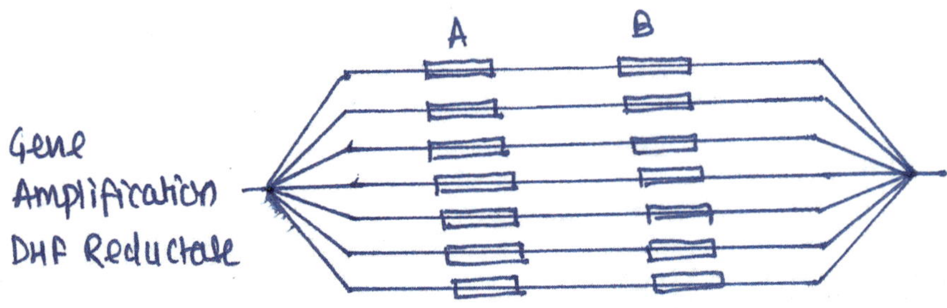

Gene Rearrangement :-

IgG V C C V

variable region of light chain of immunoglobulin IgG is coded by

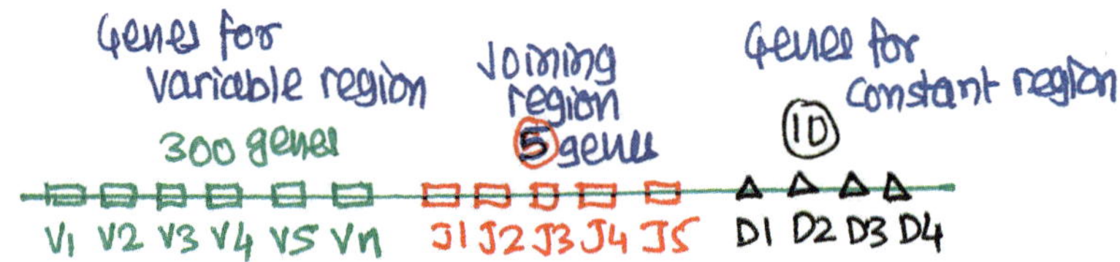

Different combination will produced different genes

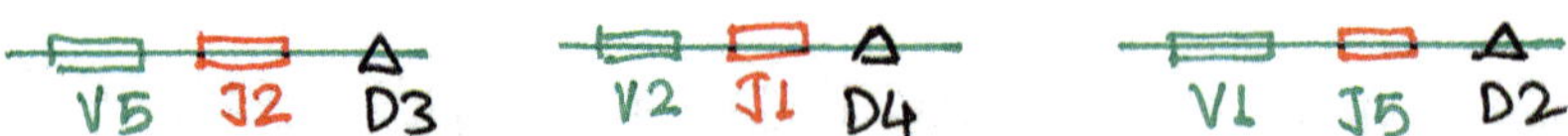

Different antibodies are produced in response to different antigens by gene rearrangement process.

Transcriptional regulation - promoter
Enhancer / Repressor elements
Transcription factors
DNA binding motifs.

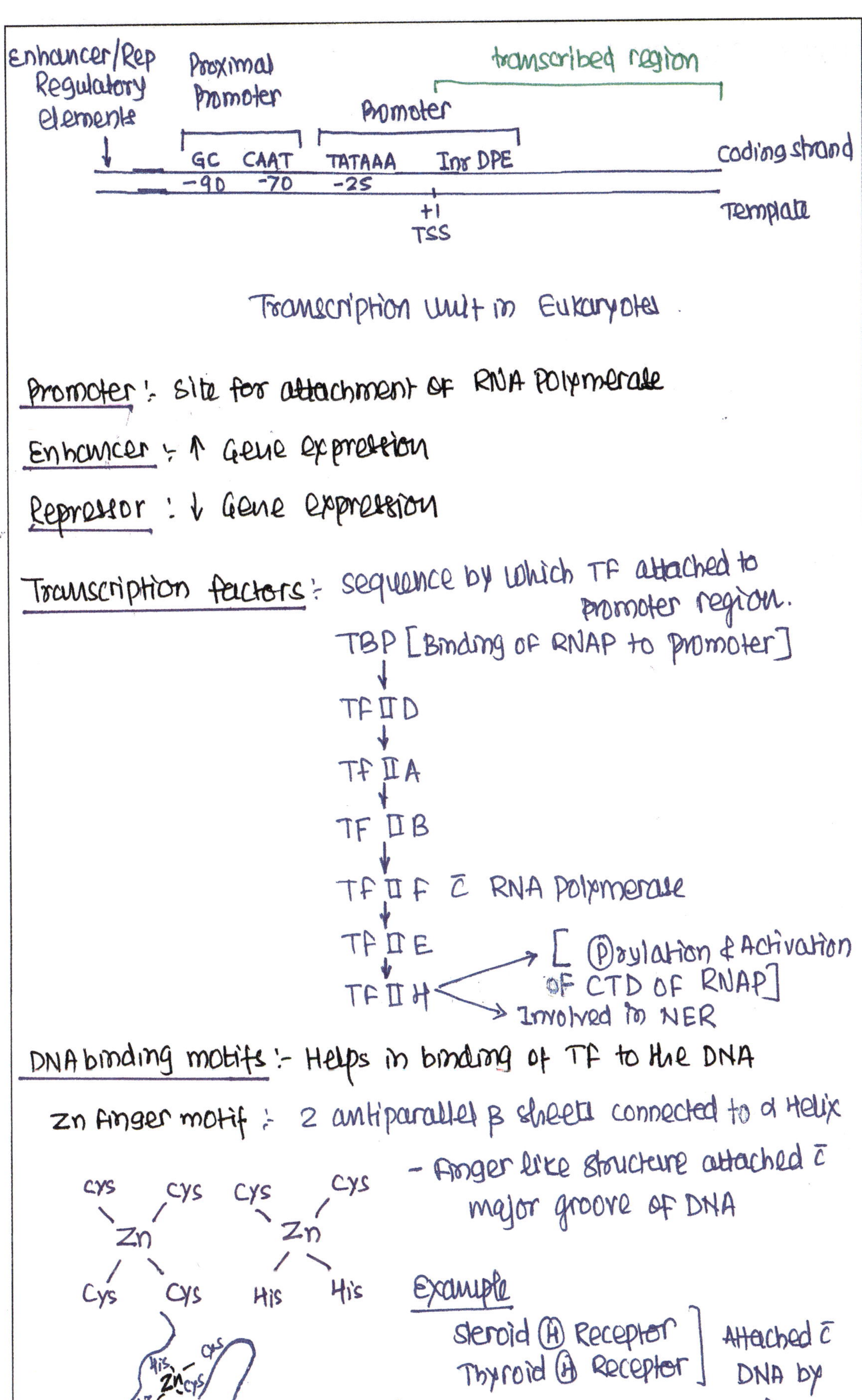
Enhancer/Rep Regulatory elements
Proximal Promoter
Promoter
transcribed region
GC CAAT TATAAA Inr DPE
-90 -70 -25
coding strand
Template
+1 TSS
Transcription unit in Eukaryotes
Promoter :- site for attachment of RNA Polymerase
Enhancer :- ↑ Gene expression
Repressor : ↓ Gene expression
Transcription factors :- sequence by which TF attached to promoter region.
TBP [Binding of RNAP to promoter]
TF II D
TF II A
TF II B
TF II F c̄ RNA Polymerase
TF II E
TF II H
[Ⓟylation & Activation of CTD of RNAP]
Involved in NER
DNA binding motifs :- Helps in binding of TF to the DNA
Zn finger motif :- 2 antiparallel β sheets connected to α Helix
- finger like structure attached c̄ major groove of DNA
Cys Cys Zn Cys Cys
Cys Cys Zn His His
His His Zn Cys Cys
Example
Steroid Ⓗ Receptor
Thyroid Ⓗ Receptor
Attached c̄ DNA by Zn finger motif

Leucine zipper motif

- 2 Alpha helices that have hydrophobic zone and basic end
- every 7th amino acid is leucine
- Positively charged aminoterminal region reacts with major and minor groove of DNA and helps in binding of Transcription factors to DNA.

C ... +N
C ... N
Major groove DNA
Leucine

- The N-terminal basic region contains Lysine and Arginine which interact with major groove in sequence specific manner.

Helix turn Helix motif

two α-helices joined by short strand of aminoacids

Helix loop helix motif

Two α helices connected by loop.
dimeric form is Active.
basic region bind with DNA.

COOH COOH
α Helix
loop
α Helix
NH_2 NH_2

Post transcriptional Regulation

1. **Splicing** :- Introns are removed and exons joined together

Two types ① self (intron) mediated

② Spliceosome mediated (SnRNA + RNP)

SnRNA → U_1, U_2, U_4, U_5, U_6 – Uracil rich small RNA.

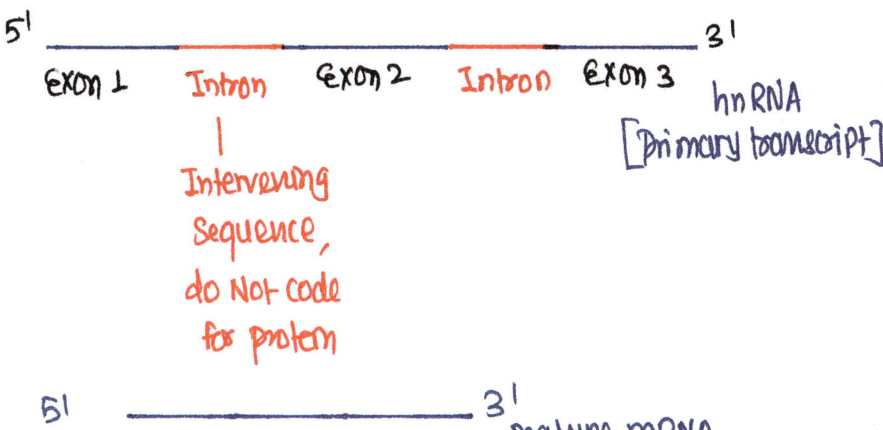

Alternative splicing :- Two different mRNA produced by single gene by combination of exons to produce different proteins

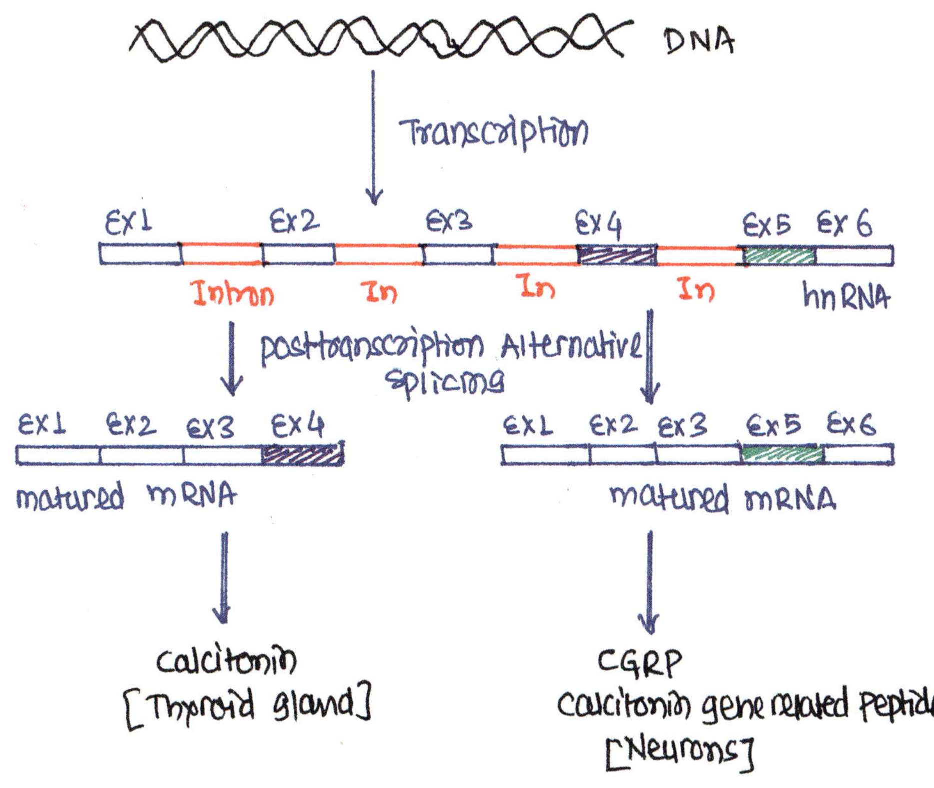

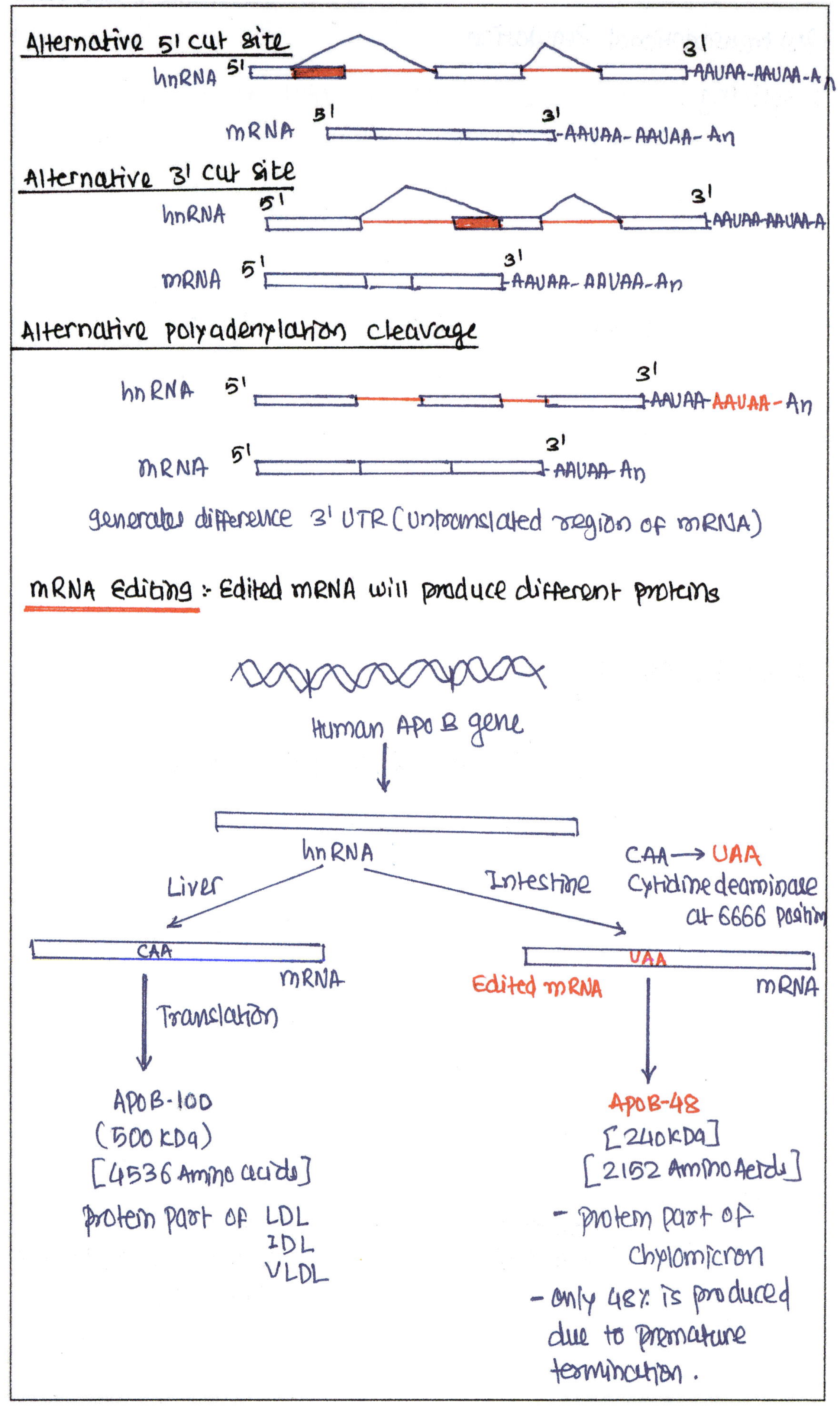

Alternative 5' cut site
hnRNA 5'
3'
AAUAA-AAUAA-An
mRNA 5'
3'
AAUAA-AAUAA-An
Alternative 3' cut site
hnRNA 5'
3'
AAUAA-AAUAA-A
mRNA 5'
3'
AAUAA-AAUAA-An
Alternative polyadenylation cleavage
hnRNA 5'
3'
AAUAA-AAUAA-An
mRNA 5'
3'
AAUAA-An
generates difference 3' UTR (untranslated region of mRNA)
mRNA Editing :- Edited mRNA will produce different proteins
Human Apo B gene
hnRNA
Liver
Intestine
CAA → UAA
Cytidine deaminase
at 6666 position
CAA
mRNA
UAA
Edited mRNA
mRNA
Translation
APO B-100
(500 kDa)
[4536 Amino acids]
Protein part of LDL
IDL
VLDL
APO B-48
[240 kDa]
[2152 Amino Acids]
- Protein part of
Chylomicron
- only 48% is produced
due to Premature
termination.

mRNA stability :-

↑ protein synthesis by making mRNA stable for longer duration of time.

Example :- Iron deficiency anemia

↓

↑ synthesis of Transferrin Receptors.

IRP Iron regulatory protein

Transferrin Receptor mRNA

Iron responsive element

5' — 3'

3' UTR

Iron regulatory protein will attach to IRE in the 3' UTR of Transferrin receptor mRNA. Now mRNA will be stable for longer duration and ↑ synthesis of Transferrin receptors.

factors imp in mRNA Stability — 5' 7mGTP cap
3' Poly A Tail
5' & 3' untranslated regions
5' & 3' stem loop structure
AU rich region

miRNA : Micro RNA :- Near about 21 - 24 NT in length - Noncoding
Produced by RNAP-II as pre-miRNA
pre mi-RNA —dicer→ miRNA
miRNA attached to 3' UTR of mRNA and induces degradation of mRNA
↓
↓ Gene expression

siRNA : small interfering RNA :- cause gene silencing by interacting with mRNA

Translation :

eIF-2	eIF 4E
$f^n \rightarrow$ Initiation of translation	$f^n \rightarrow$ Recognition of cap of mRNA
(P) eIF-2 ↓ Inactivation	(P) eIF 4E Activation.

Phosphorylation is done by kinases. Phosphorylation activates eIF-4E and inactivates eIF-2.

Posttranslational modifications :-

① proteolysis ② Covalent modifications

Recombinant DNA Technology (RDT)

Genetic engineering OR
Gene cloning OR
Gene manipulation

Very important LAQ :- Describe the steps in RDT with its applications in medicine.

Definition :- The artificial manipulation, modification and recombination of DNA OR Nucleic acid to achieve a desired goal in a predetermined way is called recombinant DNA technology.

Steps

1. Isolation of desired DNA

Cell
↓
treated c̄ Enzyme
Detergent
↓
Phenol
Protease
Ribonuclease
↓
Ultracentrifugation
↓
Chilled ethanol
↓
DNA
↓ Restriction endonuclease
Separated by Electrophoresis
↓
Identified by southern blotting technique
↓
Desired DNA c̄ sticky ends

desired DNA segment

Donor DNA

2. Isolation of Vector

- Prokaryotic vectors are selected because of rapid multiplication

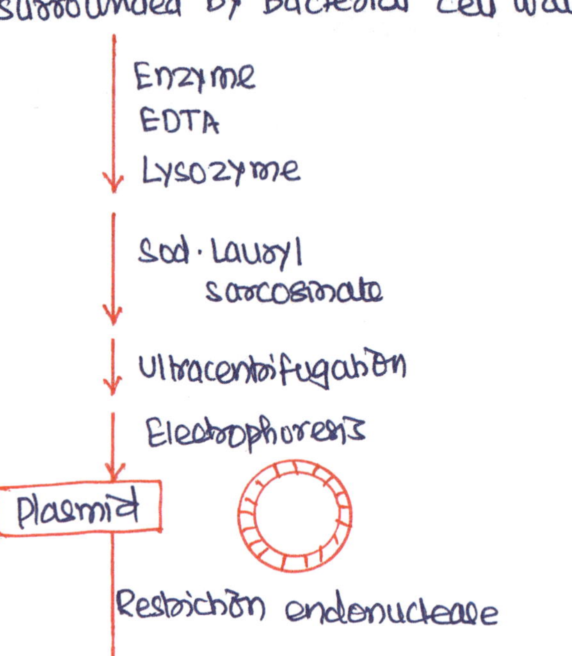

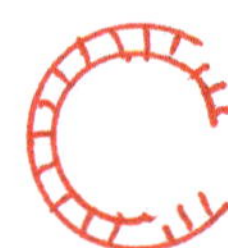

Plasmid ε sticky ends

3. formation of Hybrid DNA, Recombinant DNA/ Chimeric DNA

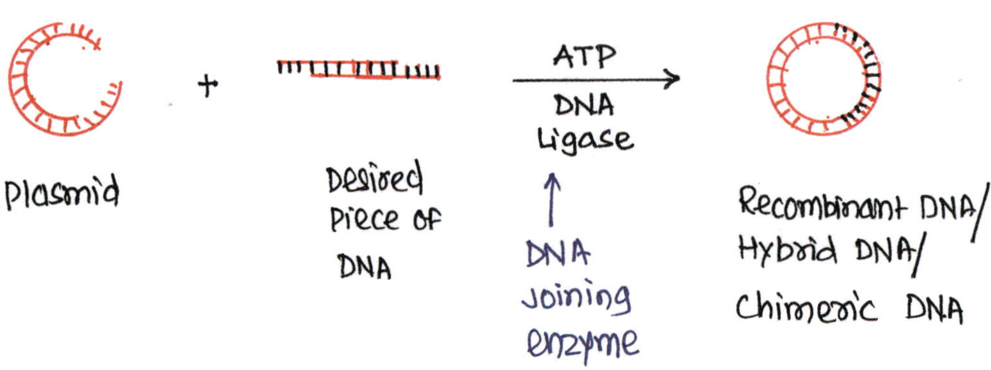

4. Introduction into Host cell

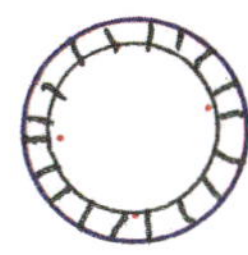

Chimeric DNA

→ Gene transfer method

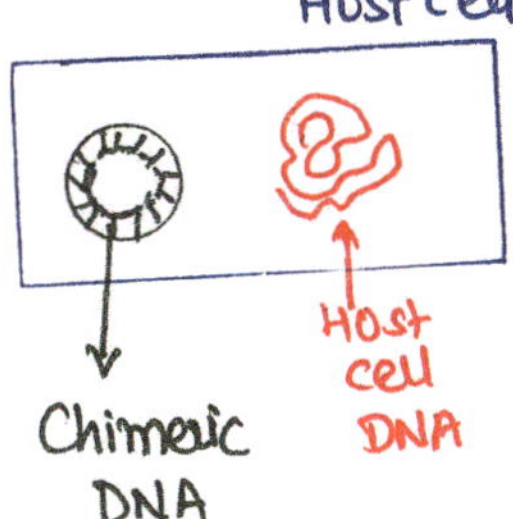

5. Multiplication into Host cell

& selection of cells that containing chimeric DNA

↓

ⓐ Hybridisation technique (using DNA probes)

ⓑ Antibiotic resistance

6. Expression of gene to produce desired product

Host cell
↓
Gene
↓
proteins

Molecular tools in RDT

1. Enzymes

Exonuclease → Act on terminal nucleotides

Endonuclease → Act on internal phosphodiester linkage

Restriction Endonuclease (Molecular scissor)

RE I → cut DNA away from restriction sequence

* RE II → cut DNA within restriction sequence

RE III → cut DNA away from restriction sequence.

Recognition sequence → 4 - 6 Nucleotide sequence which is pallindromic (Read simillar in both direction)

— G A A T T C —
— C T T A A G —

Restriction endonuclease are named according to bacteria from which they are isolated.

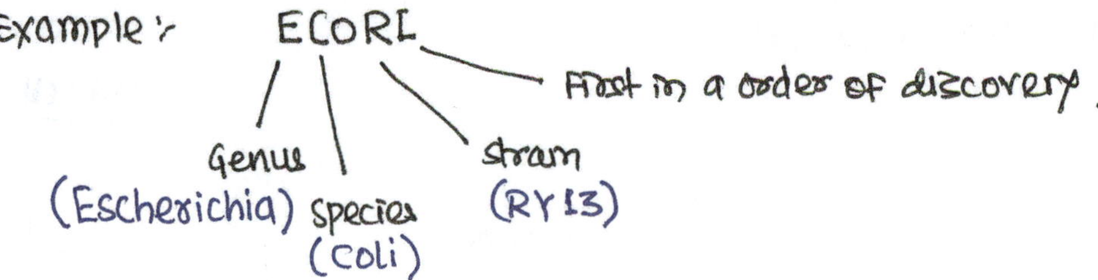

Pronounced as (ecoR one) is a restriction endonuclease enzyme isolated from E.coli.

The sequence where the enzyme cuts is

– G↓A A T T C –
– C T T A A↑G –

DNA ligase :- Molecular glue

DNA Joining enzyme
Requires ATP
Forms phosphodiester linkage

Other enzymes – DNA Polymerase
Alkaline phosphatase
Reverse transcriptase etc.

2. VECTORs

Transport vehicles used for transport of desired piece of DNA into a Host cell

	Vector	Size of insert	Characteristics.
i	Plasmid	< 10 kb	double stranded, extrachromosomal Capable of independent replication
ii	Bacteriophage	10-20 kb	Viruses that infect bacteria
iii	Cosmid	35-50 kb	Plasmid + bacteriophage
iv	YAC [Yeast artificial chromosome]	up to 1 Mb	cloning of larger size DNA
v.	MAC (Mammalian)	up to 1Mb	cloning of larger size DNA
vi.	BAC (Bacterial)	<300 kb	

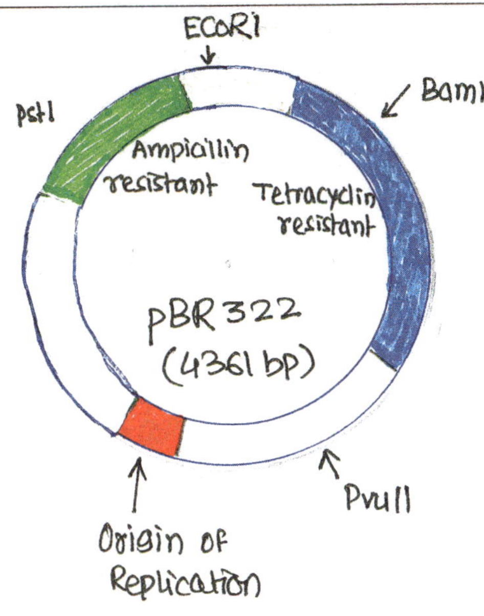

Plasmid c̄ restriction sites and two antibiotic resistance gene.

P - plasmid

BR for Bolivar and Rodriguez

It has 4361 bp

3. HOST

Prokaryotic :- E Coli
B. Subtilis

Eukaryotic :- fungi (Saccharomyces cerevisiae)
Plant cells
Animal cells (Oocytes)

Advantage

Post translation modification

4 Gene transfer techniques

Biological methods

1. Transduction using viral vectors
   - Adenoviruses
   - Adeno associated viruses
   - Retrovirus
2. Conjugation
3. Transformation

Physical methods

Electroporation
Gene gun technique
Particle bombardment
Biolistic transformation

Chemical method

Liposome mediated gene transfer
Polyethylene glycol mediated
calcium-Ⓟ mediated

Applications of Recombinant DNA Technology

1. Production of therapeutic agents

i. Insulin

For t/t of Diabetes

I P O — β-gal gene — A: Plasmid c̄ gene for chain A

I P O — β-gal gene — B: Plasmid c̄ gene for chain B

↓

Transform E coli c̄ rDNA vector and select transgenic cells by their antibiotic resistance

↓

A → (β-galactosidase–A) | B (β-galactosidase–B)

↓

β-galactosidase-Insulin A fusion protein | β-galactosidase Insulin B fusion protein

↓

Purify fusion protein

↓

Chemically remove the β-galactosidase from chain A & chain B

↓

A chain (21 AA) | B chain (30 AA)

↓

Combine polypeptide to produce functional Insulin

A, B (S–S bonds) Insulin

ii. Hepatitis B vaccine
(DNA virus)

Hepatitis B Ag producing gene + Plasmid cut c̄ Restriction endonuclease

↓

Recombinant DNA

↓

Introduced into Yeast cell

↓

Multiplication & production of Hepatitis B Ag

↓

Extraction & purification of HBV Ag

↓

Formulate into vaccine

DNA vaccine & rDNA

Gene for Ag (Immunogen) ← Plasmid cut c̄ RE

rDNA

↓

Introduced into Host cell

↓

Multiplication and production of Ag

↓

Extraction & purification

↓

Immunize c̄ Ag – B cells – Humoral immunity
T-cells | cell mediated Immunity.
MHC

iii. Erythropoietin - for t/t of anemia
iv. Filgrastim :- for t/t of Neutropenia
v. Factor VIII, IX : for t/t of hemophilia
vi. Interferons - for t/t of viral infection, cancer etc.
vii. Interleukins - for t/t of Renal cell carcinoma
vii. Epidermal growth factors - t/t of Burns
viii. Tissue plasminogen activator

Alteplace
Reteplace
Tenecteplace } Thrombolytic agent

ix. Anticoagulant - lepirudin - t/t of thrombosis

2. Production of Hormones
   FSH → t/t of infertility
   LH → Induction of ovulation
   GH → t/t of growth failure

3. Human genome project

4. Gene therapy : insertion of corrected gene - tried in SCID

5. DNA fingerprinting

6. construction of library - cDNA library

7. production of Transgenic animals c̄ desired characteristics.
   example - Dolly sheep
   Mice - Knockout mice
   transgenic mice

Dolly the sheep was cloned in 1996

Black face sheep (cytoplasmic donor) → Donor egg → Remove Nucleus ↘
White face sheep (Nuclear donor) → mammary cell ↗
Egg fused c̄ cell (↑ Direct current pulse) → embryo → Implant embryo
↓ Surrogate sheep
Black faced sheep c̄ white faced lamb → DOLLY

Polymerase Chain Reaction [PCR]

Important LAQ :- Describe the steps in PCR for identification of viral infection (SARS-COV2)

Polymerase :- DNA Polymerase
- Extention of DNA by adding NTs
- synthesis of DNA

Chain Reaction :- Chemical reaction which proceeds in exponential manner

PCR. Definition :- It is a cell free amplification technique to produce several copies of DNA.

Requirement :-

1. PCR Machine — Thermal cycler
2. Template DNA - segment of DNA
3. Primers - Oligonucleotide primer 20-30 NTs
   complementary to the ends of Target DNA
   Required for addition of NT to 3' end
   Two types - forward & Reverse primer
4. Taq DNA Polymerase - from bacillus Thermus Aquaticus
   can withstand temp. up to 90-95°c
   causes - synthesis of DNA
5. dNTP :- dATP, dGTP, dCTP, dTTP

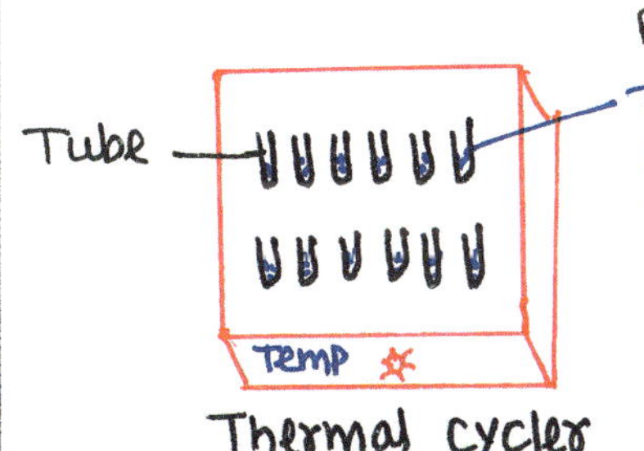

Thermal cycler

Steps

1. Denaturation
2. Annealing
3. Extension

1. Denaturation :-

separation of two strands of DNA when we heat the solution at 90-95°c for 15 second to 2 min

2. Annealing

renaturation at 56°c for 30 sec to 2 min

primer attached to separated DNA strands

3 Extension

synthesis of DNA molecule at 72°c - for 30 sec to 2 min

Total time for one cycle = 3 to 5 min

25 to 30 cycles → 3 to 6 Hrs

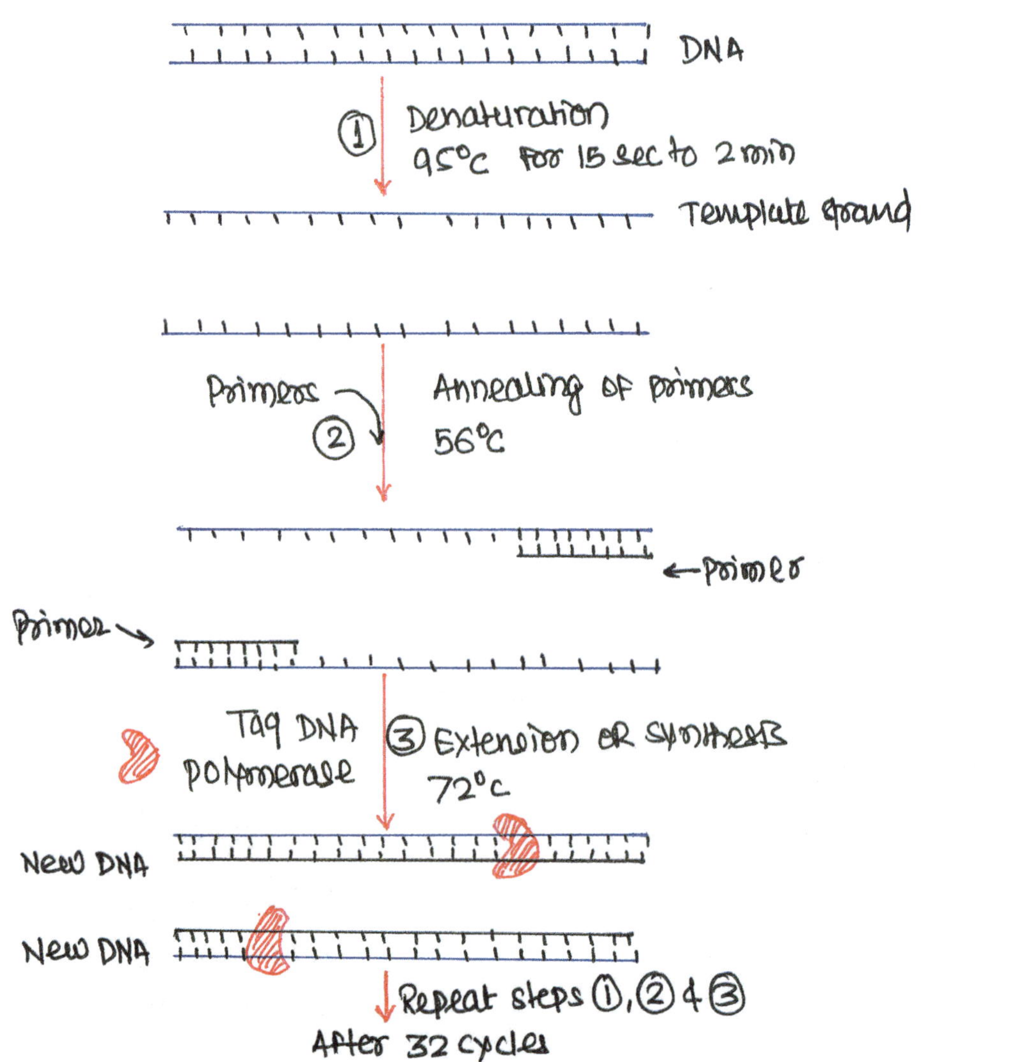

Applications

1. Detection of viral infection
2. Detection of Tuberculosis
3. Detection of cancer
4. Prenatal diagnosis (chorionic villous sampling)
   ↓
   Sickle cell disease
   Thalassemia
   Inborn errors of metabolism
5. Recombinant DNA – construction of gene library
6. Forensic Medicine – Identification of criminals, Thieves, Rapist
   paternity disputes
   (DNA fingerprinting)

SARS COV 2 Detection

Severe Acute Respiratory syndrome – coronavirus 2

Patient c̄ covid 19
↓
Nasal swab
OR
Oropharengeal swab
↓ VTM [viral transport medium]
Laboratory
↓
Nucleic Acid extraction
↓
Reverse transcriptase
RNA ⟶ cDNA
↓ PCR
Multiple copies

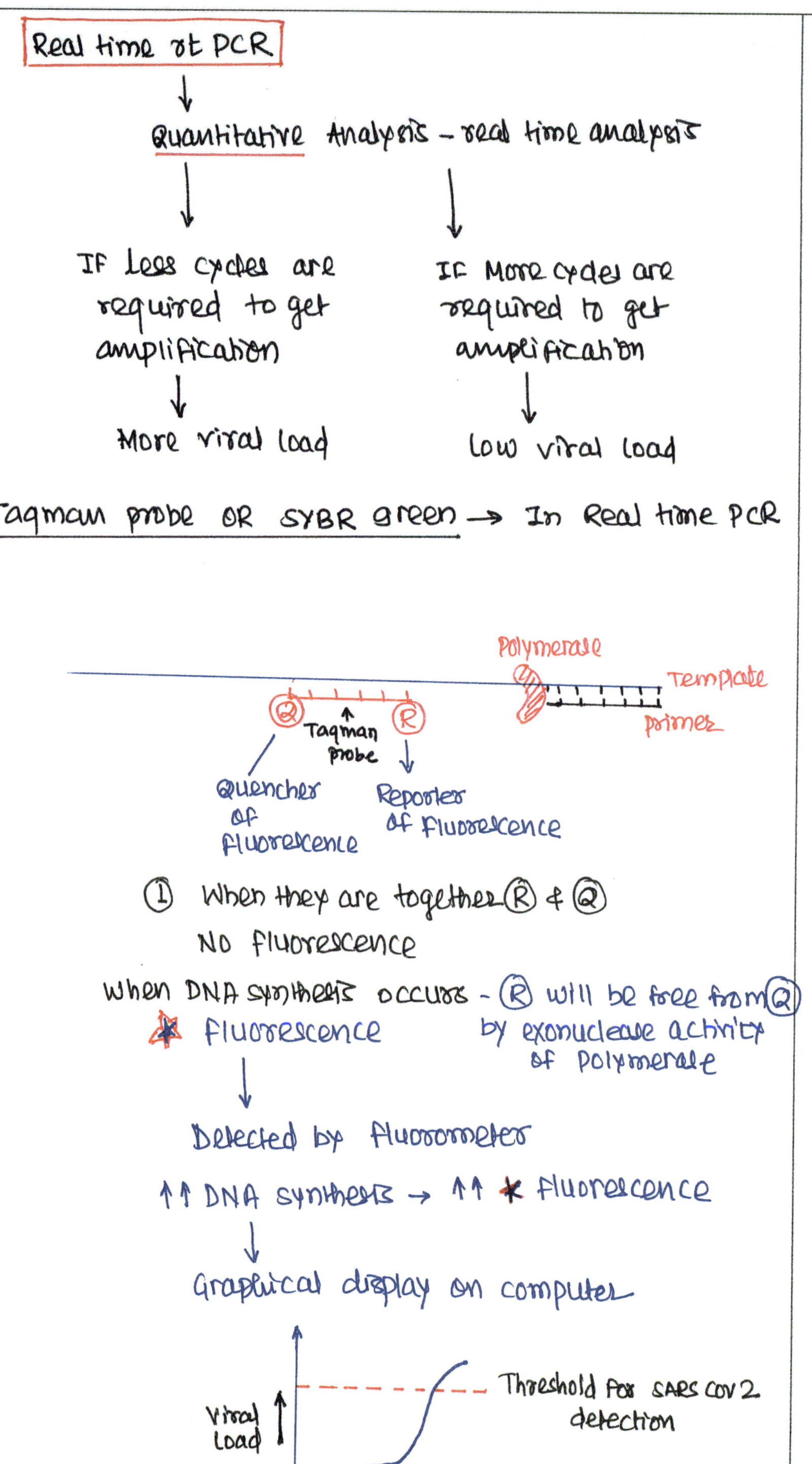
Real time rt PCR
Quantitative Analysis – real time analysis
If less cycles are required to get amplification
More viral load
If more cycles are required to get amplification
low viral load
Taqman probe OR SYBR green → In Real time PCR
Polymerase
Template
Primer
Q
R
Taqman probe
Quencher of fluorescence
Reporter of fluorescence
① When they are together Ⓡ & Ⓠ
No fluorescence
When DNA synthesis occurs – Ⓡ will be free from Ⓠ by exonuclease activity of Polymerase
fluorescence
Detected by fluorometer
↑↑ DNA synthesis → ↑↑ fluorescence
Graphical display on computer
Viral Load
Threshold for SARS cov 2 detection

DNA fingerpriting

Also known as DNA profiling

It is laboratory technique used to determine the probable identy of a person based on nucleotide sequences in certain regions of DNA that are unique to individuals.

Applications

1. criminal investigations
2. Paternity disputes
3. Immigration test cases and disputes

Aim to match two DNA fingerprints, such as a DNA sample from known person and one from unknown person.

DNA Marker's in disease diagnosis and fingerprinting

1. Restriction fragment length polymorphism (RFLP)
2. Variable number tandem repeats (VNTRs)
3. Simple tandem repeats (STRs) OR Microsatellites
4. Single nucleotide polymorphism (SNPs)

RFLP [Riflip]

A stretch of DNA exist in fragment of various length (polymorphism) derived by the action of restriction enzyme.

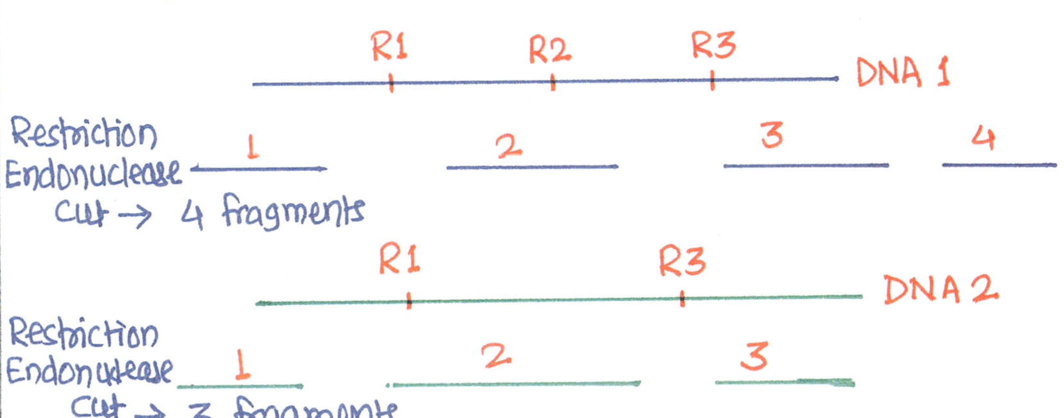

VNTRs

Variable number of tandem repeats (long NT sequences)

Short nucleotide sequences is organized as a tandem repeat and these repeats having variation in length between individuals

2 Repeats —▭—▭— DNA-1

4 Repeats —▭—▭—▭—▭— DNA-2

7 Repeats —▭—▭—▭—▭—▭—▭—▭— DNA 3

Different number of repeats in the DNA molecule.

Microsatellites (Simple tandem Repeats)

- Short repeat units usually composed of dinucleotide OR tetranucleotide units

4 Repeats CG CG CG CG DNA-1

2 Repeats CG CG DNA-2

6 Repeats CG CG CG CG CG CG DNA-3

Single Nucleotide Polymorphism (SNPs)

Most common type of genetic variation

A variation in single nucleotide in the DNA

SNPs represent the position in the genome where some individuals have one NT [G] while other will have [C]

GCAACTTGAC DNA-1

GCGACTTGAC DNA-2

GCTACTTGAC DNA-3

GCCACTTGAC DNA-4

Blotting Technique

It is a technique for transferring DNA, RNA and proteins onto a carrier so they can be separated.

Types of blotting techniques

- **Southern blot** – for detecting DNA
- **Northern blot** – for detecting RNA
- **Western blot** – for detecting protein
- **Eastern blot** – for detecting post-translational modifications in proteins

Southern blotting steps

1. Digest the DNA with Restriction enzyme

↓

2. Mixture of fragments – Gel electrophoresis to separate fragments according to size

↓

3. Fragments of DNA denatured c̄ Alkali and transferred onto Nitrocellulose filter by blotting

↓

4. Hybridization c̄ radiolabelled DNA probe

↓

5. DNA probe bound to filter is detected by Autoradiography – which reveals the DNA fragment to which the probe hybridized

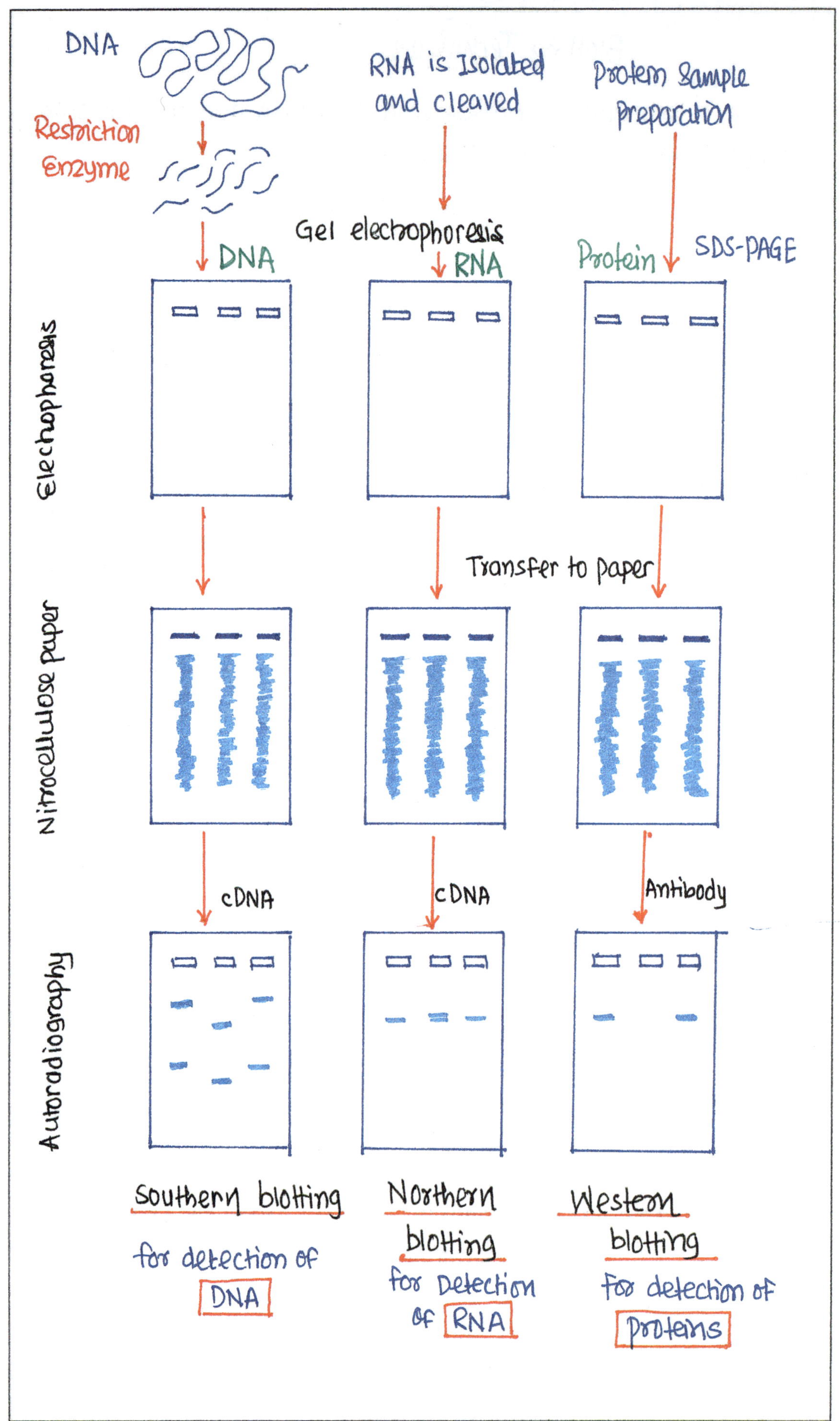
DNA
Restriction Enzyme
RNA is Isolated and cleaved
Protein Sample Preparation
Gel electrophoresis
DNA
RNA
Protein
SDS-PAGE
Electrophoresis
Transfer to paper
Nitrocellulose paper
cDNA
cDNA
Antibody
Autoradiography
Southern blotting
for detection of DNA
Northern blotting
for Detection of RNA
Western blotting
for detection of proteins

FREE RADICALS

MCI BI 7.7
SAQ

Definition:- Molecules or molecular fragments containing one or more unpaired electron/s are known as free radicals.

Characteristics - Highly reactive
- Causes generation of new free radicals
- Shorter life span
- Can cause damage to tissues.

Names of **Reactive oxygen species**.

$O_2^{\bar{\cdot}}$ → superoxide anion radical
HOO^- → Hydroxyperoxyl radical
$OH^\cdot$ → Hydroxyl radical
$NO^\cdot$ → Nitric oxide
$ONOO^\cdot$ → peroxynitrite
] Reactive Nitrogen species

H_2O_2 → Hydrogen peroxide
$^1O_2$ → Singlet oxygen
] do not have unpaired electron but follow other characteristics of free radicals

Steps in reduction of O_2 & generation of free radicals.

$$O_2 \longrightarrow O_2^{\bar{\cdot}} \longrightarrow H_2O_2 \longrightarrow OH^\cdot \longrightarrow H_2O$$

Sources :- Free Radical Sources

Endogenous
1. Electron transport chain
2. Nitric oxide synthase
   Arg ⟶ NO
3. NADPH oxidase
   Respiratory burst
4. Xanthine oxidase
5. Prostaglandin synthesis
6. Lipid peroxidation
7. Autooxidation of metals

Exogenous
1. Alcohol
2. Radiation
3. Drugs
4. Smoking

Respiratory burst

Bacteria → ◯ Macrophages

Bacteria engulfed by Macrophages

↓

↑ O_2 demand

O_2

NADPH → NADP (NADPH oxidase)

$O_2^{\bar{\cdot}}$

↓ superoxide dismutase

H_2O_2

↓

HClO

Hypochlorous acid ⟶ kills bacteria ← $O_2^{\bar{\cdot}}$

Lipid peroxidation :- Important source of free radicals.

1. **Initiation**

$$RH\ [PUFA] + OH^{\cdot} \longrightarrow R^{\cdot}\ (\text{PUFA radical}) + H_2O$$

2. **Elongation** (Propogation)

$$R^{\cdot} + O_2 \longrightarrow ROO^{\cdot}\ (\text{Lipid peroxyl radical})$$

$$ROO^{\cdot} + RH \longrightarrow ROOH\ (\text{lipide peroxide}) + R^{\cdot}$$

lipide peroxide ↓ Stimulation of lipid peroxidation

3. **Termination**

$$ROO^{\cdot} + ROO^{\cdot} \longrightarrow RO\text{-}OR + H_2O$$

When lipid peroxyl radical combine c̄ other peroxyl radical, termination occurs.

Disorders related c̄ ROS

1. Acute inflammation

2. Aging – Alzheimer's disease, Multiple sclerosis, Parkinsons ds.

3. Atherosclerosis

LDL ↑↑ $\xrightarrow{\text{free radicals}}$ oxidized LDL. ⟶ foam cell ⟶ plaque

4. Bronchoalveolar diseases (ARDS)

5. Cancer

6. Cataract, retinal injury of prematurity

7. Cardiovascular diseases – Reperfusion injury in MI

8. Diabetes Mellitus – damage to β-cells of islets of langerhans → Insulin dependent diabetes mellitus.

ANTIOXIDANTS

MCI BI 7.6

Free Radical Scavenger

Definition: Substances which inhibit / decreases the oxidation of compounds are known as antioxidants.

Antioxidants.

ENZYMATIC

1. Superoxide dismutase (SOD)
2. Catalase
3. Glutathione peroxidase (Se)
4. Glutathione reductase

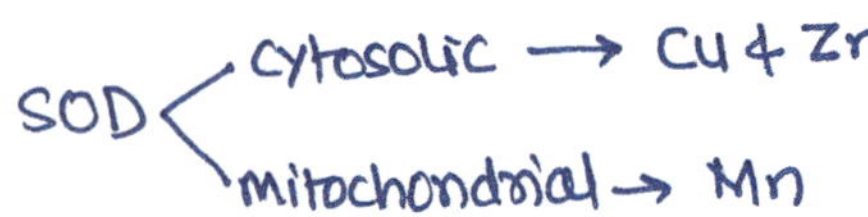

SOD
- Cytosolic → Cu + Zn
- Mitochondrial → Mn

NON-ENZYMATIC

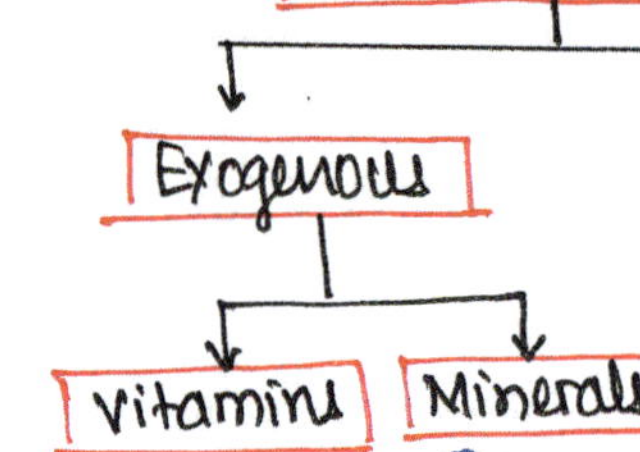

Exogenous

Vitamins
1. Vit C
2. α-Tocopherol (Vit E)
3. β-carotene

Minerals
1. Zn
2. Se
3. Cu
4. Mn

Endogenous

- Ceruloplasmin
- Uric acid
- Transferrin
- Bilirubin
- Ferritin
- Lipoic acid
- Ubiquinone
- Phytochemicals.

① $O_2^{\bar{\cdot}} + 2H^+ \xrightarrow{SOD} H_2O_2$

$H_2O_2 \xrightarrow{\text{Catalase ② peroxidase}} H_2O$

③ $2H_2O_2 \xrightarrow[\text{Glutathione Peroxidase}]{(Se)} 2H_2O + O_2$

2GSH Reduced Glutathione → GS-SG Oxidized

GS-SG → 2GSH (Glutathione reductase ④): NADPH → $NADP^+$

Role in lipid peroxidation

$ROOH + 2GSH \xrightarrow{\text{Glu. peroxidase}}$ Hydroxy FA + H_2O + Oxidized Glutathione

[Glutathione Reduced]

Vitamin E

Active form α-tocopherol

ROO (Lipid peroxyl radical) + Reduced vit E ⟶ ROOH + Oxidised Vit E

Oxidised Vit E —(Vitamin C)⟶ Reduced Vit E

Chain breaking Antioxidants (MCQ) — ① SOD
② Uric acid
③ Vit E

Vitamin C

1. Reduction of oxidized vit E
2. Ascorbate + $O_2^{\cdot -}$ ⟶ H_2O_2 + Monodehydroascorbate

   Ascorbate + $OH^{\cdot}$ ⟶ H_2O + Monodehydroascorbate
3. Along with vit E — Prevents lipid peroxidation
   Prevents cataract formation

BIOTRANSFORMATION

Detoxification
Metabolism of Xenobiotics

MCI BI 7.5

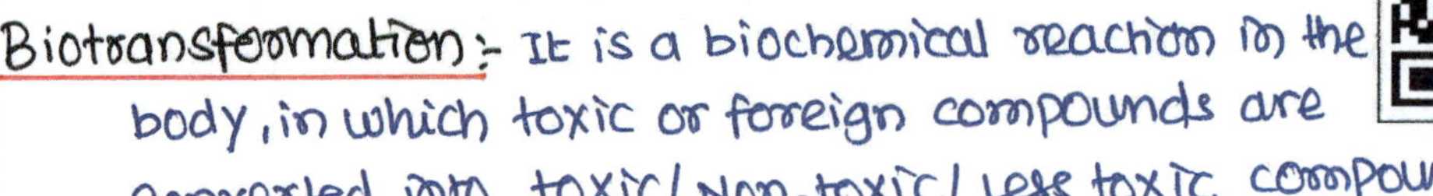

Biotransformation:- It is a biochemical reaction in the body, in which toxic or foreign compounds are converted into toxic/ Non-toxic/ Less toxic compounds.

Chemical 1 —Biotransformation→ Chemical 2

Toxic — ① Detoxification ② Entoxication → Toxic / Nontoxic / Less toxic

Detoxification:- It is a conversion of toxic compound into nontoxic or less toxic and excretable form.

Toxic ————→ Nontoxic / Less toxic } excretable form.

Entoxication:- conversion of compound into more toxic form.

Methanol (Toxic) ————→ formic acid (More toxic)

Xenobiotics:- foreign or strange compound

The compounds which need detoxification are called Xenobiotics.

Includes

- Additives
- Adulterants
- Bilirubin (toxic compound produced in the body)
- Bacterial action (produced by bact action)
    - Histamine
    - Putrescine
    - cadaverine
    - Tyramine
    - Tryptamine
- cosmetics
- Drugs
- Environmental pollutants
- Pesticides etc.

Two phases of Detoxification

Toxic compound → Phase 1 → Non toxic water soluble compound

Toxic compound → Phase 1 → Phase 2 → Non toxic water soluble compound

Benzoyl CoA → (Phase 1) → Benzoic acid → (Phase 2) → Hippuric acid

Benzene → (Phase 1) → Phenol → (Phase 2) → Phenyl sulfate

Phase 1 Reaction :- Group is added to the compound

1. Oxidation
2. Reduction
3. Hydrolysis

1. Oxidation OR Hydroxylation [SAQ]

- Cytochrome P 450 is required for hydroxylation
- Also known as mixed function oxidase
- It is Monooxygenase — only one atom is incorporated into substrate
- It requires NADPH as coenzyme
- It requires Phosphatidyl choline for its action
- 20 - 200 forms of cyt P450
- Located in microsomes of liver
- Maximum absorption at 450 nm, when exposed to CO

It includes Oxidation of – Alcohol, Aldehydes
Amines
Aromatic hydrocarbons
Sulfur compounds
Drugs.

Examples :-

Alcohol —
- Methanol ⟶ formic acid
- Ethanol ⟶ Acetic acid
- Benzoyl Alc ⟶ Benzoic acid

Aldehyde
- Benzaldehyde ⟶ Benzoic acid
- Chloral ⟶ Trichloroacetic acid

Amines
- Aliphatic amines ⟶ Aliphatic acid + Urea
- Aniline ⟶ Aminophenol

Aromatic HC → Benzene ⟶ Phenol

Sulfur ⟶ H_2SO_4

Meprobamate (Drug) ⟶ Hydroxy meprobamate

2. Reduction

Examples

Picric acid ⟶ Picramic acid

Nitrophenol ⟶ Aminophenol

Chloral ⟶ Trichloroethanol

3. Hydrolysis :- Bigger molecules are hydrolysed and converted into smaller ones.

$A/B + H_2O \longrightarrow A\text{-}OH + B\text{-}H$

examples:-

Aspirin $\xrightarrow{H_2O}$ Salicylate + Acetic acid

Atropine

Procain

Xylocain

Di-isopropyl Fluorophosphate (DFP)

Acetanilide

Phase 2 Reactions : CONJUGATION

Important SAQ

Toxic or foreign compounds are combined with compound produced in our body and converted into less toxic & water soluble form.

Toxic compound + compound produced in our body ⟶ Less toxic water soluble compound

Mnemonic – 4G SAMT

1. Glucuronic acid
2. Glycine
3. Glutathione
4. Glutamic acid
5. Sulfate
6. Acetylation
7. Methylation
8. Thiosulfate

1. **Glucuronic acid**

Active form UDP-Glucuronate

Compounds conjugated by Glucuronate B - Bilirubin
S - Steroids
P - Phenol

Bilirubin + UDP-Glucuronate $\xrightarrow[\text{transferase}]{\text{UDP-Glucuronyl}}$ Bilirubin monoglucuronide

Bilirubin monoglucuronide + UDP-Glucuronate $\xrightarrow[\text{Transferase}]{\text{UDP-Glucuronyl}}$ Bilirubin diglucuronide [Water soluble conjugated Bilirubin]

2. **Glycine**

Benzoic acid + Glycine ⟶ Hippuric acid

Cholic acid + Glycine ⟶ Glycocholic acid

Chenodeoxy cholic acid + Glycine ⟶ Chenodeoxy cholic acid

3. **Glutathione** : γ-Glutamyl-cysteinyl-glycine
↓
Important in conjugation of Alkyl
Aryl halide
Alkenes.

4. **Glutamine**

Phenylacetate ⟶ Phenylacetyl glutamine

5. **Sulfate**

Active sulfate donor ⟶ Phosphoadenosine phosphosulfate (PAPS)

Phenol + PAPS ⟶ Phenyl sulfate

6. **Acetylation** : Addition of acetyl CoA to the compound

7. **Methylation** : Active methyl donor is SAM - S Adenosyl Methionine

Epinephrine ⟶ Metanephrine

8. **Thiosulfate** :

Cyanide ⟶ Excretable form

CANCER

MCI BI 10.1–10.2

Definition:- uncontrolled, abnormal cell growth

Also known as Neoplasm / Tumor

Two main types ① Benign tumor

② Malignant tumor

Benign tumor	Malignant tumor.
Encapsulated	Not encapsulated
Does not invade local tissue	Invade the local tissues
Does not spread via blood OR Lymphatics (No metastasis)	spread to distant tissues via blood OR lymphatics [Metastasis]
No destruction of tissue of origin	Distruction of tissue of origin

Tissue	Name
from epithelium →	carcinoma
from Mesenchymal tissue →	Sarcoma
from Lymphatic →	Lymphoma
from Blood →	Leukemia

Characteristics of Cancer cell [Mnemonic - DMER IN MAP]

1. Differentiation:- resemblance of cancer cell c̄ normal cell from which it has originated.
2. Metastasis:- spread to distant organs.
3. Encapsulation:- presence of well defined capsule
4. Rate of growth:- Rapid OR slow
5. Invasion:- Invades the local tissue OR Not
6. Nuclear cytoplasmic ratio [N/C ratio]
7. Mitotic rate
8. Anaplasia – loss of morphological characteristics & orientation
9. Pleomorphism – variation in size, shape & color of the cells.
10. Angiogenesis - formation of new blood vessels.

Characteristics	Benign	Malignant
Differentiation	well differentiated	Poorly differentiated
Metastasis	No Metastasis	Metastasis
Encapsulation	well encapsulated	No capsule
Rate of growth	slow	Fast-due to loss of contact inhibition
Invasion	Not present	Invades local tissue
N/c ratio	1:4 1:6	1:1
Mitotic rate	Low	very high
Anaplasia	No	present
Pleomorphism	No	Present

Additional features of Cancer cell

1. ↑ RNA & DNA synthesis
2. ↑ Activity of ribonucleotide reductase
3. ↑ Aerobic and anaerobic glycolysis
4. ↑ Lactic acid production
5. ↑ synthesis of mutated proteins
6. ↑ Damage to cell surface due to abnormal synthesis of glycoproteins and glycolipids.

Molecular Basis of Cancer - CARCINOGENESIS

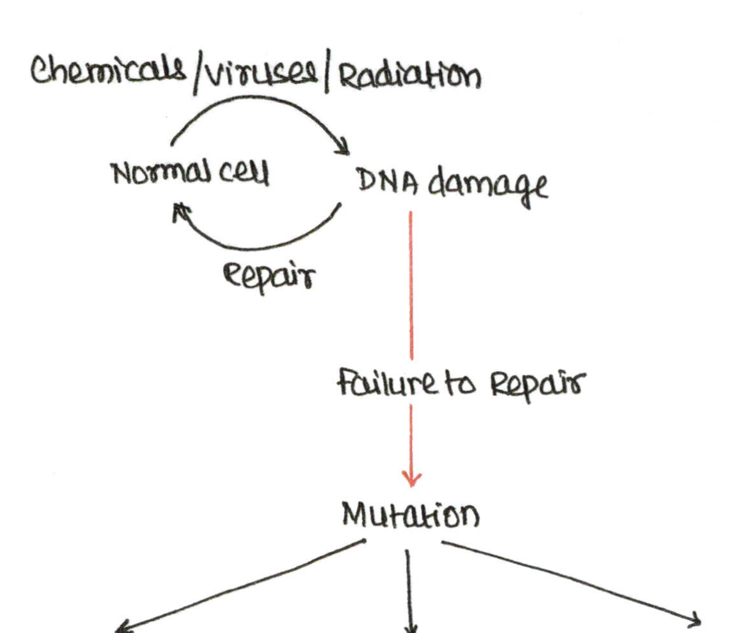

Chemical carcinogens

1. Alkylating agent :- causes spontaneous deamination, depurination
   example- cyclophosphamide — single base alteration
2. Aromatic hydrocarbon :- Benzopyrene in cigarette smoke causes cancer of lung & oral cavity
   - Tobacco & betel nuts causes cancer of oral cavity
3. Aromatic amines - Azo dye - Aniline - Bladder carcinoma
4. Nitrosocompounds - Gastric carcinoma
5. Asbestos, Nickel - Lung carcinoma
6. Aflatoxin → produced by Aspergillus flavus on peanuts
   causes - Hepatocellular carcinoma

Chemical carcinogen ⟶ Reactive electrophiles ⟶ Damage DNA
↓
① Activation of oncogenes
② Inhibition of Tu. supressor gene
③ Inhibition of Apoptosis

Viral Carcinogens

1. Human papilloma virus - 16, 18 → Cervical carcinoma
2. Herpes simplex virus (HSV -2) → Cervical carcinoma
3. HBV, HCV → Hepatic cancer
4. HIV → Kaposi's sarcoma
5. HTLV → Human T cell Lymphotropic virus → Adult T cell leukemia
6. Epstein-Barr virus (EBV) → Nasopharyngeal carcinoma
   Burkitt Lymphoma

Viruses → Insertion of promoter/enhancer
↓
Mutation
↓
Activation of oncogenes
Inhibition of Tu. supressor genes
Inhibition of Apoptosis

Radiation

Ionizing radiation → causes double strand break in DNA
↓
Prone to Precancerous / cancerous condition

UV-light → T-T dimer (Pyrimidine dimer formation)
causes xeroderma pigmentosum

X-Rays → causes skin cancer

Proto-oncogenes

Definition:- Normal genes in cell responsible for production of various Transcription factors, Growth factors, Receptor for GF, proteins that regulate Gene expression.

Proto-oncogenes

↓

1. Point Mutation
2. Chromosomal translocation
3. Gene amplification
4. Enhancer/promoter insertion

Mutation

↓

Oncogenes — Genes capable of causing cancer

↓

Mutated proteins

⊕ cell division

⊕ cell growth

⇓

CANCER

oncogenes	Location	Product produced
jun	Nucleus	Transcription factors (TF)
fos	Nucleus	Transcription factors (TF)
myc	Nucleus	DNA binding proteins
ras	cytoplasm	GTPase
raf	cytoplasm	protein Kinase
abl	cytoplasm	Tyrosine kinase
sis	cell membrane	Platelet derived growth factor

Tumor supressor Genes - Antioncogenes

Recessive genes

Protective genes → ⊖ cell division
⊖ cell growth

→ Inhibition of Tu. supressor gene causes cancer

Examples :- RB → Retinoblastoma
p53 → colon, breast cancer
WT → Wilms tumor (kidney)
FAP → familial adenomatous polyposis
BRCA → Breast cancer

① p53 - phosphoprotein
MW 53 KDa
Also known as Molecular policeman MCQ
Gatekeeper gene
Guardian of genome

Mutation
↓ Accumulation of p53

p53 induced
Transcription
↓
p21 protein

⊖ CDK
cyclin dependent kinase
↓
⊖ cell division

⊖ PCNA (Proliferating cell Nuclear Ag)
↓
⊖ Replication

② RB - Retinoblastoma gene - unphosphorylated RB bound to eTF2f

RB unphosphorylated, eTF2f bound → Eukaryotic TF2f is Not Available
↓
⊖ Transcription

Phosphorylation
cyclin D, CDK-4
↓
free → Transcription ⊕ → cell division

TUMOR MARKERS [SAQ]

Definition:- Substances produced by the tumor cell or normal cell in response to tumor tissue and detected in body fluid by various methods are known as Tumor markers.

↓

They are associated c̄ specific cancer

Samples- Blood
Urine
Body fluids - Ascitic fluid, CSF, synovial fluid etc.
Tissues

Characteristics of Ideal Marker

1. It should be specific for that cancer
2. It should be sensitive enough to be detected by various methods.
3. It should be in higher concentration than Normal individual
4. It should tell about staging of cancer.

Uses

1. **Screening** - detection of cancer before appearance of symptoms
   example - pap smear for cervical cancer
   Mammogram for breast cancer
2. **Diagnosis** - It is used in diagnosis of cancer along with imaging and clinical symptoms.
   example :- α fetoprotein — Hepatocellular carcinoma
3. **Staging of cancer** :- Extention of tumor & Metastasis
4. **Prognosis** :- course of disease
   Example :- carcinoembryonic Ag → colon cancer
5. **Monitoring** of patient in treatment
6. **Follow up**

Classification c̄ Examples

1. **Oncofetal Antigen**

Alpha fetoprotein (AFP) → Hepatocellular carcinoma
Germ cell tumor

Carcino-embryonic antigen (CEA) → Colorectal carcinoma
Breast cancer
Lung cancer

2. <u>Carbohydrate Antigen</u>

CA 15-3 → Breast carcinoma
CA 19-9 → Pancreatic cancer
CA 125 → Ovarian cancer

3. <u>Hormones</u>

β-HCG → Choriocarcinoma
Calcitonin → Medullary carcinoma of thyroid
Gastrin → Gastrinoma
Insulin → Insulinoma
ACTH → Ectopic lung cancer
Serotonin → Carcinoid tumor
Catecholamines → Pheochromocytoma

4. <u>Enzymes</u>

Alkaline phosphatase → Metastasis in liver & bone
Neuron specific enolase → Neuroblastoma
Prostatic Acid phosphatase → Prostate cancer
Lactate dehydrogenase → Testicular & germ cell tumor

5. <u>Other Markers</u>

PSA - Prostate Specific Ag → Prostate cancer
Benign hypertrophy of prostate

HER 2 - Breast carcinoma

ER & PR - Breast carcinoma

Monoclonal antibodies
Bence Jones proteins] - Multiple myeloma

EXTRACELLULAR MATRIX [ECM] MCI BI 9.1-9.2

Also Known as Connective tissue OR Ground substance

SAQ :- Describe the component of ECM with diseases.

Definition :- It is a gel like substance present in extra-cellular space in tissues.

Components of ECM

1. Structural Fibrous proteins
   - Collagen
   - Elastin
   - Fibrillin
2. Specialized glycoproteins
   - Fibronectin
   - Laminin
3. Proteoglycans [Proteins + GAGs]
   - Syndecan
   - Betaglycan
   - Serglycin
   - Perlecan
   - Aggrecan
   - Versican
   - Decorin
   - Biglycan
   - Fibromodulin

COLLAGEN

SAQ :- Structure and Functions of Collagen

Synthesis - PPTC

Preprocollagen

↓ Signal Peptidase (↑ signal peptide)

Procollagen (NH_2 — COOH)

↓ Hydroxylation, Glycosylation, disulfide bond, Triple helix formation

Tropocollagen

↓ Removal of N- and C- terminal propeptide, Assembly into Fibrillar structure

Collagen

Collagen Fibril

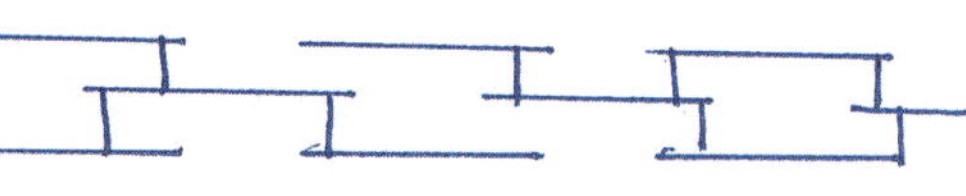

1. collagen is the most abundant protein in the animals
2. Triple helical structure
3. Gly-X-Y repeats c̄ 30% Glycine
4. Other amino acids are proline and hydroxyproline
5. proline and hydroxyproline - confers rigidity to collagen
6. stabilized by covalent cross linking by lysyl oxidase
7. Since glycine is the smallest of all the amino acids, it allows the chain to form a tight configuration and it can withstand stress

Types of Collagen

I → Most connective tissues including bones * MCQ
II → Cartilage, vitreous humor
III → Skin, lungs, vessels
IV → Basement membrane * MCQ
V → Minor component c̄ type I
VI → Most connective tissues
VII → Anchoring fibrils MCQ
VIII → Endothelium
IX → Tissue containing type II
X → Hypertrophic cartilage
XI → Tissue containing type II
XII → Tissue containing type I
XIII → Many tissues
XIV → Tissue containing type I
XV → Many tissues
XVI → Many tissues
XVII → Skin hemidesmosomes
XVIII → Many tissues
XIX → Rhabdomyosarcoma cells

Classification based on structures that they form (MCQs)

Fibril forming → 1, 2, 3, 5, 11
Network like → 4, 8, 10
FACITs' → 9, 12, 14, 16
Beaded filaments → 6
Anchoring fibril → 7
Transmembrane domain → 13, 17
Others → 15, 18

Disorders of Collagen

1. Osteogenesis imperfecta :- Mutation in the gene for type I collagen
   disorder of bone, brittle bone, blue sclera, deafness.

2. Ehler Danlos syndrome :- Mutation in gene for type I, type 3 collagen
   defect in lysyl hydroxylase

   Hypermobility of joints, Hyperextensible skin, easy bruising
   rupture of vessels.

3. Alport syndrome :- Mutation in type IV Collagen (GBM)
   Renal damage, hematuria, renal failure

4. Menkes disease :- Mutation in ATP7A gene
   Defective absorption of copper from intestine

   Neurodegeneration, kinky hair, skeletal abnormalities

5. Chondrodysplasias :- type II gene mutation - dwarfism

6. Epidermolysis bullosa - Mutation in type VII collagen
   skin breaks and blisters as a result of minor trauma.

7. Scurvy :- Vit C deficiency causes defect in prolyl & lysyl
   hydroxylase.

ELASTIN

Synthesized from tropoelastin by Post-translational modification
Simple in structure
No triple helix, No repeat Gly-X-Y
No Hydroxylysine

Function :- Confers extensibility and recoil of lung, blood vessels and ligaments.

Diseases :- ① William's syndrome :- caused by deletion of elastin gene
MR and supravalvular aortic stenosis.
② Scleroderma :- ↑ elastin
③ Pulmonary emphysema :- ↑ degradation of elastin
④ Aging of skin is associated c̄ decrease in elastin.

Fibrillin

structural component of myofibril found in various tissues
It is a large glycoprotein
Present in Lens, periosteum, aorta

Disease :- Marfan syndrome - Mutation in gene for fibrillin 1
features - ectopia lentis, long digits, tall stature
dilatation of ascending aorta

Fibronectin

Glycoprotein involved in interaction of cell c̄ ECM
dimer c̄ 7 functional domains
function :- cell adhesion and cell migration

Laminin

present in basal lamina of glomerular membrane
Has role in glomerular filtration

Proteoglycans [proteins + GAG]

Proteins that contains covalently linked GAGs
protein - core protein
carbohydrates → Glycosaminoglycans up to 95%

Names :-

Syndecan	versican
betaglycan	decorin
serglycin	biglycan
perlecan	Fibromodulin
Aggrecan	

Major GAGs →
- Hyaluronic acid - synovial fluid, vitreous humor
- Chondroitin SO_4 - cartilage, bone
- Dermatan SO_4 - skin
- Heparin — Mast cells, liver, lung, skin
- Heparan SO_4 — cell surfaces
- Keratan SO_4 — cornea, loose connective tissue

Diseases - Mucopolysaccharidoses

1. Hurler syndrome
2. Hunter syndrome
3. Sanfilippo A, B, C, D
4. Morquio A, B
5. Maroteaux - Lamy
6. Sly disease

Features of Mucopolysaccharidoses

Chronic progressive course
Multiorgan involvement
Organomegaly
Abnormal bone & cartilage
Mental retardation
Abnormal facies
Hearing & vision problem.

Functions of GAGs

1. Structural component of ECM
2. Have interaction c̄ other components
3. As polyanions bind c̄ cations
4. Act as a sieves in ECM
5. Cell migration (HA)
6. Shock absorbant (HA, CS)
7. Corneal transparency (KS)
8. Anticoagulant (Heparin)
9. Cell adhesion and cell-cell interaction (HS)

IMMUNITY

Resistance offered by the host against microorganism or any foreign substance

INNATE IMMUNITY

- Present at birth
- Immune response occurs in minutes
- Prior exposure - Not required
- Diversity is limited
- No memory
- Against Ag shared by many microbes
- Host cell receptors - Non specific

Components

1. Anatomical barriers
2. Physiological barriers
3. Phagocytes
4. Natural killer cells
5. Mast cell, Macrophages and dendritic cells
6. Complements - All
7. Fever & inflammatory responses
8. Normal resident flora
9. Cytokines
10. Acute phase reactant proteins.

ACQUIRED/ADAPTIVE IMMUNITY

Acquires during his lifetime

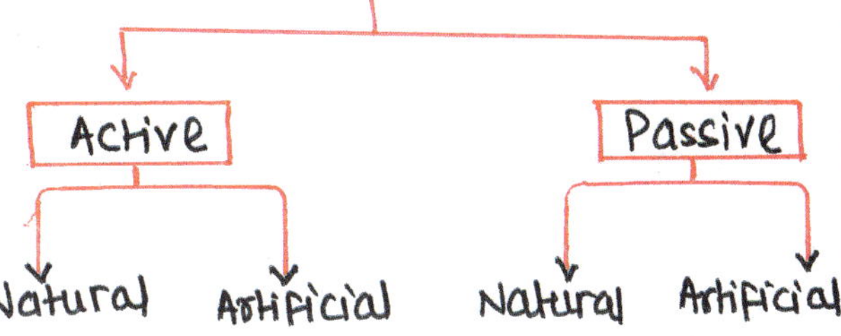

- Immune response occurs in days
- Prior exposure is needed
- More varied and specialized responses
- Memory - present
- Against Ag specific for each microbe
- Host cell receptors are specific

Components

1. T-cell
2. B-cell
3. Classical complement pathway
4. Cytokines

Active Immunity	Passive Immunity
① produced actively by host immune system	① Received passively - Immunoglobulins
② Induced by - infection (Natural) - vaccination (Artificial)	② Acquired by - Mother to fetus (IgG transfer) - Readymade antibody transfer
③ Long lasting	③ last for short time
④ Lag period present	④ No lag period
⑤ Memory present	⑤ No memory
⑥ Booster doses are useful	⑥ Less effective in booster doses
⑦ Negative phase may occur	⑦ No Negative phase
⑧ Not useful in immunodeficient individual	⑧ Useful in immunodeficient individual.

Primary Immune Response	Secondary Immune Response
① Immune response against first Ag	① Against subsequent Ag
② SSS - slow, sluggish, short lived	② PPP - Prompt, powerful, Prolonged
③ lag period - longer (4-7 days)	③ lag period - Absent or shorter
④ No negative phase	④ Negative phase may occur
⑤ Antibody produced in low titre and IgM type	⑤ Antibody produced in high titre and IgG type
⑥ Antibodies are more specific but less avid	⑥ Antibodies - less specific, more avid
⑦ Antibody producing cells - Naive B cells	⑦ Ab producing cells - Memory B cells
⑧ Both T dependent & independent Ag are processed	⑧ only T - dependent Ag are Processed

Ab conc
Initial exposure
Primary Immune resp.
Secondary exposure
secondary Immune response
Time

Mechanism of Hormone Action

LAQ: Enumerate the group I hormones and write about their mechanism of action.

LAQ: Classify the group II hormones and write about their mechanism of action.

Hormones: chemical secreted by the endocrine gland having action on target tissues.

Group I	Group II
Intracellular receptors	Receptors on cell membrane
	Act through second messenger
lipophilic (lipid soluble) Long half life (Hrs to days) require transport molecule	Hydrophilic Short half life (Min) No transport molecule required
Mediator :- Hormone-receptor complex	Mediator - Second messenger A - cAMP B - cGMP C - Ca/IP3 D - Kinase cascade

Classification on the basis of Mechanism of action

Group I

Steroid Hormones

Glucocorticoids
Mineralocorticoids
Androgen
Progestin
Estrogen
Calcitriol

Intranuclear receptors

T3, T4
Retinoic acid

Group II

Group II A. **cAMP**

Adrenaline
Glucagon
FSH
LH
MSH
ACTH
PTH
Calcitonin
Somatostatin etc.

Group II B. **cGMP**

Nitric oxide
Atrial natreuretic factor

Group II C. **Ca^{2+} / PI**

Oxytocin
GnRH
Gastrin
Cholecystokinin
~~TH~~ TRH
PDGF
Substance-P
Muscarinic ACTH
α_1 Adr. Catecholamines

Group II D. **Kinase - Phosphatase**

GH
Epidermal growth factors
Prolactin
Insulin

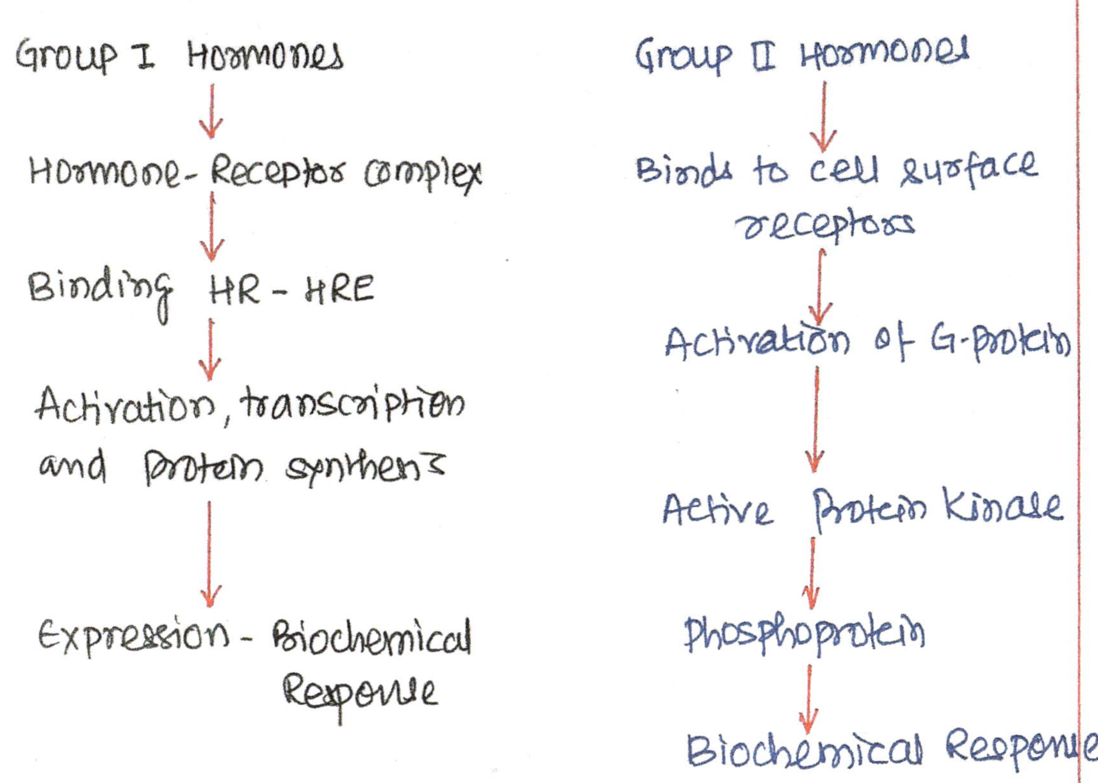

Mechanism of Action of Group I Hormones.

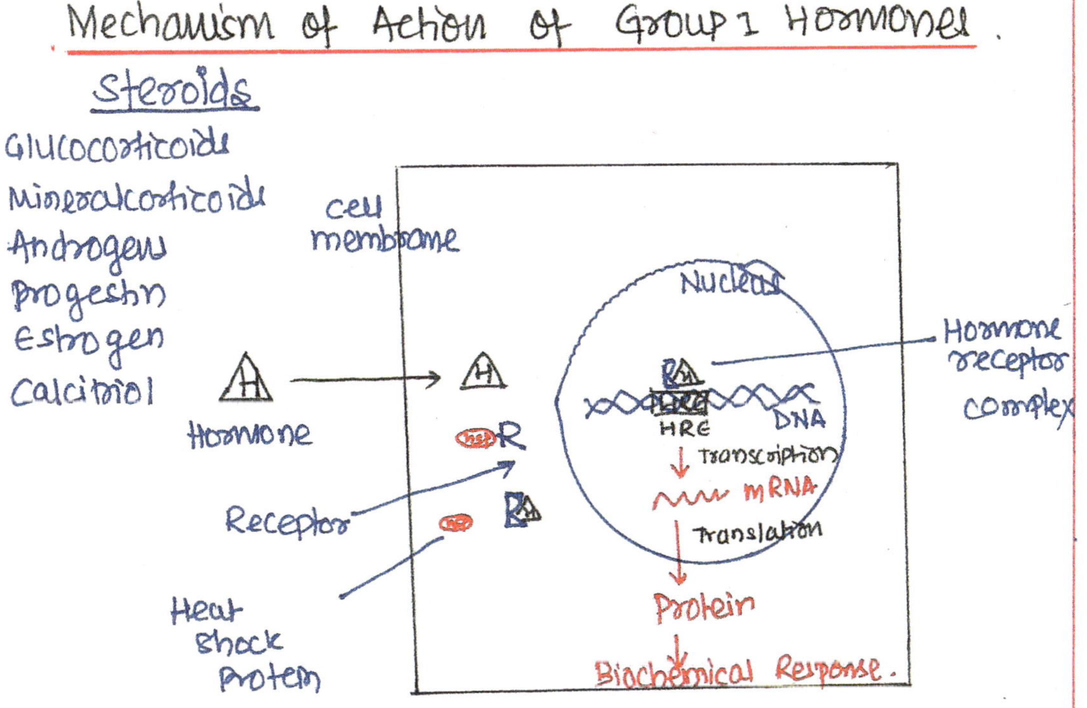

Mechanism of T_3, T_4 Retinoic acid

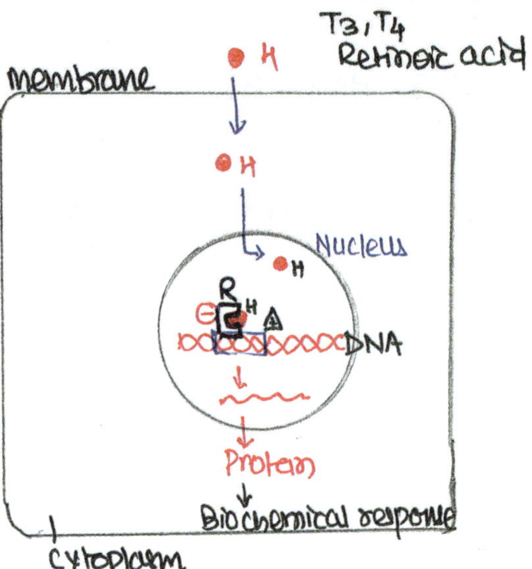

Intranuclear Receptor

- When (H) attached to receptor (R) release of negative effector ⊖ occurs.
- No. of +ve effector attached to DNA
- Transcription, translation protein
- Biochemical response.

Mechanism of Group II A - through cAMP

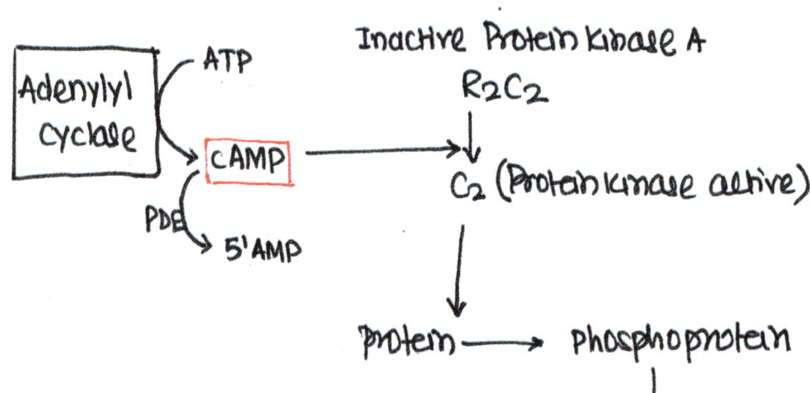

1. Adenyl cyclase converts ATP into cAMP
2. cAMP activates protein kinase A
3. protein $\xrightarrow{PKA}$ phosphoprotein
4. phosphoprotein ⟶ Biochemical Response
5. Response can be removed by phosphatase & phosphodiesterase

Examples:- Adrenaline
Glucagon
FSH, LH, MSH, ACTH
PTH
Calcitonin
Somatostatin.

Group II B - through cGMP

Examples - Nitric oxide
Atrial natriuretic factor.

Group II C - Through calcium / phosphotidyl inositol

Examples ⇒ Oxytocin, TRH, Cholecystokinin
GnRH, PDGF, Substance P
Gastrin, Muscarinic ACTH, α_1 Adr. Catecholamines.

Ca^{2+} → (Plasma membrane) → ③ Ca^{2+}

H → R → G-Protein [Phospholipase C]

PIP_2 → ② Diacylglycerol → Protein kinase

PIP_2 → Inositol triphosphate IP_3 → ER → Ca^{2+} → ① calcium-calmodulin complex → Activates various Calmodulin dependent kinases

Protein → Phosphoprotein → Biochemical Response

1. Attachment of H to Receptor activates G protein
2. G protein activates phospholipase C
3. $PIP_2 \longrightarrow IP_3 + DAG$
4. IP_3 causes release of Ca^{2+} from ER
5. Ca^{2+} forms complex c̄ calmodulin → Ca^{2+}-calmodulin complex
6. Protein $\xrightarrow{\text{Kinases}}$ Ⓟ protein ⟶ Biochemical response
   Ca^{2+}- calmodulin
   DAG, Ca^{2+}

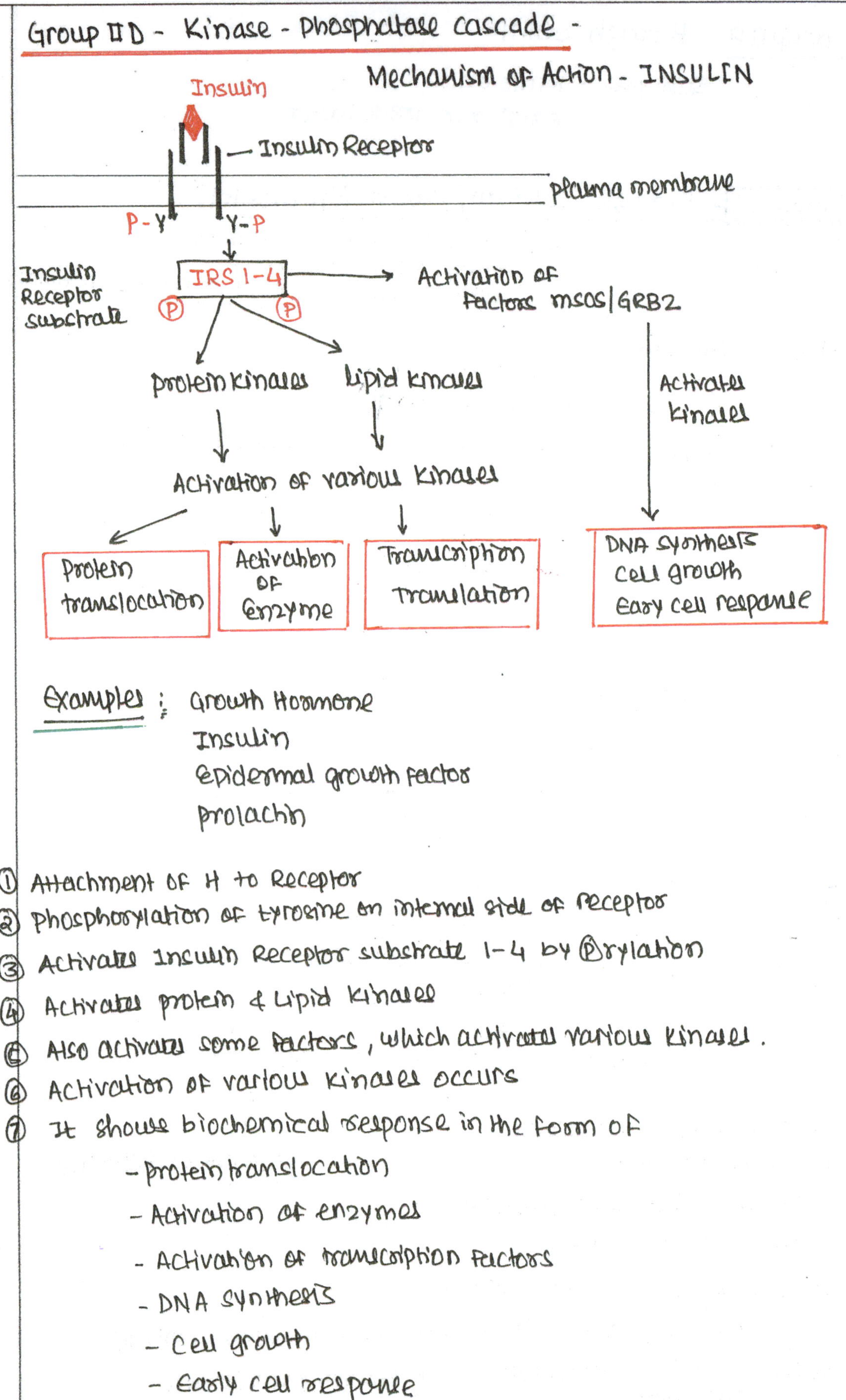

Group II D - Kinase - Phosphatase cascade -

Mechanism of Action - INSULIN

Examples : Growth Hormone
Insulin
Epidermal growth factor
Prolactin

① Attachment of H to Receptor

② Phosphorylation of tyrosine on internal side of receptor

③ Activates Insulin Receptor substrate 1-4 by Ⓟrylation

④ Activates protein & Lipid kinases

⑤ Also activates some factors, which activates various kinases.

⑥ Activation of various kinases occurs

⑦ It shows biochemical response in the form of

- Protein translocation
- Activation of enzymes
- Activation of transcription factors
- DNA synthesis
- Cell growth
- Early cell response

CHROMATOGRAPHY

Definition :- It is a technique or procedure for separation of similar compounds by its differential distribution between two phases i.e. Stationary phase & Mobile phase

If compound has more affinity towards stationary phase, it is retained on stationary phase.

If compound has more solubility in mobile phase, It will move c̄ mobile phase - less retention.

Basic principles

1. **Adsorption** :- It is a surface phenomenon in which one molecule is adsorbed on another compound
   example :- Alumina in column chromatography.

2. **Partition** :- Separation depends on solubility in mobile phase
   More soluble in mobile phase → Move fast
   Less soluble in mobile phase → Retained
   example :- Paper chromatography

3. **Molecular sieve** :- column packed c̄ Gel - has beads c̄ pores
   separation based on size & shape of particles.
   smaller particles will be retained in beads
   larger particles will elute out fast
   example :- Gel filtration chromatography

4. **Ion-exchange** :- Column is packed c̄ resins
   separation is based on charge on the molecule.
   Cation exchange Resin → Retained cations
   Anion exchange Resin → Retained Anions.

5. **Affinity** :- Columns are packed c̄ specific substrates for enzymes OR Antigen for antibody.

Paper chromatography

Principle :- Partition chromatography

Stationary phase :- Water adsorbed on cellulose by weak interaction
(liquid)

Mobile phase :- solvent system (liquid)

Types

1. **Ascending paper chromatography**

The solvent system will more in upward direction.

- Better resolution

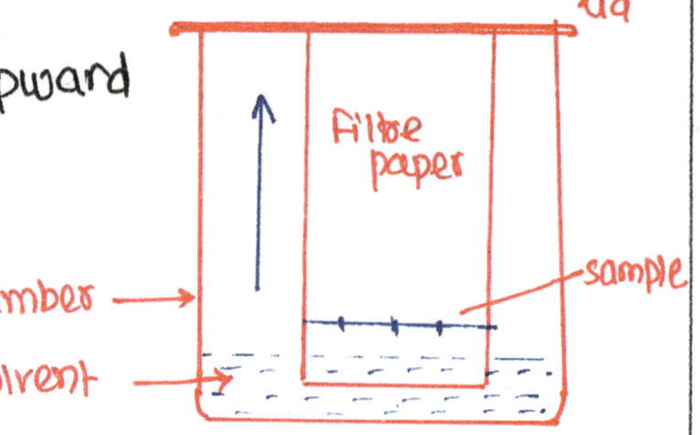

2. **Descending paper chromatography**

The solvent system will move in downward direction

- Requires less time

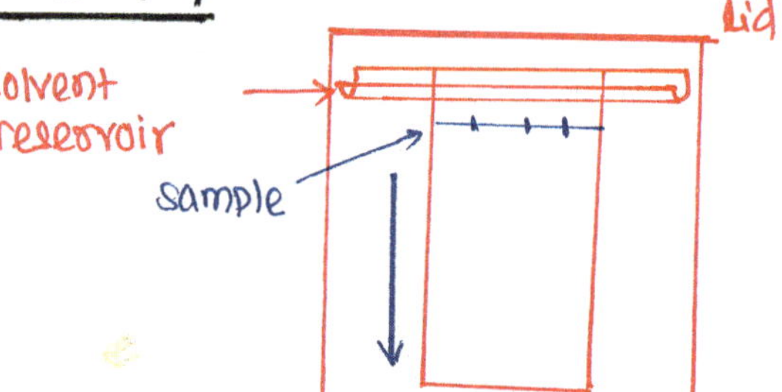

3. **Radial paper chromatography**

The solvent system move radially

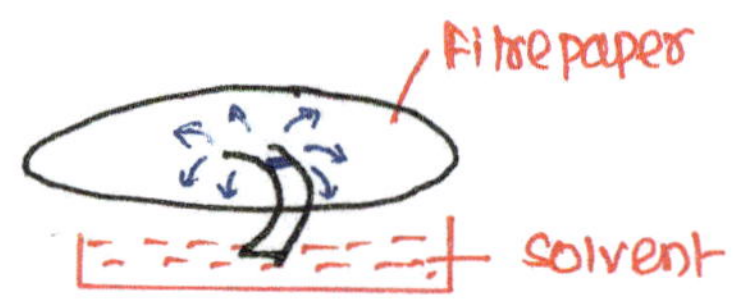

4. **2 Dimensional chromatography**

The solvent system is run and then rotated to 90°

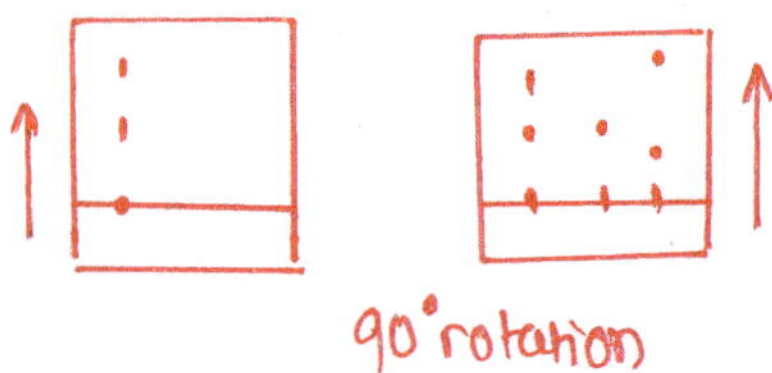

90° rotation

Steps for Aminoacid separation

1. Saturate the chamber c̄ solvent system Butanol : Aceticacid : DW
   4 : 1 : 5
2. Cut the whatman filtre paper into apropriate size
   Wash c̄ 0.1 N HCl, Dry in oven to clear impurities
3. 5cm above the lower edge, mark a line c̄ pencil
4. Apply the amino acid spots c̄ capillary tube (2-5 μl)
5. Dip the lower edge of paper into the solvent by hanging the paper in chromatography chamber
6. Allow it to run, till the solvent reaches upper edge
7. Dry and spray c̄ Ninhydrin solution

Applications

1. Separation of Amino acids – useful in detection of inborn errors of amino acid metabolism.
2. Separation and purification of – carbohydrates, proteins, lipids enzymes, Drugs, Pigments.
3. Separation of reducing sugars.
4. Identification of compound, Amino acids, reducing sugars.

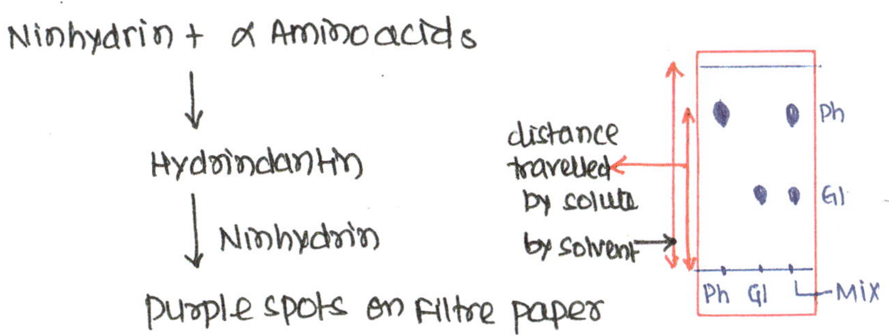

$$\text{Rf value} = \frac{\text{distance travelled by the solute}}{\text{distance travelled by the solvent}}$$

[Ratio of fronts]
Relative flow
Retardation factor

It is specific for AA in particular solvent system
Help in recognition of AA and other compounds.

THIN LAYER CHROMATOGRAPHY

Mobile phase :- solvent

Stationary phase :- glass plate coated c̄ silica gel

The separation depends on relative affinity between stationary phase and mobile phase.

The compound move over the surface of stationary phase under the influence of mobile phase - driven by capillary action. The compound c̄ greater affinity for stationary phase will travel slowly.

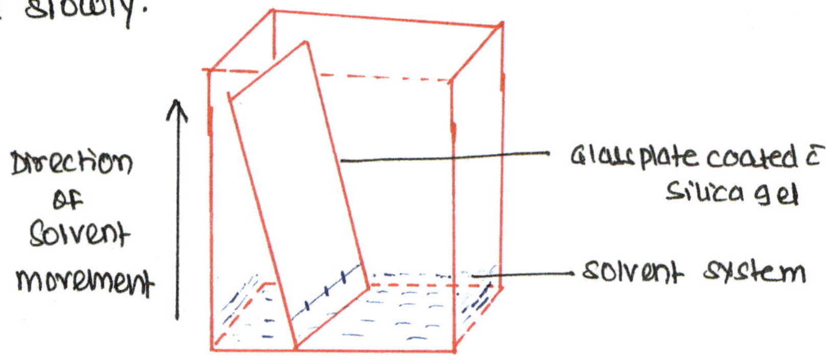

Rf value :- $\frac{\text{distance travelled by solute}}{\text{distance travelled by solvent}}$

Advantages of TLC

1. Very fast, required less time
2. Better resolution as compaired to paper chromatography
3. Less sample is required

Applications

1. Separation of AA and reducing sugars
2. Purification of various compounds - Drugs
3. Detection of inborn errors of metabolism

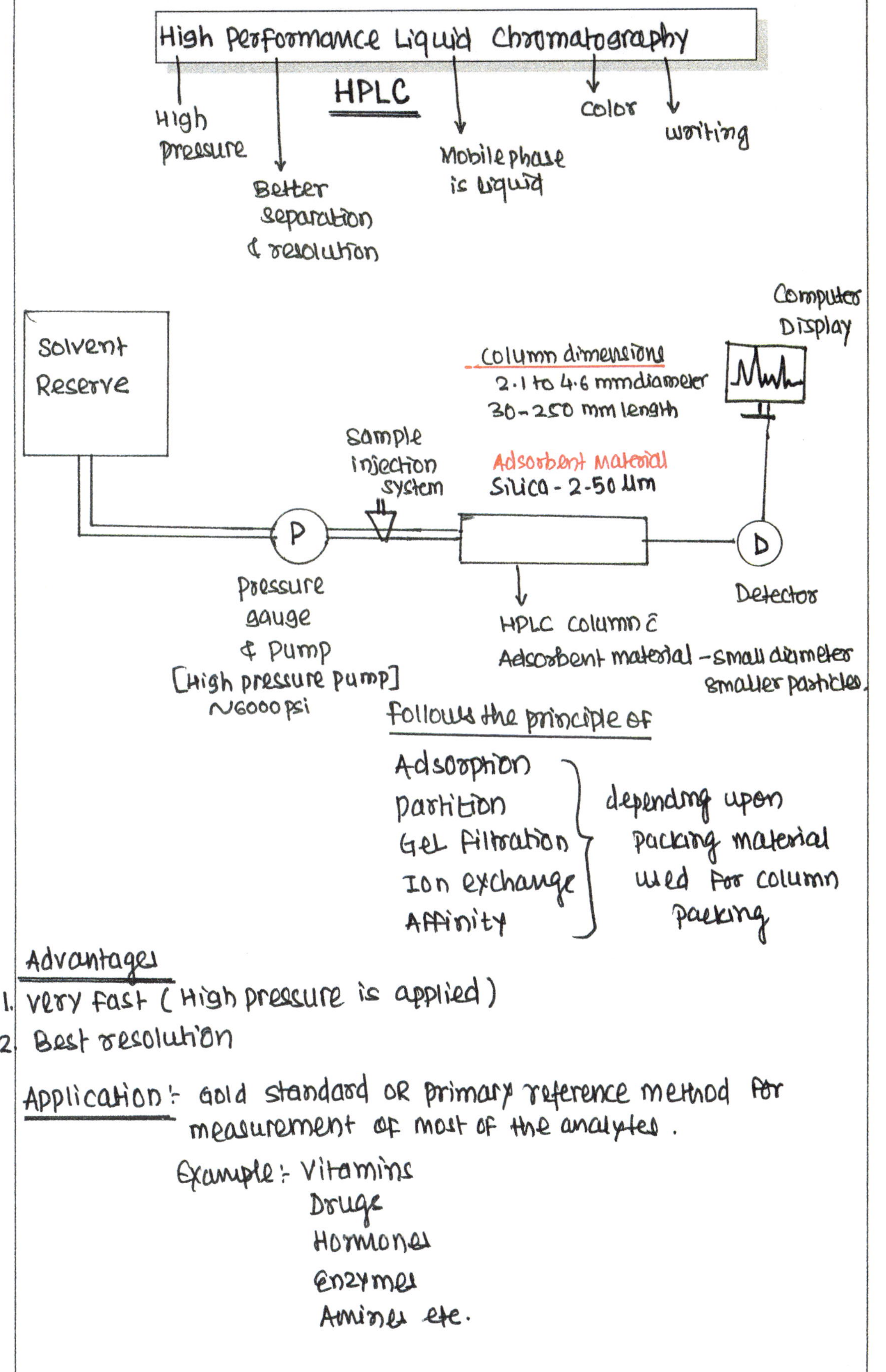

High Performance Liquid Chromatography

HPLC

- High → High pressure
- Performance → Better separation & resolution
- Liquid → Mobile phase is liquid
- Chroma → color
- tography → writing

Solvent Reserve

Pressure gauge & Pump [High pressure pump] ~6000 psi

Sample injection system

Column dimensions
2.1 to 4.6 mm diameter
30–250 mm length

Adsorbent Material
Silica - 2-50 µm

HPLC column c̄ Adsorbent material - small diameter smaller particles.

Detector

Computer Display

Follows the principle of

- Adsorption
- Partition
- Gel Filtration
- Ion exchange
- Affinity

} depending upon packing material used for column packing

Advantages

1. very fast (High pressure is applied)
2. Best resolution

Application:- Gold standard OR primary reference method for measurement of most of the analytes.

Example:- Vitamins
Drugs
Hormones
Enzymes
Amines etc.

ELECTROPHORESIS

Definition :- Migration of the charged particles through a solution under the influence of an external electrical field.

Arne Tiselius :- Nobel prize for electrophoresis in 1948

Principle :- An ampholyte becomes positively charged in acidic medium and migrate to cathode, whereas in alkaline medium they becomes negatively charged and migrate to anode.

pH above pI ⟶ Negative charge ⟶ move toward +ve electrode

pH below pI ⟶ Positive charge ⟶ move toward −ve electrode

factors affecting electrophoresis (Migration)

1. Net charge on molecule – directly proportional
2. size + shape of particles - inversely proportional to size
3. pH of the buffer
4. strength of electric field
5. properties of supporting medium
6. Temp. of system

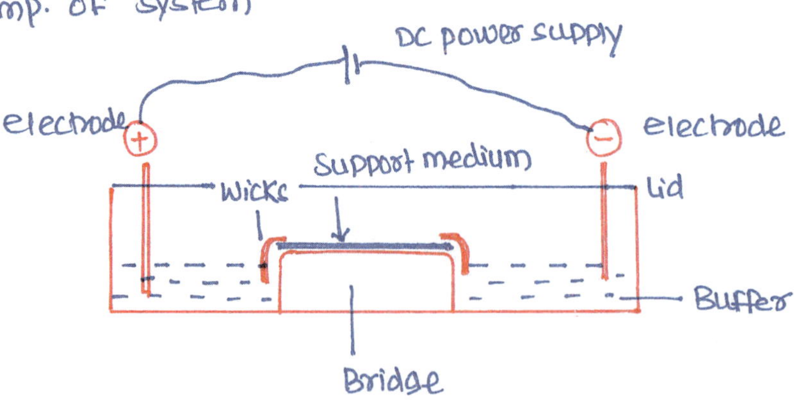

Horizontal electrophoresis Apparatus

Support Medium -
1. Filter paper
2. Cellulose Acetate membrane
3. Agar OR Agarose Gel
4. Starch
5. Polyacrylamide Gel

Steps

1. The sample to be separated is applied to a supporting medium
2. Electrophoresis is carried out at constant voltage or current
3. pH of 8.6 is used in electrophoresis – Buffer solution
4. Fix the separated bands by 7% Acetic acid
5. Visualized by staining – Bromophenol blue, Amido black } for proteins
   Sudan black → for lipoproteins
6. Quantification is done by densitometry OR elution followed by spectrophotometry.

	Type of electrophoresis	Applications
1.	Paper electrophoresis support :- Whatman paper No. 1	for separation of peptides, nucleic acid nucleotides and charged carbohydrates
2.	Cellulose Acetate Membrane electrophoresis	Separation of Lipoproteins Glycoproteins Hemoglobins.
3.	Agarose Gel Electrophoresis Separation based on charge & size	Separation of plasma proteins Lipoproteins Isoenzymes – LDH (1 to 5) CPK (1 to 3) ALP Used in immunodiffusion immunoelectrophoresis
4	Starch Gel electrophoresis	Separation of Isoenzymes (ALP, LDH)
5.	Polyacrylamide Gel electrophoresis Separation based on Molecular weight	Separation of plasma proteins Enzymes, Isoenzymes Lipoproteins

Abnormal electrophoretic Pattern

1. Multiple Myeloma - Prominent M-Band
2. Cirrhosis of liver - decreased Albumin, increased γ-globulins β-γ-bridging.
3. Nephrotic syndrome. All proteins lost in urine except α_2 prominent α_2 band, due to retained α_2 macroglobulin, because of larger size.

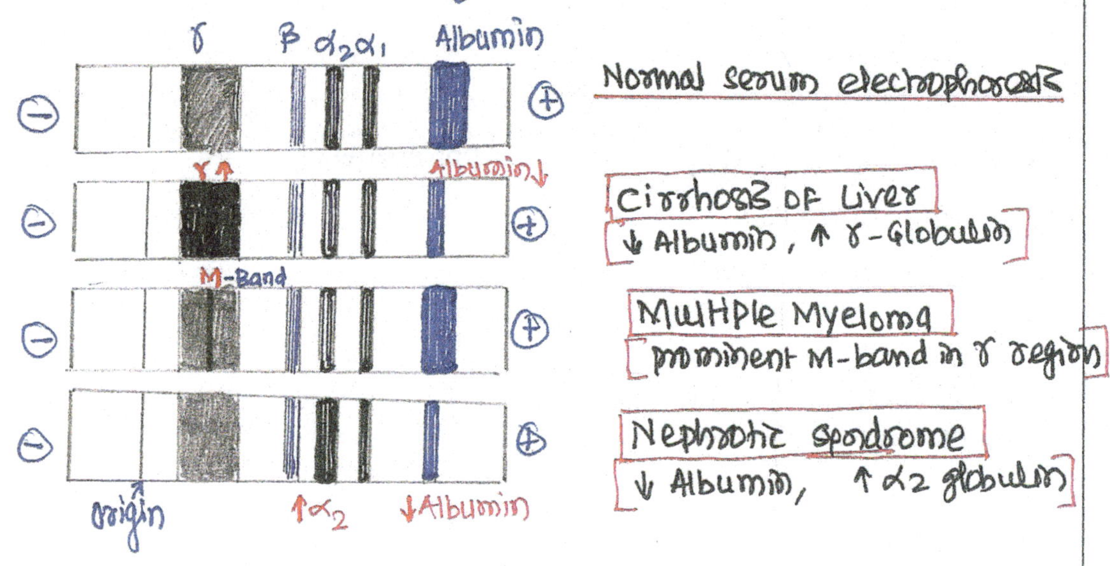

SDS - PAGE

Sodium dodecyl SO_4 - Polyacrylamide Gel electrophoresis

Loading buffer :- SDS [Sodium dodecyl SO_4]
+
B- Mercaptoethanol [to cleave S-S bond]
+
Bromophenol blue [Tracer dye]
+
Glycerol [↑ density & settle sample to the bottom of well]

SDS - Anionic surfactant - causes denaturation of proteins
breaks H bonds, hydrophobic interaction, Ionic interaction

Linear peptide c̄ Negative charge
- Gives Negative charge to peptide
- Now proteins are linear molecule c̄ Net Negative charge.

Acrylamide
↓ Free Radicals [Ammonium persulfate - free radical initiator; TEMED - catalyst - Accelerate free radical formation]
Activated monomers
↓
Polymerisation

Bisacrylamide :- cross links the acrylamide polymers

⊖

Sample well

Stacking Gel - All proteins should arrive at same time in separating gel
PH 6.8

MW KDa
86
42
18 KDa
7 KDa

Separating Gel
PH 8.6

⊕

Running Buffer - Tris Glycine - chloride

In stacking Gel - Protein → -ve charge
Cl^- → -ve charge
Glycine → Neutral charge
[PI 5.97]

At PH 6.8 in stacking Gel Glycine carries No Net charge and lag behind Cl^- & proteins - Proteins concentrated into Narrow zone

Separating Gel :- Glycine carries –ve charge
[PH 8·6]

High Molecular weight peptides ⟶ Move slowly
Low Molecular weight Peptides ⟶ Moves fast

Stop the run when tracer dye reaches the end.
Fix the bands by using Acetic acid
Stained c̄ coomassie blue (Anionic dye)

Separation is based on Molecular sieving effect and molecular weight of peptides (polypeptide chain length)

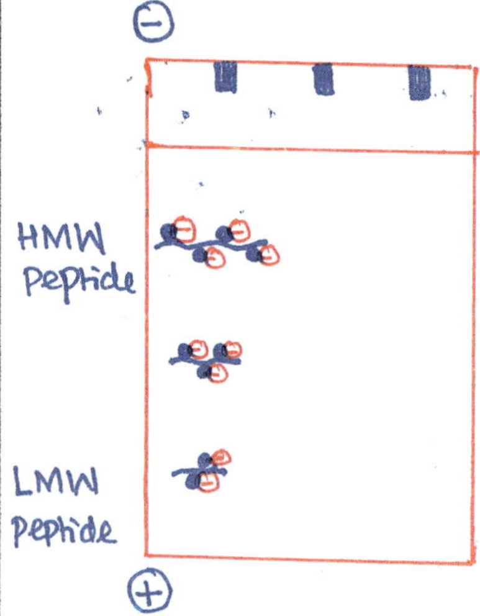

Applications

1. To measure the molecular weight of the molecules by using MW Markers
2. To estimate the size of protein
3. For mapping of peptide
4. To analyze size and number of polypeptide subunits.
5. Used in combination c̄ western blot, for determination of presence of certain proteins.
6. For the analysis of post-translational modifications.
7. To evaluate the levels of various plasma proteins and diagnosis of Multiple myeloma, Nephrotic syndrome, cirrhosis of liver, chronic infection etc.

COLORIMETER

Principle :- The concentration of compound is measured by the intensity of color. The intensity of color is directly proportional to the concentration of analyte.

Beer's law :- When a monochromatic light passes through a light absorbing material, the intensity of transmitted light decreases exponentially as the concentration of light absorbing material increases.

Lambert's Law :- The intensity of transmitted light decreases exponentially as the length of light path through the light absorbing material increases.

We keep the length of light path constant, then it follows Beer's law.

Instrumentation

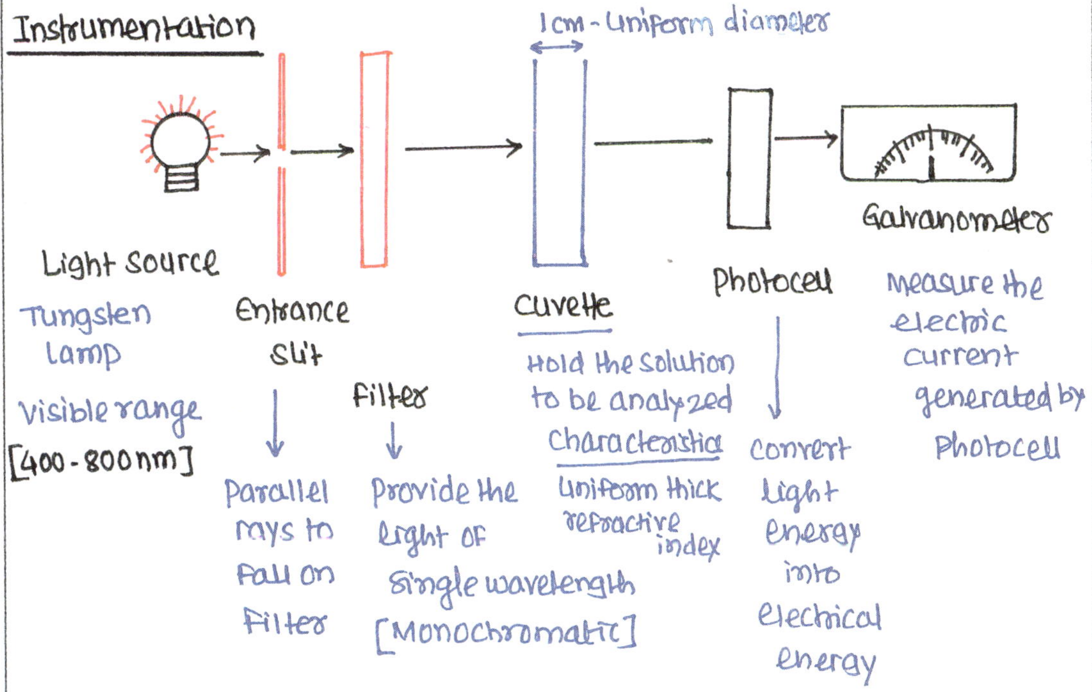

Procedure :- Label 3 test tubes as Test, Blank & Standard

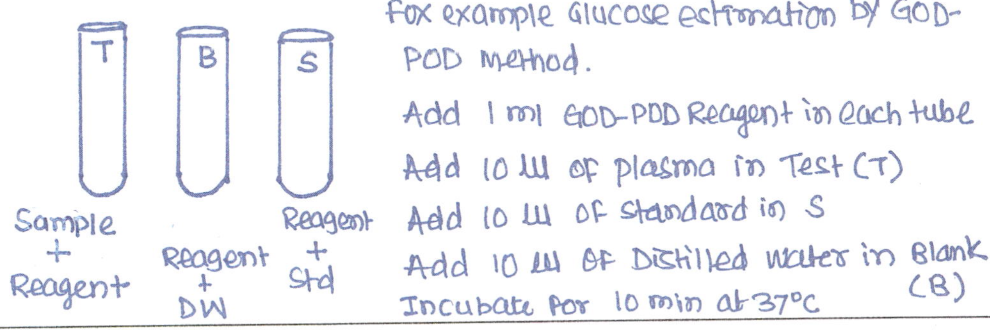

For example glucose estimation by GOD-POD method.

Add 1 ml GOD-POD Reagent in each tube

Add 10 µl of plasma in Test (T)

Add 10 µl of standard in S

Add 10 µl of Distilled water in Blank (B)

Incubate for 10 min at 37°C

Reagent blank - some amount of light is absorbed by reagent itself. to minus the color/effect/OD of reagent.

Standard: The solution of known concentration, If both concentration & OD of standard is known, the concentration of unknown can be calculated.

Operational steps- ① Proper filter (complementary color filter) is selected

② set the instrument OD to zero

③ OD of Test, standard & Blank is Read.

Formula

$$\text{concentration of unknown mg/dl} = \frac{OD_T - OD_B}{OD_S - OD_B} \times \text{conc. of std.}$$

Application

It is used for estimation of various analytes in samples like plasma, serum, CSF, urine and other body fluid

It is used for measuring optical density for estimation of plasma Glucose

serum urea, creatinine, uric acid, cholesterol

serum enzymes - SGOT, SGPT, ALP

serum minerals - calcium, phosphorus

serum Total protein, Albumin, bilirubin etc.

Filter (nm)	Filter color	Color of solution to be Analyzed
405 nm	violet	Yellow green
450 nm	Blue	Yellow
505 nm	bluish green	Red
546 nm	Green	Red violet
578 nm	Yellow	Violet
630 nm	orange	Greenish blue
670 nm	Red	Blue green

ENZYME LINKED IMMUNOSORBENT ASSAY [ELISA]

↓

Immuno

↓

Ag-Ab Reaction

Commonely used for detection of
- T_3, T_4, TSH
- HIV-Ab
- SARS COV2-Ab

Ag – proteins, Peptides, Hormones, bacterial Ag, viral Ag.

Ab – Against various infectious particles.

RIA - Radio immunoassay

ELISA

CLIA - Chemiluminescence immunoassay

Disadvantages of RIA

Radiation hazards
Shorter shelf life of reagents
] overcome by ELISA

ELISA

1. No Radiation hazard
2. Longer shelf life of reagents
3. Easy automation
4. equally sensitive and specific

Definition:- It is an immunochemical reaction based on Ag-Ab reaction in which one of the immunoreagent is immobilised on solid phase support through passive adsorption and other is detected via enzyme linked reaction.

Enzymes used in ELISA
- Horse radish peroxidase
- ALP
- β-Glucosidase
- Glucose oxidase

Linking substances
- Glutaraldehyde
- Periodate
- Maleimide

Sorbent - Adsorbed on solid phase support via passive adsorption
- Polyvinyl chloride
- Polystyrene

Types of ELISA

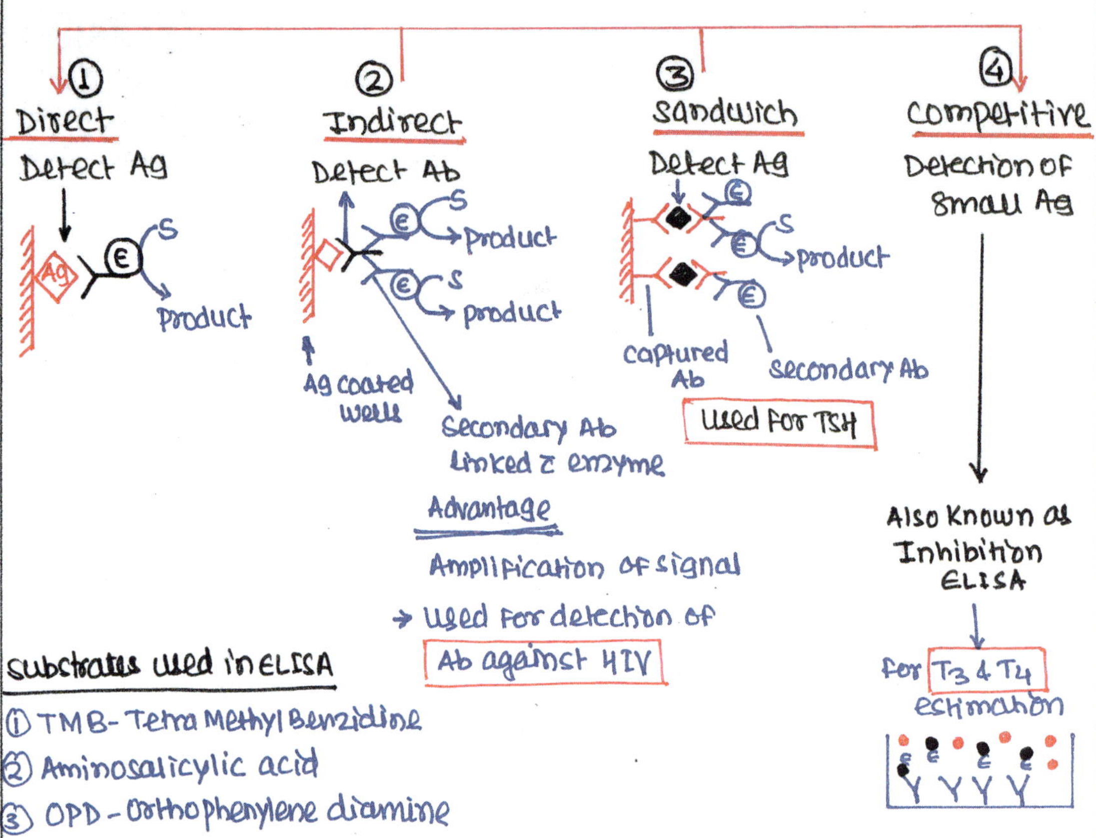

Substrates used in ELISA

① TMB - Tetra Methyl Benzidine
② Aminosalicylic acid
③ OPD - Orthophenylene diamine
④ Azino didiammonium salt

Competition betn serum Ag ● and enzyme labelled Ag ●E

Conc. of $T_3 \propto \frac{1}{\text{color}}$

Avidin-Biotin Chemistry

In place of Ab conjugated c̄ Enzyme Biotin labelled Ab are used then Avidin-Ⓔ complex is added

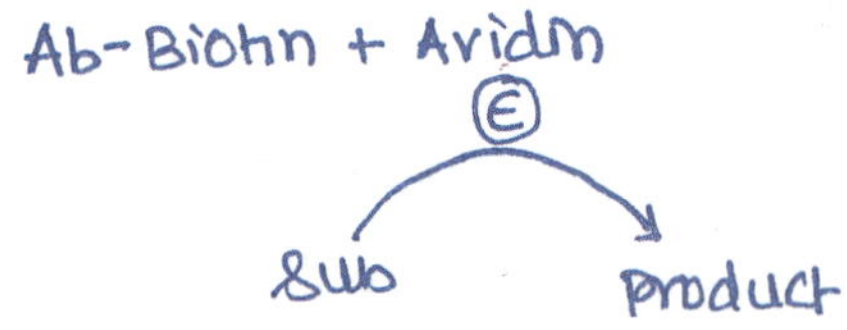

The color produced in ELISA is measured spectrophotometrically.

Quality Control in laboratories

Definition:- study of all sources of variation and procedures used to recognize and minimize them, which arises within laboratory from receipt of specimen to dispach of report.

Internal QC

For continuous and immediate monitoring of laboratory work, to decide whether the results are reliable enough to be released

Done daily in the laboratory

External QC

Periodic & retrospective assessment of lab values, to improve the quality of laboratory work

Evaluation by outside agency

compare the lab work c̄ other labs involved in QC program

INTERNAL Quality Control

Control material :- ① In house patient sera (Pooled clinical samples)
② Randox, Biorad OR other commercially available IQC material

Frequency :- Daily in the laboratory for all test parameters
It can be per batch / per shift

Results available :- Immediately to take appropriate action

- Values are plotted in Levey Jenning chart c̄ ±3SD
- Westgard Rule is applied to detect any error
- control material should be in normal range as well as pathological range

Levey Jenning Chart

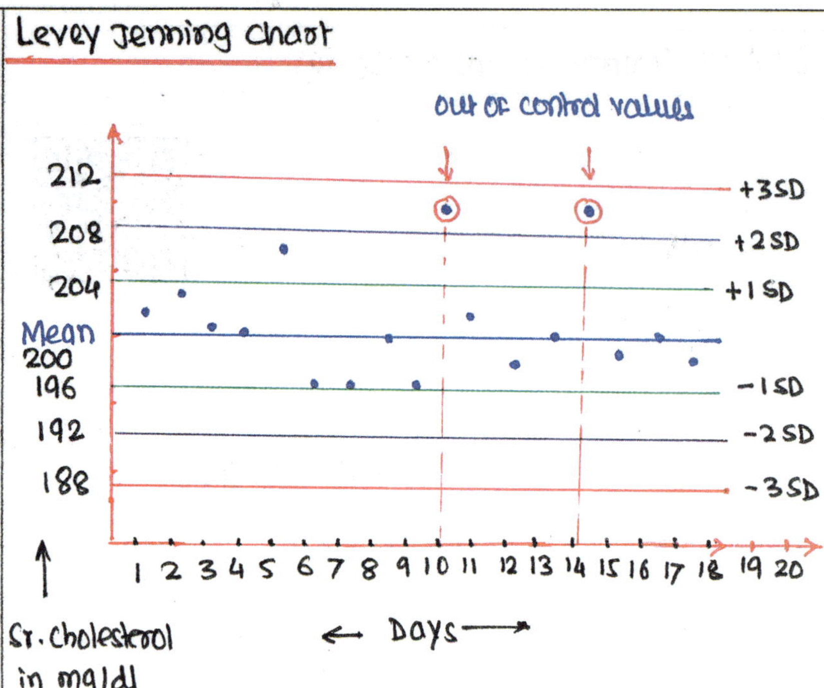

- Put up the IQC control material for atleast 20 or more days and record the values.
- Commercial preparation - mean & SD is already provided c̄ material
- Make a plot c̄ assay run on x-axis and control values on Y axis
- Every time the control is run, the value is plotted on LJ chart.
- As long as control value are between ± 2SD lines on LJ chart the control values are "in control"
- See the observation on 10th & 14th day, values are outside ± 2SD, Therefore on these days values are "out of control".

Westgard Rules

1 2s Rule :- one value outside ± 2SD - suspected Inaccuracy/Imprecision
Warning Rule

1 3s Rule :- one value outside ± 3SD - Inaccuracy/Imprecision
Rejection rule

2 2s Rule :- 2 consecutive values are outside 2 SD - Inaccuracy - Rejection Rule

R 4s Rule :- In same run, difference of 4 SD, means one value above +2SD & other below -2SD - Imprecision - Rejection Rule

4 1s Rule :- 4 consecutive values outside 1 SD - Inaccuracy - Rejection

10 x Rule :- 10 consecutive values on same side of mean - Inaccuracy Rejection

External Quality Control [Proficiency Testing]

The common control material is used by the different lab enrolled in the program.

This give idea about the accuracy of analysis

Results are tabulate monthly and sent to the main laboratory on monthly basis.

Mean of all peer group calculated = Target value.

Target value compared $\bar{c}$ individual lab value by SDI

$$\underset{(\text{z score})}{\text{SDI}} = \frac{\text{Lab Result} - \text{Group mean}}{\text{Group SD}}$$

Standard deviation index > 2 indicate that laboratory <u>NOT</u> in agreement $\bar{c}$ rest of the lab in program.

Most of the laboratories are enrolled $\bar{c}$ CMC, Vellore QC program

- CMC, Vellore lab sent the sample in dried form to different labs across the India
- Individual Lab Analyse the sample for different parameters
- Sent the result to CMC, Vellore website online
- They analyze the data and calculate SDI and VIS

$$\text{Variance Index Score (VIS)} = \frac{\%\ \text{variation}}{\text{CCV}} \times 100$$

$$\%\ \text{variation} = \frac{\text{Diff. betn lab value \& designated value (mean)}}{\text{Designated value (mean)}} \times 100$$

CCV - chosen coefficient of variation (given for Analyte)

VIS <100 → very good
100-150 → Good
150-200 → Satisfactory
>200 → Not Acceptable

Accuracy :- The closeness of a measured value to a target value.

Precision :- closeness between repeatative meaurement.

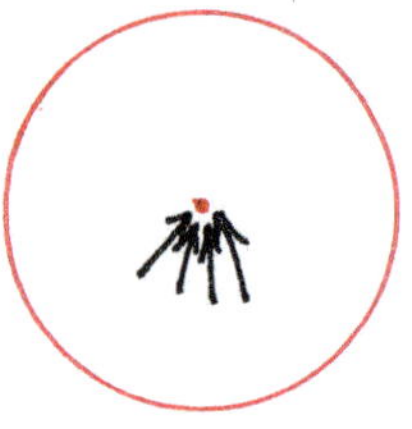

Accurate

Precise but Not accurate

Accurate & Precise

Random error :- Error varies in an unpredictable manner under effectively identical condition.

- Because of lack of Reproducibility (repeatability)

Examples :- Errors in pipetting
changes in incubation period & temp.
The mixing of sample & reagent

can be minimised by training, supervision and adherent to SOP (standard operating procedures)

Systematic error :- Error varies according to definite law or Pattern in effectively identical conditions.

- Due to calibration problem

examples :- Impure/ unstable/ contaminated calibration material
Improper preparation of calibrator
Inadequate calibration techniques
unstable reagent blank

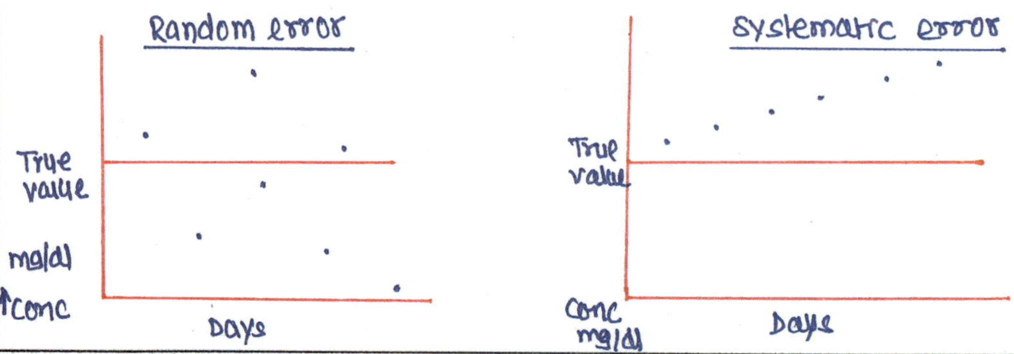

Automated Analyzer - Biochemistry

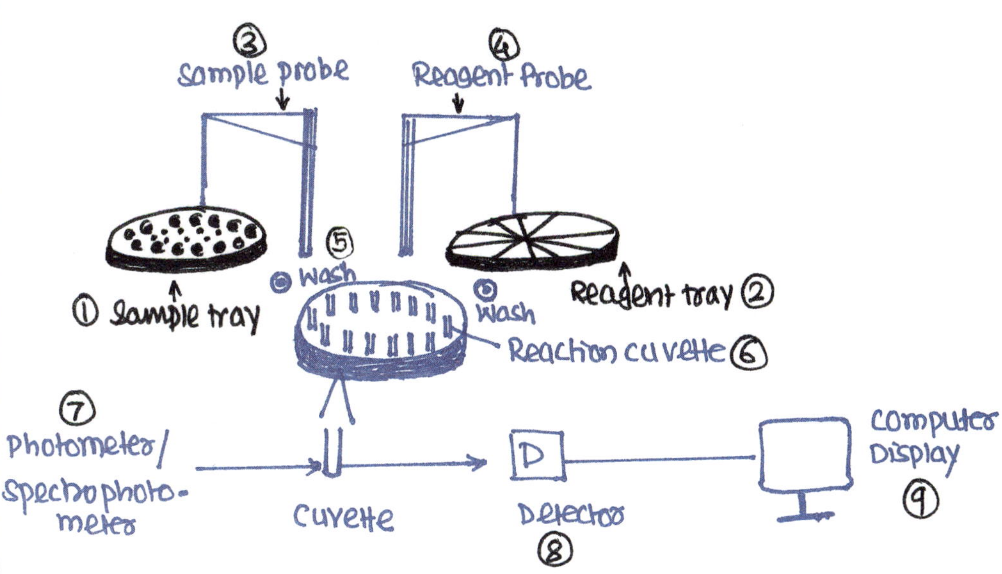

Instrumentation

1. Sample tray → for holding sample - example serum
2. Reagent tray → Reagent holding → cool compartment
3. Sample probe → for taking samples in uL
4. Reagent probe → for taking reagent in uL
5. Wash station → Washing of probes
6. Reaction cuvettes → Actual reaction where sample & reagent mixed & incubated
7. Photometer / spectrophotometer → to read absorbance (nm)
8. Detector → converts light into electrical energy
9. computer display of result → to give display of Result.

Advantages [It is a Heart of Biochemistry Laboratory]

1. Analyzer can handle much larger workload according its throughput example :- XL 640 → 640 tests/hour
2. Decreases Manual errors
3. Results c̄ accuracy and reproducibility
4. Easy to operate, training required
5. Multiple tests can be done on single serum sample by using different reaction cuvettes.
6. Less reagent volume is required, Reduces cost per test
7. Digital display and printed Results
8. Computer based self monitoring of quality control programme
9. Reduces requirement of laboratory staff
10. Very less time is required for processing the samples.
11. Can be connected c̄ lab information system

Uses of Automated Analyzers

① Estimation of normal routine parameters – Sr. Bilirubin < T, D
Sr. Urea
Sr. Creatinine
Sr. Uric acid
Sr. Cholesterol
Sr. Triglyceride
Sr. HDL, LDL
Sr. Total protein
Serum Albumin
Plasma Glucose
CSF-Glucose
CSF-proteins.

② Estimation of special parameters
- Sr. ferritin
- Ceruloplasmin
- D-dimer
- C-Reactive protein
- Cystatin C

③ Estimation of Enzymes – SGOT
SGPT
ALP
LDH
Lipase
Amylase
CK-MB

VITAMINS REVISION CHART

FAT Soluble Vitamins 

Vitamin	Active forms	RDA	Sources	Functions	Deficiency manifestations
Vitamin A	Retinal, Retinol, Retinoic acid, Beta Carotene	Male-600 µg, Female 600 µg, Pregnancy 800 µg, Lactation- 950 µg	Fish liver oil, Cod liver oil, liver, meat, Carrots, Spinach, Mango, etc.	Vision: Walds visual cycle (Rhodopsin), Cell growth and differentiation, Glycoprotein synthesis, Antioxidant, Cholesterol synthesis, Maintenance of germinal epithelium	Night blindness, Conjunctival xerosis, Bitots spot, Corneal xerosis, Keratomalacia, Growth retardation, **Phrynoderma**: Rough dry skin and hyperkeratinisation (Toad skin), Sterility in males
Vitamin D	Calcitriol (1,25-dihydroxy-cholecalciferol)	400 IU in Children, 200 IU in adults	Sunlight, Sea foods, Egg yolk (Cholesterol)	Bone: Mineralization, Intestine: Increases absorption of calcium, Renal tubules: Reabsorption of calcium	**Rickets**: Craniotabes, frontal bossing, widening of fontanelles, bowlegs, widening of wrist, prominent costochondral junctions, increased ALP, **Osteomalacia** in adults: Easy fractures, decreased bone mineral density, increased ALP
Vitamin E	Alpha: Tocopherol, Tocotrienol	10 mg in males, 8 mg in females	Vegetable oils (Wheat germ oil, cottonseed oil, Safflower oil, Sunflower oil, etc.)	Antioxidant, Prevent lipid peroxidation, Protect RBCs from hemolysis, Maintain germinal epithelium	Vitamin in search of disease, Anemia, sterility
Vitamin K	K1: Phylloquinone, K2: Menaquinone, K3: Menadione	70-140 microgram	K1: Green vegetables, Spinach, K2: Produced by intestinal bacteria, K3: Synthetic, water soluble	Clotting: Post translational modification of clotting factors (II, VII, IX, X), Activation of osteocalcin	Bleeding disorders, Increased clotting time

WATER SOLUBLE VITAMINS

Vitamin	Active form	RDA	Sources	Functions	Deficiency
Vitamin C	Ascorbic acid	60-75 mg	Indian Gooseberry (Amla), Oranges, Lemon, Guava	Collagen formation: Hydroxylation of proline and lysine, Antioxidant, steroid synthesis, Bone formation, Tyrosine and tryptophan metabolism, prevention of cataract	**Scurvy**: Bleeding spongy gums, loose teeth, delayed wound healing, hemorrhagic spots on skin and mucous membrane, pseudoarthritis
B1 (Thiamine)	TPP	1-1.5 mg	Outer alerone layer of whole grains, cereals, pulses, Polishing removes 80% thiamine	PDH, KGDH, Transketolase, Branched chain alpha ketoacid DH.	Dry **beriberi** (neuritis), wet beriberi (CVS), Infantile beriberi, Wernicke's Korsakoff syndrome
B2 (Riboflavin)	FAD, FMN	1.5 mg	Cereals, milk, fish	SDH, Acyl Co-A DH, PDH, KGDH	Glossitis, Chelosis, Angular stomatitis, pink eye, seborrheic dermatitis
B3 (Niacin)	NAD, NADP, NADH	20 mg	Whole grains, yeast, liver, meat	Glyceraldehyde 3 (p) DH, LDH, PDH, Isocitrate DH, B.OH Acyl co-A DH, HMG co-A reductase, Ketoacyl reductase	**Pellagra**: Diarrhea, dermatitis (Casal's necklace), Dementia
B5 (Pantothenic Acid)	Coenzyme A	10 mg	Egg, liver, milk, yeast, meat, plant sources	Serves as a carrier of active acyl groups, PDH, alpha KGDH, FAS complex, Succinyl co-A, ACYL Co-A	Burning feet syndrome (pain and Numbness in legs)
B6 (Pyridoxine)	PLP	1-2 mg	Egg, fish, milk, corn, wheat	Transamination, Glycogen phosphorylase, ALA Synthase, decarboxylation and synthesis of Neurotransmitters	Neurological symptoms: seizures, convulsions, neuropathy, anemia, xanthurenate in urine, affect Niacin synthesis

B7 (Biotin)	Biotin	30-100 µg	Milk, egg, meat, grains	**Carboxylation** reaction: Acetyl Co-A carboxylase, Propionyl Co-A carboxylase, Pyruvate carboxylase	Rash, Alopecia, Anemia, loss of appetite, depression, Hallucinations, muscle pain
B9 (Folic Acid)	THF	200 µg	Green leafy vegetables,	Involved in **one C-Metabolism**, synthesis C-2, 8 of Purines, TMP Synthesis, Glycine, serine, ethanolamine, choline synthesis, Initiator N formyl methionine of protein synthesis	Megaloblastic anemia, FIGLU excretion on histidine load test, defect in pregnancy causes neural tube defect in child
B12 (Cobalamin)	Adenosyl B12, Methyl B12	1-2 µg	Liver, meat, fish, curd (Animal origin)	Methyl malonyl co-A →Succinyl co-A, 1) Methyl malonyl co-A mutase, 2) methionine synthase, Homocysteine → Methionine	Megaloblastic anemia, Pernicious anemia, subacute combined degeneration of cord, Methyl malonic acid in urine

MINERALS REVISION CHART

Minerals—Major Elements

Minerals	Serum level	RDA	Sources	Functions	Deficiency
Calcium	9-11 mg/dL	800 mg in adult, Pregnancy, lactation and after menopause—1,500 mg	Milk and milk products, Eggs, fish, vegetables, cereals, pulses	Formation of hydroxyapatite and mineralisation of bones, Muscle contraction Coagulation Integrity of cell membrane Second messenger Enzyme activation Nerve impulse Transmission Heart contraction Secretory Processes	**CATS**-Covulsions, Arrythmia, Tetany, Stridor, Spasm-**Carpopedal spasm**
Phosphorus	3-4 mg/dL	800 mg/day	Milk and milk products, Eggs, fish, vegetables, cereals, pulses	Bone and Teeth formation Nucleoproteins- DNA and RNA Phospholipid, Phosphoproteins Enzyme regulation by phosphorylation and dephosphorylation, Synthesis of ATP, Creatine P Intracellular buffer Coenzymes	Bone deformity, Muscle weakness, arrythmias
Magnesium	1.5-2.2 mg/dL	350 mg in males, 300 mg in females	Green vegetables, milk, meat, cereals	Bone and Teeth formation Activation of enzymes—GK, HK, PFK, Enolase, Glucose 6 P-ase, Adenylate cyclase	Neuromuscular irritability, Muscle spasm, convulsions
Sodium	135-145 mmol/L	5 grams	Table salt Cereals Legumes Eggs Milk meat	Plasma osmolality Acid-base balance Water-electrolyte balance Na^+-K^+ Pump Symport of Glu, Gal. AA Nerve Muscle function	Dehydration, Muscle cramps, decreased blood pressure, headache

Potassium	3.5-5.0 mmol/L	3-4 grams	Bananas, Coconut, Fruits	Plasma osmolality Acid-base balance Water-electrolyte balance Na^+-K^+ Pump Resting membrane potential Depolarization and contraction of heart Secretion of HCl in gastric juice Nerve Muscle function	Muscle weakness, paralysis, skipped heartbeat or irregular heart rhythm
Chloride	96-106 mmol/L	1-5 grams	Table salt Cereals Legumes Eggs Milk meat	Plasma osmolality Acid-base balance Water-electrolyte balance	

Trace Elements

Mineral	RDA	Sources	Functions	Deficiency
Iron	Males—10 mg Females—20 mg Pregnancy and lactation—40 mg	Green leafy vegetables (20 mg/100 g), Pulses, Cereals, Meat, Jaggery, Dry fruits, Beetroot	**Iron containing proteins**: Hemoglobin, Mb, Cytochrome, Cyt P450, Try Pyrrolase, Catalase, NOS synthase Xanthine oxidase, Aconitase, Cytochromes	**Iron deficiency anemia**: Fatigue, pallor, Koilonychia—spoon shaped nails, Decrease MCV, MCH, Microcytic RBCs, Decreased ferritin, Increased TIBC
Copper	2 mg	Organ meat, green leafy veg, Fruits, Nuts, Shellfish	**Copper enzymes**: Catalase, Cytochrome oxidase, Ceruloplasmin, ALA synthase, Superoxide dismutase, Tyrosinase, Lysyl Oxidase	Anemia CVS manifestations Menkes disease: MR and kinky steal hairs Hypopigmentation

Zinc	10 mg	Meat, Milk, Eggs, Shellfish, Grains, Nuts, Beans	**Zn containing enzymes**: Carbonic anhydrase, SOD, RNA Polymerase, ALA dehydratase, Carboxypeptidase, Alcohol DH, Alkaline Phosphatase Storage and release of insulin Maintenance of Vitamin A in serum Reproduction Wound healing Gusten: Zinc metalloprotein: Taste sensation	**Acrodermatitis Enteropathica**: Rash around mouth, nose, fingers, toes and perineal area Poor wound healing Alopecia Loss of taste sensation Impaired spermatogenesis
Iodine	150-200 μg	Drinking water Iodized Salt Vegetables grown on seashore, Sea foods	Thyroid Hormone (T3, T4) synthesis	**Simple endemic Goitre**/Iodine deficiency goitre—enlarged thyroid, fatigue, cold intolerance, Hair loss, menstrual irregularity
Fluoride	Less than 2 ppm	Drinking water	Development of Bone and Teeth	Dental Caries if less than 0.5 ppm for prolonged period
Selenium	50-200 μg	Meat Seafood Liver	Glutathione Peroxidase: Antioxidant Thioredoxin reductase Deiodinase	**Keshan disease**: Cardiomyopathy
Manganese	5-6 mg	Meat, Fish, Pulses, Vegetables, Cereals	Mitochondrial Superoxide dismutase, RNA polymerase, Arginase, Enolase, Kinase, Carboxylase	Impaired growth and development, Rash

WESTGARD RULES: QUALITY CONTROL CHARTS

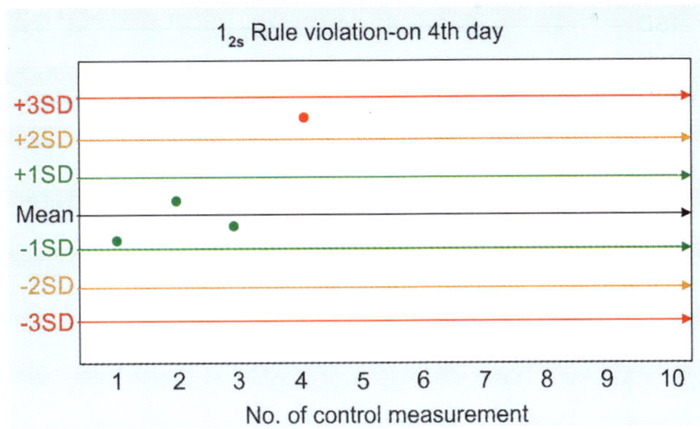

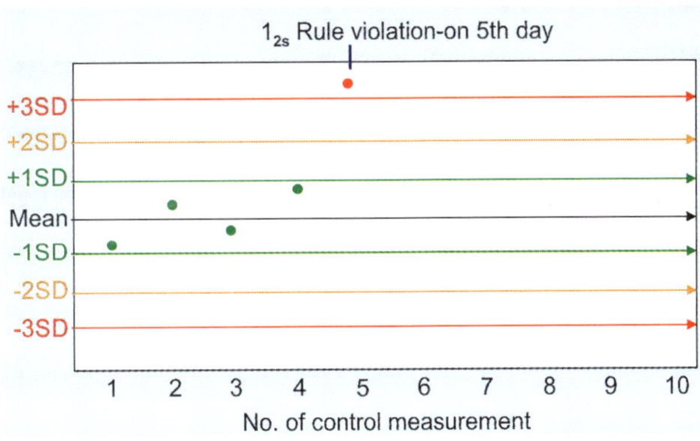

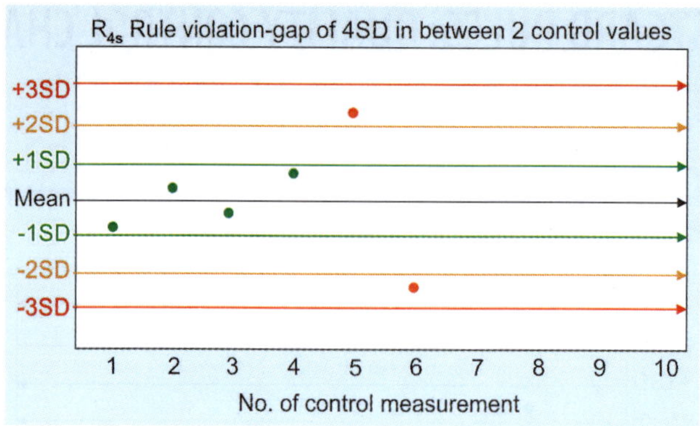
R_{4s} Rule violation-gap of 4SD in between 2 control values
+3SD
+2SD
+1SD
Mean
-1SD
-2SD
-3SD
1 2 3 4 5 6 7 8 9 10
No. of control measurement

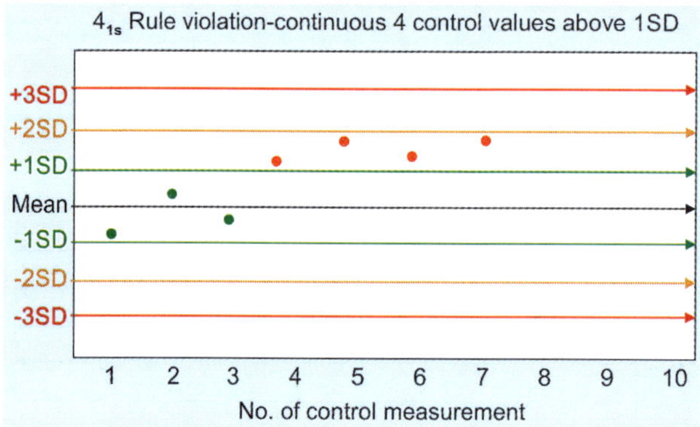
4_{1s} Rule violation-continuous 4 control values above 1SD
+3SD
+2SD
+1SD
Mean
-1SD
-2SD
-3SD
1 2 3 4 5 6 7 8 9 10
No. of control measurement

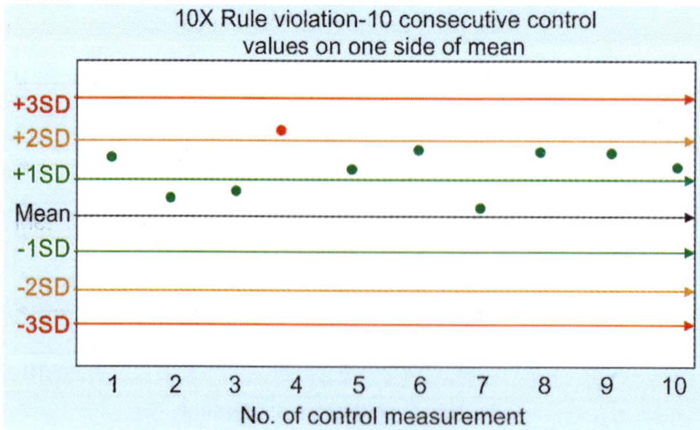
10X Rule violation-10 consecutive control values on one side of mean
+3SD
+2SD
+1SD
Mean
-1SD
-2SD
-3SD
1 2 3 4 5 6 7 8 9 10
No. of control measurement